Selected Titles in This Series

26 **J. Harnad, G. Sabidussi, and P. Winternitz, Editors,** Integrable systems: From classical to quantum, 2000

25 **Decio Levi and Orlando Ragnisco, Editors,** SIDE III—Symmetries and integrability of difference equations, 2000

24 **B. Brent Gordon, James D. Lewis, Stefan Müller-Stach, Shuji Saito, and Noriko Yui, Editors,** The arithmetic and geometry of algebraic cycles, 2000

23 **Pierre Hansen and Odile Marcotte, Editors,** Graph colouring and applications, 1999

22 **Jan Felipe van Diejen and Luc Vinet, Editors,** Algebraic methods and q-special functions, 1999

21 **Michel Fortin, Editor,** Plates and shells, 1999

20 **Katie Coughlin, Editor,** Semi-analytic methods for the Navier-Stokes equations, 1999

19 **Rajiv Gupta and Kenneth S. Williams, Editors,** Number theory, 1999

18 **Serge Dubuc and Gilles Deslauriers, Editors,** Spline functions and the theory of wavelets, 1999

17 **Olga Kharlampovich, Editor,** Summer school in group theory in Banff, 1996, 1998

16 **Alain Vincent, Editor,** Numerical methods in fluid mechanics, 1998

15 **François Lalonde, Editor,** Geometry, topology, and dynamics, 1998

14 **John Harnad and Alex Kasman, Editors,** The bispectral problem, 1998

13 **Michel C. Delfour, Editor,** Boundaries, interfaces, and transitions, 1998

12 **Peter C. Greiner, Victor Ivrii, Luis A. Seco, and Catherine Sulem, Editors,** Partial differential equations and their applications, 1997

11 **Luc Vinet, Editor,** Advances in mathematical sciences: CRM's 25 years, 1997

10 **Donald E. Knuth,** Stable marriage and its relation to other combinatorial problems: An introduction to the mathematical analysis of algorithms, 1997

9 **D. Levi, L. Vinet, and P. Winternitz, Editors,** Symmetries and integrability of difference equations, 1996

8 **J. Feldman, R. Froese, and L. M. Rosen, Editors,** Mathematical quantum theory II: Schrödinger operators, 1995

7 **J. Feldman, R. Froese, and L. M. Rosen, Editors,** Mathematical quantum theory I: Field theory and many-body theory, 1994

6 **Guido Mislin, Editor,** The Hilton Symposium 1993: Topics in topology and group theory, 1994

5 **D. A. Dawson, Editor,** Measure-valued processes, stochastic partial differential equations, and interacting systems, 1994

4 **Hershy Kisilevsky and M. Ram Murty, Editors,** Elliptic curves and related topics, 1994

3 **Rémi Vaillancourt and Andrei L. Smirnov, Editors,** Asymptotic methods in mechanics, 1993

2 **Philip D. Loewen,** Optimal control via nonsmooth analysis, 1993

1 **M. Ram Murty, Editor,** Theta functions: From the classical to the modern, 1993

Volume 26

CRM PROCEEDINGS & LECTURE NOTES

Centre de Recherches Mathématiques
Université de Montréal

Integrable Systems: From Classical to Quantum

Proceedings of the 38th Session of the
Séminaire de mathématiques supérieures
July 26–August 6, 1999
Montréal, Québec, Canada

J. Harnad
G. Sabidussi
P. Winternitz
Editors

The Centre de Recherches Mathématiques (CRM) of the Université de Montréal was created in 1968 to promote research in pure and applied mathematics and related disciplines. Among its activities are special theme years, summer schools, workshops, postdoctoral programs, and publishing. The CRM is supported by the Université de Montréal, the Province of Québec (FCAR), and the Natural Sciences and Engineering Research Council of Canada. It is affiliated with the Institut des Sciences Mathématiques (ISM) of Montréal, whose constituent members are Concordia University, McGill University, the Université de Montréal, the Université du Québec à Montréal, and the Ecole Polytechnique. The CRM may be reached on the Web at www.crm.umontreal.ca.

American Mathematical Society
Providence, Rhode Island USA

The production of this volume was supported in part by the Fonds pour la Formation de Chercheurs et l'Aide à la Recherche (Fonds FCAR) and the Natural Sciences and Engineering Research Council of Canada (NSERC).

2000 *Mathematics Subject Classification.* Primary 81–02, 81–06, 82–02, 82–06, 33–02, 33–06, 35Qxx; Secondary 37J15, 37J35, 37K10, 37K15, 17N37, 17B80, 70H06, 81R12, 81R50, 82B20, 82B23.

Library of Congress Cataloging-in-Publication Data

Integrable systems : from classical to quantum : proceedings of the 38th session of the séminaire de mathématiques supérieures, July 26–August 6, 1999 Montréal, Québec, Canada / J. Harnad, G. Sabidussi, P. Winternitz, editors.

p. cm. — (CRM proceedings & lecture notes ; v. 26)

Includes bibliographical references.

ISBN 0-8218-2093-1 (alk. paper)

1. Hamiltonian systems—Congresses. I. Harnad, J. P. (John P.), 1946– II. Sabidussi, Gert. III. Winternitz, Pavel. IV. Series.

QA614.83 I655 2000

514′.74—dc21 00-064611

∞ The paper used in this book is acid-free and falls within the guidelines
established to ensure permanence and durability.
This volume was typeset using $\mathcal{A}_{\mathcal{M}}\mathcal{S}$-TeX,
the American Mathematical Society's TeX macro system,
and submitted to the American Mathematical Society in camera ready
form by the Centre de Recherches Mathématiques.
Visit the AMS home page at URL: http://www.ams.org/

10 9 8 7 6 5 4 3 2 1 05 04 03 02 01 00

Contents

Preface

The 1999 Séminaire de mathématiques supérieures concerned the modern theory of integrable systems, both classical and quantum. This subject has been in a remarkably active state of development throughout the past three decades, finding renewed motivation for its study in some quite surprisingly new directions. Besides the intrinsic interest in the study of integrable models of many-particle systems, spin chains, lattice and field theory models at both the classical and the quantum level, and completely solvable models in statistical mechanics, there have been new applications in relation to a number of other fields of current interest in theoretical physics and pure mathematics, such as: the Seiberg-Wittten approach to supersymmetric Yang-Mills theory, the spectral theory of random matrices, topological models of quantum gravity, conformal field theory, mirror symmetry, quantum cohomology, etc.

The series of lectures at the 1999 SMS gave a nice cross-section of the current state of work in the area of integrable systems, presented by some of the leading researchers active in this area. There were a total of 65 participants, coming from a dozen different countries in Europe, North America and the Far East. The principal lecture series each consisted of a sequence of between three and five hour-long lectures, and these were supplemented by a number of more specialized presentations by some of the participants. Three of these, given by L. Fehér, D. Korotkin and J.-M. Maillet have been included in the present volume.

The topics covered by the principal lectures in the present volume include: universality of distribution functions in random matrix theory (C. A. Tracy and H. Widom), the classical R-matrix approach to isospectral flows and their algebro-geometric linearization (J. Harnad), the extension of the latter to the elliptic R-matrix case and Hitchin systems, and the link between the latter and Calogero-Moser systems (J. Hurtubise), integrability and symmetry of the XXZ spin chain model (T. Miwa), the relation of Bäcklund transformations to the Baxter Q-operator (E. Sklyanin), characteristic systems on Poisson Lie groups (N. Reshetikhin), multi-separability and superintegrability of classical and quantum systems (W. Miller, Jr.), determinant representations of form factors (V. Korepin) and the developments in special function theory related to quantum integrable systems of Calogero-Moser type (S. Ruijsenaars). The supplementary topics include: the solution of the quantum inverse scattering problem for spin chains and its application to the computation of correlation functions (J.-M. Maillet), algebro-geometric solutions of isomonodromic deformation equations (D. Korotkin), and Poisson-Lie groupoids in relation to the chiral WZNW phase space (J. Balog, L. Fehér, and L. Palla).

Unfortunately, not every lecture sequence was available in time for inclusion, and some that are included are presented in a somewhat reduced form. Among the excellent series which unfortunately could not be included here are the one by A. Its, on "The Riemann-Hilbert Approach to Exactly Solvable Quantum Field Theory and Statistical Mechanical Models," the one by A. Polychronakos on "Calogero-Like Systems; Physics and Mathematics" and the one by P. Winternitz, on "Integrable Systems, Symmetries and Lie Algebra Contractions".

The series that are presented in this volume in a somewhat shortened form include the ones by T. Miwa, whose original title was "Algebraic Analysis of Solvable Lattice Models"; by C. A. Tracy, whose title was "Random Matrix Models and Integrable Systems"; by V. Korepin, whose original title was "Determinant Representations for Quantum Correlation Functions for Exactly Solvable Models"; by N. Reshetikhin, whose title was "Quantization of Integrable Systems", and by J. Harnad, which originally included an additional lecture covering the transition from classical to quantum systems solvable by separation of variables techniques.

The scope and quality of the lectures that are included will, we hope, make this a useful resource for all those interested in an up-to-date introduction to and overview of many of the main currents in the theory of integrable systems at the end of the second millennium.

The Editors

Proceedings of the 38th session of the Séminaire de mathématiques supérieures, held at the Département de mathématiques et de statistique of the Université de Montréal, July 26–August 6, 1999.

1 2 3 4 5 6 7 8 9 10 11 12 13 14 15 16 17 18 19 20 21
22 23 24 25 26 27 28 29
30 31 32 33 34 35 36 37 38 39 40
41 42 43 44 45 46 47

Key to Group Picture

1. T. Sasamoto
2. Y. Takeyama
3. M. Légaré
4. S. Tremblay
6. K. Styrkas
7. D. Richter
8. M. Koelling
9. V. Itskov
10. M. Yakimov
11. P. Desrosiers
12. H. Sati
13. O. Ragnisco
14. A. Degenhard
15. D. Gomez-Ullate
16. Z. Thomova
17. P. Bracken
18. O. Yermolayeva
19. R. Mohammadlikhani
20. A. Tovbis
21. L. Fehér
22. G. David
23. V. Fokin
24. M. Foursov
25. T. Kimura
26. Y. Ohyama
27. R. Smirnov
28. O. Stoyanov
29. Y. Prykarpatsky
30. A. Daigneault
31. S. Ruijsenaars
32. A. Oblomkov
33. A. Penskoi
34. M. Gekhtman
35. P. Casati
36. B. Springborn
37. M. Bertola
38. V. Terras
39. J.-M. Maillet
40. S. Abenda
41. E. Sklyanin
42. P. Winternitz
43. J. Harnad
44. V. Korepin
45. A. Polychronakos
46. N. Reshetikhin
47. C. A. Tracy

List of Participants

Abenda, Simonetta, Universitá di Bologna, Italy.
Alexeyev, Alexander, Peoples' Friendship University, Russia.
Bertola, Marco, International School for Advanced Studies, SISSA, Italy.
Bourama, Toni, Universidad Autonoma del Estado de Morelos, Mexico.
Bracken, Paul, CRM, Université de Montréal, Canada.
Bruce, Aaron, University of Waterloo, Canada.
Butler, Leo, Queen's University, Canada.
Casati, Paolo, Universitá di Milano, Italy.
Degenhard, Andreas, Universität Bielefeld, Germany.
Desrosiers, Patrick, Université Laval, Canada.
Fehér, Laszlo, Joszef Attila University of Szeged, Hungary.
Fokin, Vladimir, Indiana University-Purdue University, USA.
Foursov, Mikhail, University of Minnesota Twin Cities, USA.
Gauthier, Paul, Université de Montréal, Canada.
Gekhtman, Michael, University of Notre Dame, USA.
Gomez-Ullate, David, Universidad Computense Madrid, Spain.
Grunland, Michel, CRM, Université de Montréal and Université du Québec à Trois-Rivières, Canada.
Guimond, Louis-Sébastien, CRM, Université de Montréal, Canada.
Guzzetti, David, International School for Advanced Studies, SISSA, Italy.
Harnad, John, Concordia University and CRM, Université de Montréal, Canada.
Its, Alexander, Indiana University-Purdue University, USA.
Its, Elisabeth, Indiana University-Purdue University, USA.
Itskov, Vladimir, University of Minnesota Twin Cities, USA.
Kang, Seok-Jin, Seoul National University, Korea.
Kedem, Rinat, University of Massachussets, USA.
Kimura, Takashi, Boston University, USA.
Koelling, Melinda, University of Michigan, USA.
Korepin, Vladimir, SUNY-York, Stony Brook, USA.
Korotkin, Dmitrii, Max-Planck Institute for Gravitational Physics, Germany.
Lafortune, Stéphane, CRM, Université de Montréal, Canada.
Légaré, Martin, University of Alberta, Canada.
Lewis, Marc-André, Université Pierre et Marie Curie—Paris VI, France.
Maillet, Jean-Michel, École Normale Supérieure de Lyon, France.
Major, Olivier, CRM, Université de Montréal, Canada.
Miller, Willard, IMA, University of Minnesota, USA.

Miwa, Tetsuji, Kyoto University, Japan.
Mohammadalikhani, R., University of Toronto, Canada.
Polychronakos, Alexios, Uppsala University and University of Ioanninna, Greece.
Oblomkov, Alexei, Moscow State University and Independent University of Moscow, Russia.
Ohyama, Yousuke, Osaka University, Japan.
Patera, Jiri, CRM, Université de Montréal, Canada.
Penskoi, Alexei, Université de Montréal, Canada.
Prykarpatsky, Yarema, National Academy of Sciences, Ukraine.
Ragnisco, Orlando, Universitá degli Studi Roma III, Italy.
Reshetikhin, Nicolai, University of California, Berkeley, USA.
Richter, David, McGill University, Canada.
Ruijsenaars, Simon, Ctr. Wiskunde & Informatica, The Netherlands.
Saint-Aubin, Yvan, CRM, Université de Montréal, Canada.
Sasamoto, Tomohiro, University of Tokyo, Japan.
Sati, Hisham, Texas A & M University, USA.
Sklyanin, Evgueni, St. Petersburg University, Russia.
Smirnov, Roman, University of Waterloo, Canada.
Springborn, Boris, Technische Universität Berlin, Germany.
Stiévenart, Nathalie, Concordia University, Canada.
Stoyanov, Ognyan, Rutgers University, USA.
Styrkas, Konstantin, Brandeis University, USA.
Takeyama, Yoshihiro, Kyoto University, Japan.
Terras, Véronique, École Normale Supérieure de Lyon, France.
Thomova, Zora, SUNY—Institute of Technology, USA.
Tovbis, Alexander, University of Central Florida, USA.
Tracy, Craig A., University of California, Davis, USA.
Tremblay, Sébastien, CRM, Université de Montréal, Canada.
Winternitz, Pavel, CRM, Université de Montréal, Canada.
Yakimov, Milen, University of California, Berkeley, USA.
Yamilov, Ravil, Russian Academy of Sciences—Ufa, Russia.
Yermolayeva, Oksana, Donetsk State University, Ukraine.

Authors' Addresses

Balog, J., Research Institute for Nuclear and Particle Physics, Hungarian Academy of Sciences, H-1525 Budapest 114, P.O.B. 49, Hungary; `balog@rmki.kfki.hu`

Fehér, L., Institute for Theoretical Physics, József Attila University, H-6726 Szeged, Tisza Lajos krt 84–86, Hungary; `lfeher@sol.cc.u-szeged.hu`

Harnad, J., Department of Mathematics and Statistics, Concordia University, 7141 Sherbrooke W., Montréal, Québec, Canada H4B 1R6, and Centre de recherches mathématiques, Université de Montréal, C. P. 6128, succ. centre-ville, Montréal, Québec, Canada H3C 3J7; `harnad@crm.umontreal.ca`

Hurtubise, J. C., Department of Mathematics and Statistics, McGill University, 805 Sherbrooke W., Montréal, Québec, Canada H3A 2K6, and Centre de recherches mathématiques, Université de Montréal, C. P. 6128, succ. centre-ville, Montréal, Québec, Canada H3C 3J7; `hurtubis@crm.umontreal.ca`

Korepin, V. E., Institute for Theoretical Physics, State University of New York at Stony Brook, Stony Brook, NY 11794-3840, USA; `korepin@insti.physics.sunysb.edu`

Korotkin, D. A., Max-Planck-Institut für Gravitationsphysik, Am Mühlenberg 1, D-14476 Golm, Germany;
Current address: Department of Mathematics and Statistics, Concordia University, 7141 Sherbrooke W., Montréal, Qué., H4B 1R6, Canada; `korotkin@discrete.concordia.ca`

Maillet, J.-M., Laboratoire de Physique, Groupe de Physique Théorique, ENS Lyon, 46 allée d'Italie, 69364 Lyon Cedex 07, France; `jean-michel.maillet@ens-lyon.fr`

Miller, Jr., W., Institute for Mathematics and its Applications University of Minnesota, Minneapolis, MN 55455, USA; `miller@ima.umn.edu`

Miwa, Tetsuji, Department of Mathematics, Faculty of Science, Kyoto University, Kyoto 606, Japan; `tetsuji@kusm.kyoto-u.ac.jp`

Palla, L., Institute for Theoretical Physics, Roland Eötvös University, H-1117, Budapest, Pázmány P. sétány 1 A-ép, Hungary; `palla@ludens.elte.hu`

Reshetikhin, Nicolai, Department of Mathematics, University of California at Berkeley, Berkeley, CA 94720-3840, USA; `reshetik@math.berkeley.edu`

Ruijsenaars, S. N. M., Centre for Mathematics and Computer Science, P.O. Box 94079, 1090 GB Amsterdam, The Netherlands; siru@wxs.nl

Sklyanin, E. K., Steklov Mathematical Institute at St. Petersburg, Fontanka 27, St. Petersburg 191011, Russia; sklyanin@euclid.pdmi.ras.ru

Tracy, Craig A., Department of Mathematics, Institute of Theoretical Dynamics, University of California, Davis, CA 95616, USA; tracy@itd.ucdavis.edu

Widom, Harold, Department of Mathematics, University of California, Santa Cruz, CA 95064, USA; widom@cats.ucsc.edu

Centre de Recherches Mathématiques
CRM Proceedings and Lecture Notes
Volume **26**, 2000

On the Chiral WZNW Phase Space, Exchange r-Matrices and Poisson-Lie Groupoids

J. Balog, L. Fehér, and L. Palla

ABSTRACT. This is a review of recent work on the chiral extensions of the WZNW phase space describing both the extensions based on fields with generic monodromy as well as those using Bloch waves with diagonal monodromy. The symplectic form on the extended phase space is inverted in both cases and the chiral WZNW fields are found to satisfy quadratic Poisson bracket relations characterized by monodromy dependent exchange r-matrices. Explicit expressions for the exchange r-matrices in terms of the arbitrary monodromy dependent 2-form appearing in the chiral WZNW symplectic form are given. The exchange r-matrices in the general case are shown to satisfy a new dynamical generalization of the classical modified Yang-Baxter (YB) equation and Poisson-Lie (PL) groupoids are constructed that encode this equation analogously as PL groups encode the classical YB equation. For an arbitrary simple Lie group G, exchange r-matrices are exhibited that are in one-to-one correspondence with the possible PL structures on G and admit them as PL symmetries.

1. Introduction

The Wess-Zumino-Novikov-Witten (WZNW) model [**33**] of conformal field theory has proved to be the source of interesting structures that play an increasingly important rôle in theoretical physics and in mathematics [**14, 16**]. One of the fascinating aspects of the model is that in addition to its built-in affine Kac-Moody symmetry it also exhibits certain quantum group properties [**26**]. The quantum group properties were originally discovered in the quantized model, which raised the question to find their classical analogues. The studies at the beginning of the nineties led to the consensus that the origin of these quantum group properties lies in the Poisson-Lie symmetries of the so called chiral WZNW phase

2000 *Mathematics Subject Classification.* Primary: 81T40, secondary: 37K30, 17B80, 53D17.

First-named author supported in part by the Hungarian National Science Fund (OTKA) under grants T019917 and T030099.

Second-named author supported in part by the Hungarian National Science Fund (OTKA) under grants T025120 and T030099, and by the Ministry of Education under grant FKFP 0596/1999.

Third-named author supported in part by the Hungarian National Science Fund (OTKA) under grants T019917 and T030099, and by the Ministry of Education under grant FKFP 0178/1999.

This is the final form of the paper.

space that emerges after splitting the left- and right-moving degrees of freedom [**5, 10, 19, 2, 7, 13, 6, 24, 28, 21, 20, 3**]. The chiral separation arises from the product form of the solution of the WZNW field equation given by $g(x_L, x_R) = g_L(x_L)g_R^{-1}(x_R)$, where x_C ($C = L, R$) are lightcone coordinates and the g_C are quasiperiodic group valued fields with equal monodromies, $g_C(x + 2\pi) = g_C(x)M$ for some M in the WZNW group G. The chiral WZNW Poisson structures found in the literature have the form

$$\{g_C(x) \overset{\otimes}{,} g_C(y)\} = \frac{1}{\kappa_C}\big(g_C(x) \otimes g_C(y)\big)\left(\hat{r} + \frac{1}{2}\hat{I}\,\mathrm{sign}(y - x)\right), \qquad 0 < x, y < 2\pi, \tag{1.1}$$

where $\hat{I}$ is given by the quadratic Casimir of the simple Lie algebra, $\mathcal{G}$, of the WZNW group, G, and the interesting object is the 'exchange r-matrix' $\hat{r}$. These classical 'exchange algebras' can be regarded as fundamental since the current algebra follows as their consequence, and help to better understand the quantum group properties of the model by means of canonical quantization [**12, 23, 11**]. However, the choice of the chiral Poisson structure is highly non-unique due to the fact that the g_C are determined by the physical field g only up to the gauge freedom $g_C \mapsto g_C h$ for any constant $h \in G$.

There are two qualitatively different cases that correspond to building the WZNW field out of chiral fields with diagonal monodromy ('Bloch waves') or out of fields with generic monodromy. For Bloch waves [**7, 13, 6**], the Poisson structure is essentially unique and the associated r-matrix is a solution of the so-called classical dynamical Yang-Baxter (CDYB) equation, which has recently received a lot of attention [**17**]. For chiral fields with generic monodromy, it has been argued in [**24, 20**] that the possible exchange r-matrices should correspond to certain local differential 2-forms ρ on open domains $\check{G} \subset G$, whose exterior derivative is the 3-form that occurs in the WZNW action. Until recently, the precise connection between ρ and $\hat{r}$ has not been elaborated, and in most papers dealing with generic monodromy actually only those very special cases were considered for which $\hat{r}$ is a monodromy independent constant.

We here review the main results obtained in our recent papers [**8, 9**], where a detailed analysis of the chiral extensions of the WZNW phase space was undertaken. The next section contains an outline of the background to the problem. The subsequent two sections describe the chiral WZNW Hamiltonian structures in detail. The final section is devoted to an interpretation of these structures in terms of Poisson-Lie groupoids.

2. Chiral Extensions of the WZNW Phase Space

We below describe the WZNW Hamiltonian system and the chiral extension of its solution space following the spirit of [**24, 20**].

We consider a simple, real or complex, Lie algebra, $\mathcal{G}$, with a corresponding connected Lie group, G, and identify the phase space of the WZNW model associated with the group G as

$$\mathcal{M} = T^*\widetilde{G} = \{(g, J_L) \mid g \in \widetilde{G},\ J_L \in \widetilde{\mathcal{G}}\}, \tag{2.1}$$

where $\widetilde{G} = C^\infty(S^1, G)$ is the loop group and $\widetilde{\mathcal{G}} = C^\infty(S^1, \mathcal{G})$ is its Lie algebra. The isomorphism of the cotangent bundle $T^*\widetilde{G}$ with $\widetilde{G} \times \widetilde{\mathcal{G}}$ is established by means of right-translations on $\widetilde{G}$. The elements $g \in \widetilde{G}$ (resp. $J_L \in \widetilde{\mathcal{G}}$) are modeled as 2π-periodic G-valued (resp. $\mathcal{G}$-valued) functions on the real line $\mathbf{R}$. The phase space is equipped with the symplectic form

$$\Omega^\kappa = d\int_0^{2\pi} d\sigma\ \mathrm{Tr}(J_L dgg^{-1}) + \frac{\kappa}{2}\int_0^{2\pi} d\sigma\ \mathrm{Tr}(dgg^{-1}) \wedge (dgg^{-1})' \tag{2.2}$$

with some constant κ. Here prime denotes derivative with respect to the space variable, $\sigma \in \mathbf{R}$, and for any $A, B \in \mathcal{G}$, $\mathrm{Tr}(AB)$ denotes a fixed multiple of the Cartan-Killing form on $\mathcal{G}$. If T_α and T^α ($\alpha = 1, \ldots, \dim \mathcal{G}$) are dual bases of $\mathcal{G}$, $\mathrm{Tr}(T_\alpha T^\beta) = \delta_\alpha^\beta$, then $\mathrm{Tr}(AB) = A_\alpha B^\alpha$ with $A_\alpha = \mathrm{Tr}(AT_\alpha)$, $B^\alpha = \mathrm{Tr}(BT^\alpha)$ and the usual summation convention in force. For the wedge product we use the conventions in [**1**].

Although the expression of Ω^κ appears rather formal at first sight, it can be used to unambiguously associate Hamiltonian vector fields and Poisson brackets (PBs) with a set of admissible functions, which include, for example, the Fourier components of the WZNW field g, the 'left-current' J_L and the 'right-current' J_R given by

$$J_R = -g^{-1}J_L g + \kappa g^{-1}g'. \tag{2.3}$$

The currents are the momentum maps that generate two commuting actions of $\widetilde{G}$ on $\mathcal{M}$ that correspond respectively to left- and right-translations on $\widetilde{G}$. This means that the following local PB relations are valid:

$$\begin{aligned}
\{\mathrm{Tr}(T_\alpha J_L)(\sigma), \mathrm{Tr}(T_\beta J_L)(\bar\sigma)\}_{WZ} &= \mathrm{Tr}([T_\alpha, T_\beta]J_L)(\sigma)\delta + \kappa\ \mathrm{Tr}(T_\alpha T_\beta)\,\delta' \\
\{\mathrm{Tr}(T_\alpha J_R)(\sigma), \mathrm{Tr}(T_\beta J_R)(\bar\sigma)\}_{WZ} &= \mathrm{Tr}([T_\alpha, T_\beta]J_R)(\sigma)\delta - \kappa\ \mathrm{Tr}(T_\alpha T_\beta)\,\delta' \\
\{g(\sigma), \mathrm{Tr}(T_\alpha J_L)(\bar\sigma)\}_{WZ} &= T_\alpha g(\sigma)\delta \\
\{g(\sigma), \mathrm{Tr}(T_\alpha J_R)(\bar\sigma)\}_{WZ} &= -g(\sigma)T_\alpha \delta,
\end{aligned} \tag{2.4}$$

together with $\{J_L(\sigma), J_R(\bar\sigma)\}_{WZ} = 0$, where $\delta := \delta(\sigma - \bar\sigma) = (2\pi)^{-1}\sum_{n\in\mathbf{Z}} e^{in(\sigma-\bar\sigma)}$. These PBs can be derived from the symplectic form Ω^κ, whose precise meaning is explained in several papers (see, e.g., [**27**]). Thus we need not dwell on this point, but note that in the case of a complex Lie algebra the admissible functions depend holomorphically on the matrix elements of g, J_L, J_R in the finite dimensional irreducible representations of G, and $\widetilde{G} \times \widetilde{\mathcal{G}}$ is then a model of the holomorphic cotangent bundle.

The phase space $\mathcal{M}$ represents the initial data for the WZNW system, whose dynamics is generated by the Hamiltonian

$$H_{WZ} = \frac{1}{2\kappa}\int_0^{2\pi} d\sigma\ \mathrm{Tr}(J_L^2 + J_R^2). \tag{2.5}$$

Denoting time by τ and introducing lightcone coordinates as

$$\begin{gathered}
x_L := \sigma + \tau, \quad x_R := \sigma - \tau, \\
\partial_L = \frac{\partial}{\partial x_L} = \frac{1}{2}(\partial_\sigma + \partial_\tau), \quad \partial_R = \frac{\partial}{\partial x_R} = \frac{1}{2}(\partial_\sigma - \partial_\tau),
\end{gathered} \tag{2.6}$$

Hamilton's equation can be written in the alternative forms [**33**]

$$\kappa\partial_L g = J_L g,\ \partial_R J_L = 0 \iff \kappa\partial_R g = gJ_R,\ \partial_L J_R = 0. \tag{2.7}$$

Let $\mathcal{M}^{\rm sol}$ be the space of solutions of the WZNW system. $\mathcal{M}^{\rm sol}$ consists of the smooth G-valued functions $g(\sigma,\tau)$ which are 2π-periodic in σ and satisfy $\partial_R(\partial_L g\, g^{-1})=0$. The general solution of this evolution equation can be written as

$$g(\sigma,\tau)=g_L(x_L)g_R^{-1}(x_R), \tag{2.8}$$

where (g_L,g_R) is any pair of G-valued, smooth, quasiperiodic functions on $\mathbf{R}$ with *equal monodromies*, i.e., for $C=L,R$ one has $g_C(x_C+2\pi)=g_C(x_C)M$ with some C-independent $M\in G$. To elaborate this representation of the solutions in more detail, we define the space $\widehat{\mathcal{M}}$:

$$\widehat{\mathcal{M}}:=\{(g_L,g_R)|g_{L,R}\in C^\infty(\mathbf{R},G),\ g_{L,R}(x+2\pi)=g_{L,R}(x)M,\ M\in G\}. \tag{2.9}$$

There is a free right-action of G on $\widehat{\mathcal{M}}$ given by

$$G\ni h:(g_L,g_R)\mapsto(g_Lh,g_Rh). \tag{2.10}$$

Notice that $\widehat{\mathcal{M}}$ is a *principal fibre bundle* over $\mathcal{M}^{\rm sol}$ with respect to the above action of G. The projection of this bundle, $\vartheta:\widehat{\mathcal{M}}\to\mathcal{M}^{\rm sol}$, is given by

$$\vartheta:(g_L,g_R)\mapsto g=g_Lg_R^{-1},\quad \text{i.e., } g(\sigma,\tau)=g_L(x_L)g_R^{-1}(x_R). \tag{2.11}$$

We can identify $\mathcal{M}$ with $\mathcal{M}^{\rm sol}$ by associating the elements of the solution space with their initial data at $\tau=0$. Formally, this is described by the map $\iota:\mathcal{M}^{\rm sol}\to\mathcal{M}$,

$$\iota:\mathcal{M}^{\rm sol}\ni g(\sigma,\tau)\mapsto\big(g(\sigma,0),J_L(\sigma)=(\kappa\partial_L g\,g^{-1})(\sigma,0)\big)\in\mathcal{M}. \tag{2.12}$$

Obviously, $\iota^*(\Omega^\kappa)$ is then the natural symplectic form on the solution space. Explicitly,

$$(\iota^*\Omega^\kappa)(g)= -\kappa\Big(d\int_0^{2\pi}d\sigma\,{\rm Tr}(g^{-1}\partial_R g\,g^{-1}dg)+\frac{1}{2}\int_0^{2\pi}d\sigma\ {\rm Tr}(g^{-1}dg)\wedge\partial_\sigma(g^{-1}dg)\Big)\Big|_{\tau=0}. \tag{2.13}$$

Regarding now $\mathcal{M}^{\rm sol}$ as the base of the bundle $\vartheta:\widehat{\mathcal{M}}\to\mathcal{M}^{\rm sol}$, we obtain a closed 2-form, $\widehat{\Omega}^\kappa$, on $\widehat{\mathcal{M}}$, $\widehat{\Omega}^\kappa:=\vartheta^*(\iota^*\Omega^\kappa)$. By substituting the explicit formula (2.11) of ϑ, one finds

$$\widehat{\Omega}^\kappa(g_L,g_R)=\kappa_L\Omega_{\rm chir}(g_L)+\kappa_R\Omega_{\rm chir}(g_R)\quad \text{with } \kappa_L:=\kappa,\kappa_R:=-\kappa, \tag{2.14}$$

where $\Omega_{\rm chir}$ is the so called chiral WZNW 2-form:

$$\begin{aligned}\Omega_{\rm chir}(g_C)=&-\frac{1}{2}\int_0^{2\pi}dx_C\,{\rm Tr}(g_C^{-1}dg_C)\wedge(g_C^{-1}dg_C)'\\ &-\frac{1}{2}{\rm Tr}\big((g_C^{-1}dg_C)(0)\wedge dM_CM_C^{-1}\big),\quad M_C=g_C^{-1}(x)g_C(x+2\pi).\end{aligned} \tag{2.15}$$

This crucial formula of $\widehat{\Omega}^\kappa$ was first obtained by Gawędzki [**24**].

It is clear from its definition that $d\widehat{\Omega}^\kappa=0$, but $\widehat{\Omega}^\kappa$ is not a symplectic form on $\widehat{\mathcal{M}}$, since it is degenerate. Of course, its restriction to any (local) section of the bundle $\vartheta:\widehat{\mathcal{M}}\to\mathcal{M}^{\rm sol}$ is a symplectic form, since such sections yield (local) models of $\mathcal{M}^{\rm sol}$. On the other hand, one can check that $\Omega_{\rm chir}$ has a non-vanishing exterior derivative [**24**]:

$$d\Omega_{\rm chir}(g_C)=-\frac{1}{6}{\rm Tr}(M_C^{-1}dM_C\wedge M_C^{-1}dM_C\wedge M_C^{-1}dM_C). \tag{2.16}$$

Although this cancels from $d\widehat{\Omega}^\kappa$, since $M_L = M_R$ for the elements of $\widehat{\mathcal{M}}$, it makes the chiral separation of the WZNW degrees of freedom a very non-trivial problem.

The idea of the chiral separation arises from the observation that the currents J_C *almost* completely determine the chiral WZNW fields g_C, and thus also $g = g_L g_R^{-1}$, by means of the differential equations

$$\kappa_C \partial_C g_C = J_C g_C \quad \text{for } C = L, R. \tag{2.17}$$

Thus it appears an interesting possibility to construct the WZNW model as a reduction of a simpler model, in which the left- and right-moving degrees of freedom would be separated in terms of *completely independent* chiral fields g_L and g_R, regarded as fundamental variables. It is clear that the solution space of such a chirally extended model must be a direct product of two identical but independent spaces, i.e., it must have the form

$$\widehat{\mathcal{M}}^{\text{ext}} := \mathcal{M}_L \times \mathcal{M}_R, \tag{2.18}$$

$$\mathcal{M}_C := \{g_C \mid g_C \in C^\infty(\mathbf{R}, G),\ g_C(x + 2\pi) = g_C(x) M_C,\ M_C \in G\}. \tag{2.19}$$

Ideally, one would like to endow the space $\widehat{\mathcal{M}}^{\text{ext}}$ with a symplectic structure, $\widehat{\Omega}^\kappa_{\text{ext}}$, that reduces to $\widehat{\Omega}^\kappa$ on the submanifold $\widehat{\mathcal{M}} \subset \widehat{\mathcal{M}}^{\text{ext}}$ defined by the periodicity constraint $M_L = M_R$. It is easy to see that these requirements force $\widehat{\Omega}^\kappa_{\text{ext}}$ to have the following form:

$$\widehat{\Omega}^\kappa_{\text{ext}}(g_L, g_R) = \kappa_L \Omega^\rho_{\text{chir}}(g_L) + \kappa_R \Omega^\rho_{\text{chir}}(g_R), \tag{2.20}$$

$$\Omega^\rho_{\text{chir}}(g_C) = \Omega_{\text{chir}}(g_C) + \rho(M_C) \tag{2.21}$$

with some 2-form ρ depending *only* on the monodromy of g_C. Since in the extended model the factors $(\mathcal{M}_C, \kappa_C \Omega^\rho_{\text{chir}})$ should be *symplectic* manifolds *separately*, the condition

$$d\Omega^\rho_{\text{chir}} = -\frac{1}{6}\,\text{Tr}(M_C^{-1} dM_C \wedge M_C^{-1} dM_C \wedge M_C^{-1} dM_C) + d\rho(M_C) = 0 \tag{2.22}$$

arises. But then we have to face the problem that no globally defined smooth 2-form exists on G that would satisfy this condition for all $M_C \in G$.

There are two rather different ways out of the above difficulty [**20**]. The first is to restrict the possible domain of the monodromy matrix M_C to some open submanifold in G on which an appropriate 2-form ρ may be found. We refer to a choice of such a domain and 2-form ρ as *a chiral extension of the WZNW system* with generic monodromy.

The second possibility is to restrict the domain of the allowed monodromy matrices much more drastically from the beginning, in such a way that after the restriction $d\Omega_{\text{chir}}$ vanishes, whereby the difficulty disappears. For example, one may achieve this by restricting the monodromy matrices to vary in a fixed maximal torus of G, which amounts to constructing (a subset of) the solutions of the WZNW field equation in terms of chiral 'Bloch waves'. This second possibility is especially natural in the case of compact or complex Lie groups, for which there is only one maximal torus up to conjugation. Geometrically, the restriction to Bloch waves corresponds to taking a (local and partial) gauge fixing of the bundle $\vartheta : \widehat{\mathcal{M}} \to \mathcal{M}^{\text{sol}}$.

3. Hamiltonian Structures for Generic Monodromy

We here investigate the chiral WZNW phase space $\mathcal{M}_C$ introduced above. The analysis is the same for both chiralities, $C = L, R$, and we simplify our notation by putting $\mathcal{M}_{\rm chir}$ for $\mathcal{M}_C$ and g, M, κ for g_C, M_C, κ_C, respectively. Thus $\mathcal{M}_{\rm chir}$ is parametrized by the G-valued, smooth, quasiperiodic field $g(x)$ satisfying the monodromy condition

$$g(x+2\pi) = g(x)M, \quad M \in G. \tag{3.1}$$

The corresponding chiral current, $J(x) = \kappa g'(x)g^{-1}(x) \in \mathcal{G}$, is a smooth, 2π-periodic function of x. We then consider a 2-form ρ on a domain $\check{G} \subset G$, for which we assume that $d\rho(M) = 1/6\,{\rm Tr}(M^{-1}dM)^{3\wedge}$ and let $\check{\mathcal{M}}_{\rm chir} \subset \mathcal{M}_{\rm chir}$ be the set of chiral WZNW fields whose monodromy matrix lies in $\check{G}$. It turns out that $\kappa\Omega^{\rho}_{\rm chir}$ defines a symplectic structure on $\check{\mathcal{M}}_{\rm chir}$ if a further condition holds for the pair $(\check{G}, \rho)$. In order to describe this condition let us introduce the parametrization

$$\rho(M) = \frac{1}{2} q^{\alpha\beta}(M)\,{\rm Tr}\big(T_\alpha M^{-1}dM\big) \wedge {\rm Tr}\big(T_\beta M^{-1}dM\big), \quad q^{\alpha\beta} = -q^{\beta\alpha}, \tag{3.2}$$

with T_α denoting a basis of $\mathcal{G}$, whose dual basis with respect to Tr is T^α. For any $M \in \check{G}$, then also introduce the linear operator $q(M) : \mathcal{G} \to \mathcal{G}$ by

$$q(M) : T^\beta \mapsto q^{\alpha\beta}(M)T_\alpha, \tag{3.3}$$

as well as its shifts, $q_\pm(M) := q(M) \pm \frac{1}{2}I$, by the identity operator I. The further condition that we need is that

$$\det(q_+(M) - q_-(M) \circ {\rm Ad}\, M^{-1}) \neq 0 \quad \forall M \in \check{G}. \tag{3.4}$$

This condition will guarantee the (weak) non-degeneracy of $\kappa\Omega^{\rho}_{\rm chir}$ on $\check{\mathcal{M}}_{\rm chir}$.

To use $\kappa\Omega^{\rho}_{\rm chir}$ in practice we need to establish some notation for tangent vectors $X[g]$ at $g \in \mathcal{M}_{\rm chir}$ and vector fields X over the chiral phase space. To this end we consider smooth curves on $\mathcal{M}_{\rm chir}$ described by functions $\gamma(x,t) \in G$ satisfying

$$\gamma(x+2\pi, t) = \gamma(x,t)M(t) \quad M(t) \in G; \qquad \gamma(x,0) = g(x). \tag{3.5}$$

$X[g]$ is obtained as the velocity to a curve at $t = 0$, encoded by the $\mathcal{G}$-valued, smooth function

$$\xi(x) := \frac{d}{dt} g^{-1}(x)\gamma(x,t)\Big|_{t=0}. \tag{3.6}$$

The monodromy properties of $\xi(x)$ can be derived by taking the derivative of the first equation in (3.5): $\xi'(x+2\pi) = M^{-1}\xi'(x)M$, and this can be solved in terms of a $\mathcal{G}$-valued, smooth, 2π-periodic function, $X_J \in \widetilde{\mathcal{G}}$, and a constant Lie algebra element, ξ_0, as follows:

$$\xi(x) = \xi_0 + \int_0^x dy\, g^{-1}(y)X_J(y)g(y). \tag{3.7}$$

A vector field X on $\mathcal{M}_{\rm chir}$ is an assignment, $g \mapsto X[g]$, of a vector to every point $g \in \mathcal{M}_{\rm chir}$. Thus it can be specified by the assignments $g \mapsto \xi_0[g] \in \mathcal{G}$ and $g \mapsto X_J[g] \in \widetilde{\mathcal{G}}$. Using any curve that defines $X[g]$, X acts on a differentiable function, $g \mapsto F[g]$, on $\mathcal{M}_{\rm chir}$ as

$$X(F)[g] = \frac{d}{dt} F[g_t]\Big|_{t=0} \qquad g_t(x) = \gamma(x,t). \tag{3.8}$$

Note that the evaluation functions $F^x[g] := g(x)$ and $\mathcal{F}^x[g] := J(x)$ are differentiable with respect to any vector field, and their derivatives are given by

$$X\big(g(x)\big) = g(x)\xi(x) \quad \text{and} \quad X\big(J(x)\big) = \kappa X_J(x). \tag{3.9}$$

This clarifies the meaning of X_J as well. It is also obvious from its definition that the monodromy matrix yields a G-valued differentiable function on $\mathcal{M}_{\rm chir}$, $g \mapsto M = g^{-1}(x)g(x+2\pi)$, whose derivative is characterized by the $\mathcal{G}$-valued function

$$X(M)M^{-1} = M\xi(x+2\pi)M^{-1} - \xi(x). \tag{3.10}$$

Having defined vector fields, one can also introduce differential forms as usual. We only remark that by (3.9) evaluation 1-forms like $dg(x)$, $dJ(x)$ or $(g^{-1}dg)'(x)$ are perfectly well-defined: e.g. $dg(x)(X) = X\big(g(x)\big) = g(x)\xi(x)$.

Let us now show that $\Omega^\rho_{\rm chir}$ is weakly non-degenerate, that is $\Omega^\rho_{\rm chir}(X,Y) = 0$ $\forall X$ only for $Y = 0$, if and only if (3.4) holds. In order to compute

$$\Omega^\rho_{\rm chir}(X,Y) = \Omega_{\rm chir}(X,Y) + \rho(X,Y) \tag{3.11}$$

for two arbitrary vector fields, we take X to be parametrized by $\xi(x)$ and further by the pair $\big(\xi_0, X_J(x)\big)$, while the analogous parametrization for Y is given by $\eta(x)$ and the pair $\big(\eta_0, Y_J(x)\big)$. Then a straightforward calculation gives that

$$\begin{aligned}\Omega^\rho_{\rm chir}(X,Y) = \int_0^{2\pi} dx\ \mathrm{Tr}\Big(X_J(x)g(x)\Big(\eta(x) + q_-(M)\big(M^{-1}Y(M)\big)\Big)g^{-1}(x)\Big)\\ + \mathrm{Tr}\Big(\xi_0\big(q_-(M) - \mathrm{Ad}\, M\circ q_+(M)\big)\big(M^{-1}Y(M)\big)\Big).\end{aligned} \tag{3.12}$$

This vanishes for every X, that is for arbitrary $X_J \in \widetilde{\mathcal{G}}$ and $\xi_0 \in \mathcal{G}$, if and only if

$$\begin{aligned}\big(q_-(M) - \mathrm{Ad}\, M\circ q_+(M)\big)\big(M^{-1}Y(M)\big) = 0,\\ \eta(x) + q_-(M)\big(M^{-1}Y(M)\big) = 0.\end{aligned} \tag{3.13}$$

Since the transpose with respect to the scalar product on $\mathcal{G}$ satisfies

$$\big(q_-(M) - \mathrm{Ad}\, M\circ q_+(M)\big)^T = \big(q_-(M)\circ \mathrm{Ad}\, M^{-1} - q_+(M)\big), \tag{3.14}$$

if (3.4) holds then it follows from (3.13) that $\eta(x)$ must vanish, that is $Y = 0$. Thus we proved that (3.4) implies the non-degeneracy of $\Omega^\rho_{\rm chir}$. The converse statement is also easy to establish, since if the determinant in (3.4) vanished say at M^0, then there would exist a non-zero $A \in \mathcal{G}$ such that $\big(q_-(M^0) - \mathrm{Ad}\, M^0\circ q_+(M^0)\big)(A) = 0$. We could hence define a tangent vector Y^0 at a corresponding point in $\mathcal{M}_{\rm chir}$ by $\eta^0(x) = -q_-(M^0)(A)$, and this vector would annihilate $\Omega^\rho_{\rm chir}$. (The definition of Y^0 is consistent since $\eta^0(x+2\pi) - (M^0)^{-1}\eta^0(x)M^0 = A$ holds.)

Now we turn to our main problem: For a differentiable (scalar) function F on the phase space $\check{\mathcal{M}}_{\rm chir}$, we wish to find a corresponding vector field, Y^F, satisfying

$$X(F) = \kappa\Omega^\rho_{\rm chir}(X, Y^F) \tag{3.15}$$

for all vector fields X. Notice that Y^F does not necessarily exist for a given F. We say that F is an element of the set of *admissible Hamiltonians*, denoted as $\mathtt{H}$, if the corresponding Hamiltonian vector field, Y^F, exists. On account of the non-degeneracy of $\Omega^\rho_{\rm chir}$, if Y^F exists then it is uniquely determined.

We may use the formula (3.12) for $Y := Y^F$ to establish the following three necessary and sufficient conditions that F must obey to guarantee that Y^F exists:

- There must exist a *smooth* $\mathcal{G}$-valued function on $\mathbf{R}$, $A^F(x)$, and a constant Lie algebra element, a^F, such that for any vector field X

$$X(F) = \kappa \int_0^{2\pi} dx\, \mathrm{Tr}\big(X_J(x)A^F(x)\big) + \kappa\, \mathrm{Tr}\big(\xi_0 a^F\big). \tag{3.16}$$

  (This means that $F \in \mathtt{H}$ must have an exterior derivative parametrized by the assignments $g \mapsto A^F(x)[g]$ and $g \mapsto a^F[g]$. The restriction of $A^F(x)$ to $x \in [0, 2\pi]$ and a^F are uniquely determined by (3.16), and $A^F(x)$ is made a unique function on $\mathbf{R}$ by the next requirement.)
- The expression

$$\big[A^F(x), J(x)\big] + \kappa \frac{dA^F(x)}{dx} \tag{3.17}$$

  must define a smooth 2π-*periodic* function on $\mathbf{R}$.
- $A^F(x)$ and a^F must be related by

$$a^F = g^{-1}(0)\big[A^F(0) - A^F(2\pi)\big]g(0). \tag{3.18}$$

If these conditions are satisfied, then Y^F is in fact given by

$$g^{-1}(x)Y^F\big(g(x)\big) = g^{-1}(x)A^F(x)g(x) - \frac{1}{2}a^F + r(M)\big(a^F\big), \tag{3.19}$$

where $r(M)$ is the linear operator on $\mathcal{G}$ defined by

$$r(M) = \frac{1}{2}\big(q_+(M) - q_-(M)\circ \mathrm{Ad}(M^{-1})\big)^{-1} \circ \big(q_+(M) + q_-(M)\circ \mathrm{Ad}(M^{-1})\big). \tag{3.20}$$

Note that the matrix $r^{\alpha\beta}(M)$ of $r(M)$ is antisymmetric. Later we shall also use the operators $r_\pm(M) := r(M) \pm \frac{1}{2}I$ and the corresponding $\mathcal{G}\otimes\mathcal{G}$-valued functions on $\check{G}$:

$$\hat{r}(M) := r^{\alpha\beta}(M)T_\alpha \otimes T_\beta, \quad \hat{r}_\pm(M) = \hat{r}(M) \pm \frac{1}{2}\hat{I}, \quad \hat{I} = T^\alpha \otimes T_\alpha. \tag{3.21}$$

The above explicit description of the Hamiltonian map $F \mapsto Y^F$ induced by $\kappa\Omega^\rho_{\mathrm{chir}}$ on $\check{\mathcal{M}}_{\mathrm{chir}}$ is one of our main results [**8**]. Its proof can be sketched as follows. First, by assuming that (3.15) holds we see from (3.12) for $Y = Y^F$ that, at every point in the phase space, $X(F)$ has the form (3.16) with

$$g^{-1}(x)A^F(x)g(x) = \eta(x) + q_-(M)\big(M^{-1}Y^F(M)\big), \tag{3.22}$$

$$a^F = \big(q_-(M) - \mathrm{Ad}\, M \circ q_+(M)\big)\big(M^{-1}Y^F(M)\big). \tag{3.23}$$

Since by the meaning of tangent vectors we must have $\eta'(x) = g^{-1}(x)Y^F_J(x)g(x)$, by taking the derivative of (3.22) we immediately get that

$$Y^F\big(J(x)\big) = \big[A^F(x), J(x)\big] + \kappa\partial_x A^F(x) \tag{3.24}$$

must hold, which in particular means that the right hand side must define a smooth, 2π-periodic function on $\mathbf{R}$. As for equation (3.18), this is a direct consequence of (3.22) and (3.23) by taking into account that as a tangent vector Y^F satisfies

$$M^{-1}Y^F(M) = \eta(2\pi) - M^{-1}\eta(0)M. \tag{3.25}$$

This proves that the elements of H indeed meet the conditions (3.16), (3.17), (3.18). Moreover, if the Hamiltonian vector field exists then by combining (3.22) and (3.23) we obtain

$$\begin{aligned} g^{-1}(x)Y^F\big(g(x)\big) \\ = g^{-1}(x)A^F(x)g(x) - q_-(M)\circ\big(q_-(M) - \mathrm{Ad}\, M\circ q_+(M)\big)^{-1}(a^F), \end{aligned} \tag{3.26}$$

which is equivalent to (3.19), since for the operator $r(M)$ defined by (3.20)

$$r_-(M) = -q_-(M)\circ\big(q_-(M) - \mathrm{Ad}\, M\circ q_+(M)\big)^{-1} \tag{3.27}$$

is an identity. To complete the proof, one checks that if the expression in (3.17) is periodic with period 2π, then (3.19) gives a well-defined vector field (since $g^{-1}(x)Y^F\big(g(x+2\pi)\big)M^{-1} - g^{-1}(x)Y^F\big(g(x)\big)$ is independent of x), which satisfies (3.15) if (3.16) and (3.18) hold.

Now we elaborate the Hamiltonian vector field for some particular elements in H. First note that the matrix elements of the evaluation functions $\mathcal{F}^x$ and F^x fail to satisfy the first condition, thus they are not in H. However, their smeared out versions

$$\mathcal{F}_\mu := \int_0^{2\pi} dx\ \mathrm{Tr}\big(\mu(x)J(x)\big),\quad F_\phi[g] := \int_0^{2\pi} dx\ \mathrm{Tr}\big(\phi(x)g^\Lambda(x)\big), \tag{3.28}$$

(where in defining F_ϕ we use a representation[1] $\Lambda : G\to GL(V)$ of G with $g^\Lambda = \Lambda(g)$ and a smooth test function $\phi : \mathbf{R}\to \mathrm{End}(V)$) can be shown to belong to H, if $\mu(x)$ is a $\mathcal{G}$-valued, smooth, 2π-periodic test function, and ϕ satisfies $\phi^{(k)}(0) = \phi^{(k)}(2\pi) = 0$ for every integer $k\geq 0$. The corresponding Hamiltonian vector fields obtained from (3.19) satisfy

$$\begin{aligned} Y^{\mathcal{F}_\mu}\big(g(x)\big) &= \mu(x)g(x),\\ Y^{\mathcal{F}_\mu}\big(J(x)\big) &= [\mu(x), J(x)] + \kappa\mu'(x),\\ Y^{\mathcal{F}_\mu}(M) &= 0, \end{aligned} \tag{3.29}$$

and, for $x\in[0,2\pi]$,

$$g^{-1}(x)Y^{F_\phi}\big(g(x)\big) = \frac{1}{\kappa}T^\alpha \int_x^{2\pi} dy\, \mathrm{Tr}\big(T^\Lambda_\alpha \phi(y) g^\Lambda(y)\big) - \frac{1}{2}a^{F_\phi} + r(M)(a^{F_\phi}). \tag{3.30}$$

Eq. (3.29) shows that the $\mathcal{F}_\mu$ generate an infinitesimal action of the loop group on the phase space with respect to which $g(x)$ is an affine Kac-Moody primary field, and the current $J(x)$ transforms according to the co-adjoint action of the centrally extended loop group. The matrix elements M^Λ_{kl} of the monodromy matrix in representation Λ also belong to H. The action of $Y^{M^\Lambda_{kl}}$ on $g^\Lambda_{ij}(x)$ and on M^Λ_{ij} can be written in tensorial form as

$$Y^{M^\Lambda_{kl}}\big(g^\Lambda_{ij}(x)\big) = \frac{1}{\kappa}\big(g(x)\otimes M\,\hat\Theta(M)\big)^\Lambda_{ik,jl}, \tag{3.31}$$

$$Y^{M^\Lambda_{kl}}(M^\Lambda_{ij}) = \frac{1}{\kappa}\big((M\otimes M)\hat\Delta(M)\big)^\Lambda_{ik,jl}, \tag{3.32}$$

---

[1] We also use the notation $\mathrm{Tr} = c_\Lambda \mathrm{tr}_\Lambda$, where tr_Λ is the trace over the representation Λ and c_Λ is a normalization factor that makes $c_\Lambda \mathrm{tr}(A^\Lambda B^\Lambda)$ independent of Λ for $A, B\in\mathcal{G}$.

where our tensor product notation is $(K \otimes L)_{ik,jl} = K_{ij} L_{kl}$, and

$$(3.33)\qquad \hat{\Theta}(M) = \hat{r}_+(M) - M_2^{-1}\hat{r}_-(M)M_2, \quad \hat{\Delta}(M) = \hat{\Theta}(M) - M_1^{-1}\hat{\Theta}(M)M_1$$

with $M_1 = M \otimes 1$, $M_2 = 1 \otimes M$.

We now wish to rewrite the above Hamiltonian vector fields in a symbolic notation of Poisson brackets. Recall that the PB of two smooth functions F_1 and F_2 on a *finite dimensional* smooth symplectic manifold is defined by

$$(3.34)\qquad \{F_1, F_2\} = Y^{F_2}(F_1) = -Y^{F_1}(F_2) = \Omega(Y^{F_2}, Y^{F_1}),$$

where Y^{F_i} is the Hamiltonian vector field associated with F_i by the symplectic form Ω. One may formally apply the same formula in the infinite dimensional case to the 'smooth enough' admissible functions. However, it is a non-trivial problem to precisely specify the set of functions that form a closed Poisson algebra. Setting this question aside, it is clear from (3.29) and (3.32) that the admissible functions of J and those of M will form two closed Poisson subalgebras that centralize each other. Furthermore, we may use the perfectly well-defined expression

$$(3.35)\qquad \{F_\chi, F_\phi\} := Y^{F_\phi}(F_\chi)$$

for the PB of two admissible Hamiltonians of type F in eq. (3.28) to define the ('distribution valued') PB of the evaluation functions $g(x)$ by the equality:

$$(3.36)\qquad \{F_\chi, F_\phi\} := \int_0^{2\pi}\int_0^{2\pi} dx\, dy\, \mathrm{Tr}_{12}(\chi(x)\otimes\phi(y)\{g^\Lambda(x) \overset{\otimes}{,} g^\Lambda(y)\}),$$

where Tr_{12} is the (normalized) trace over $V \otimes V$ and $\{g^\Lambda(x) \overset{\otimes}{,} g^\Lambda(y)\}_{ik,jl} = \{g^\Lambda_{ij}(x), g^\Lambda_{kl}(y)\}$. With these definitions, our explicit formula of the Hamiltonian vector field Y^{F_ϕ} in (3.30) is equivalent to the following quadratic 'exchange algebra' type PB for the chiral field $g(x)$:

$$(3.37)\qquad \{g^\Lambda(x) \overset{\otimes}{,} g^\Lambda(y)\} = \frac{1}{\kappa}\big(g^\Lambda(x)\otimes g^\Lambda(y)\big)\big(\hat{r}(M) + \frac{1}{2}\hat{I}\,\mathrm{sign}(y-x)\big)^\Lambda, \qquad 0 < x, y < 2\pi.$$

Proceeding in the same way with the $\{F_\phi, M^\Lambda_{kl}\}$ PB as we did with the $\{F_\chi, F_\phi\}$ one, we conclude that the right hand side of (3.31) should be interpreted as the expression of the $\{g^\Lambda_{ij}(x), M^\Lambda_{kl}\}$ PB, and similarly for (3.32).

It is an open question if the admissible Hamiltonians of type $\mathcal{F}_\mu$, F_ϕ and M^Λ_{kl} together generate a closed Poisson algebra. Leaving this for a future study, we here only remark that the Jacobi identity for three functions of type F_ϕ is in fact equivalent to the following equation for $\hat{r}(M)$:

$$(3.38)\qquad \big[\hat{r}_{12}(M), \hat{r}_{23}(M)\big] + \Theta_{\alpha\beta}(M)T_1^\alpha \mathcal{R}^\beta \hat{r}_{23}(M) + \text{cycl. perm.} = -\frac{1}{4}\hat{f},$$

where $\hat{f}$ is defined by

$$(3.39)\qquad \hat{f} := f^\gamma_{\alpha\beta} T^\alpha \otimes T^\beta \otimes T_\gamma, \quad [T_\alpha, T_\beta] = f^\gamma_{\alpha\beta} T_\gamma,$$

and the cyclic permutation is over the 3 tensorial factors with $\hat{r}_{23} = r^{\alpha\beta}(1 \otimes T_\alpha \otimes T_\beta)$, $T_1^\alpha = T^\alpha \otimes 1 \otimes 1$, and so on. We use the components of $\hat{\Theta} = \Theta_{\alpha\beta} T^\alpha \otimes T^\beta$ given by

(3.33), and the left-invariant differential operators $\mathcal{R}^\beta$ that act on a function ψ of M by

$$(\mathcal{R}^\beta\psi)(M) := \frac{d}{dt}\psi(Me^{tT^\beta})\Big|_{t=0}. \tag{3.40}$$

Eq. (3.38) can be viewed as a dynamical generalization of the classical modified Yang-Baxter equation, to which it reduces if the r-matrix is a monodromy independent constant. As a consequence of $d\Omega^\rho_{\rm chir} = 0$, (3.38) is satisfied for any $\hat{r}(M)$ given by (3.20).

What are the Poisson-Lie (PL) symmetries of the chiral WZNW phase space? To make this question more definite, we equip the group $G = \{h\}$ with a PL structure by means of the Sklyanin bracket

$$\{h \stackrel{\otimes}{,} h\}_{\widehat{R}} = \frac{1}{\kappa}[h\otimes h, \widehat{R}], \tag{3.41}$$

where $\widehat{R} = R^{\alpha\beta}T_\alpha \otimes T_\beta \in \mathcal{G}\wedge\mathcal{G}$ is a *constant* r-matrix satisfying

$$[\widehat{R}_{12}, \widehat{R}_{23}] + \text{cycl. perm.} = -\nu^2 \hat{f} \tag{3.42}$$

for some constant ν. We then seek the conditions on $\hat{r}(M)$ and $\widehat{R}$ that guarantee the standard right action[2] of G on $\check{\mathcal{M}}_{\rm chir}$,

$$\check{\mathcal{M}}_{\rm chir} \times G \ni (g,h) \mapsto gh \in \check{\mathcal{M}}_{\rm chir}, \tag{3.43}$$

to be a PL action. This leads to the requirement

$$\hat{r}(h^{-1}Mh) - \widehat{R} = (h\otimes h)^{-1}\big(\hat{r}(M) - \widehat{R}\big)(h\otimes h), \tag{3.44}$$

i.e., right multiplication is a PL symmetry if and only if the exchange r-matrix $\hat{r}(M)$ is such a solution of (3.38) that the difference $(\hat{r}(M) - \widehat{R})$ is equivariant. We can provide such solutions explicitly in association with *any* given solution of (3.42). These solutions are obtained by assuming the validity of the exponential parametrization for $M \in \check{G}$:

$$M := e^{2\pi\Gamma} \text{ for } \Gamma \in \check{\mathcal{G}} \subset \mathcal{G} \quad\text{and}\quad \mathcal{Y} := 2\pi(\operatorname{ad}\Gamma). \tag{3.45}$$

Any odd analytic function of $\mathcal{Y}$ is equivariant, and it is possible to prove [**8**] that the r-matrix corresponding to the linear operator

$$r(M) = \frac{1}{2}\coth\frac{\mathcal{Y}}{2} - \nu\coth(\nu\mathcal{Y}) + R \tag{3.46}$$

solves both (3.44) and (3.38) (on the domain where its power series converges).

We end this section with some remarks on the above formula. First, note that for $\nu = 0$ (3.46) is understood as the limit of the corresponding complex analytic function. Thus for $\nu = 0$ and $R = 0$ it yields $r_0 = \frac{1}{2}\coth\frac{\mathcal{Y}}{2} - \frac{1}{\mathcal{Y}}$. If the PB (3.37) on $\check{\mathcal{M}}_{\rm chir}$ is defined by r_0, then (3.43) is a *classical* G*-symmetry.* Second, if $\nu = 1/2$ then $r = R$, which is the case of the constant exchange r-matrices [**20**]. Third, it is worth stressing that for a compact Lie algebra $\mathcal{G}$ constant exchange r-matrices do not exist, because of the negative sign on the right hand side of (3.38), but our formula (3.46) gives explicit solutions of (3.38) also in this case using a purely imaginary ν in (3.42). Finally, we remark that in the $\nu = 1/2$ case the construction of the 2-form ρ that corresponds to the r-matrix in (3.46) is presented in [**20**], while

[2]Since $M \mapsto h^{-1}Mh$, we here have to assume that $\check{G} \subset G$ is invariant under the adjoint action of G, or should restrict our attention to the corresponding $\mathcal{G}$-action.

in general a suitable local 2-form can be obtained by solving (3.20) for q. Further comments are contained in [**8**].

4. Hamiltonian Structures for Diagonal Monodromy

We now describe the Hamiltonian structure that results by restricting the symplectic form $\Omega^\rho_{\rm chir}$ to a submanifold $\mathcal{M}_{\rm Bloch} \subset \mathcal{M}_{\rm chir}$ consisting of chiral WZNW fields with diagonal monodromy. The corresponding exchange algebra PB turns out to contain the classical dynamical r-matrix (4.32).

In this section, let $\mathcal{G}$ be either a complex simple Lie algebra or its normal real form, and G a corresponding Lie group. Choose a Cartan subalgebra $\mathcal{H} \subset \mathcal{G}$ that admits the root space decomposition

$$\mathcal{G} = \mathcal{H} \oplus \sum_{\alpha\in\Phi} \mathcal{G}_\alpha, \tag{4.1}$$

and an associated basis $H_k \in \mathcal{H}$, $E_\alpha \in \mathcal{G}_\alpha$ normalized by ${\rm Tr}(E_\alpha E_{-\alpha}) = 2/|\alpha|^2$. By using this basis any $A \in \mathcal{G}$ can be decomposed as

$$A = A^0 + A^r \quad \text{with } A^0 \in \mathcal{H},\ A^r = \sum_{\alpha\in\Phi} E_\alpha\, {\rm Tr}(E^\alpha A),\ E^\alpha := \frac{1}{2}|\alpha|^2 E_{-\alpha}. \tag{4.2}$$

Fix an open domain $\mathcal{A} \subset \mathcal{H}$ which has the properties that $\alpha(\omega) \notin i2\pi\mathbf{Z}$ for any root, $\alpha \in \Phi \subset \mathcal{H}^*$, and the map $\mathcal{A} \ni \omega \mapsto e^\omega \in G$ is injective.

Then define $\mathcal{M}_{\rm Bloch} \subset \mathcal{M}_{\rm chir}$ by

$$\mathcal{M}_{\rm Bloch} := \{b \in C^\infty(\mathbf{R}, G) \mid b(x+2\pi) = b(x)e^\omega, \quad \omega \in \mathcal{A} \subset \mathcal{H}\}. \tag{4.3}$$

Let $\mathcal{M}_{\rm Bloch}$ be equipped with the 2-form $\kappa\Omega^{\rho_B}_{\rm Bloch}$, where

$$\Omega^{\rho_B}_{\rm Bloch}(b) := -\frac{1}{2}\int_0^{2\pi} dx\ {\rm Tr}(b^{-1}db) \wedge (b^{-1}db)' - \frac{1}{2}{\rm Tr}\big((b^{-1}db)(0) \wedge d\omega\big) + \rho_B(\omega) \tag{4.4}$$

with an arbitrary *closed* 2-form ρ_B on $\mathcal{A}$. Clearly, $\Omega^{\rho_B}_{\rm Bloch}$ could be obtained from $\Omega^\rho_{\rm chir}$ (2.21) upon imposing the constraint $M = e^\omega$. Now ρ_B is parametrized as

$$\rho_B(\omega) = \frac{1}{2} q_B^{kl}(\omega)\, {\rm Tr}(H_k d\omega) \wedge {\rm Tr}(H_l d\omega), \quad q_B^{kl} = -q_B^{lk}, \tag{4.5}$$

and a corresponding linear operator $q_B(\omega)$ on $\mathcal{H}$ is defined by

$$q_B(\omega)(C) = H_k q_B^{kl}(\omega)\, {\rm Tr}(H_l C) \quad \forall C \in \mathcal{H}. \tag{4.6}$$

To show that $\Omega^{\rho_B}_{\rm Bloch}$ is symplectic, it will be convenient to parametrize $b \in \mathcal{M}_{\rm Bloch}$ as

$$b(x) = h(x)\exp(x\bar\omega), \quad \bar\omega := \frac{\omega}{2\pi}, \tag{4.7}$$

where $\omega \in \mathcal{A}$ and $h \in \widetilde{G}$. This one-to-one parametrization yields the identification

$$\mathcal{M}_{\rm Bloch} = \widetilde{G} \times \mathcal{A} = \{(h, \omega)\}. \tag{4.8}$$

Correspondingly, a vector field X on $\mathcal{M}_{\rm Bloch}$ is parametrized by

$$X = (X_h, X_\omega) \quad X_h \in T_h\widetilde{G},\ X_\omega \in T_\omega\mathcal{A} \simeq \mathcal{H} \tag{4.9}$$

with $h^{-1}X_h \in T_e\widetilde{G} \simeq \widetilde{\mathcal{G}}$. By regarding ω and h as evaluation functions on $\mathcal{M}_{\rm Bloch}$, we may write $X_\omega = X(\omega)$ and $X_h(x) = X\big(h(x)\big)$. Equivalently, X can be characterized by its action on $b(x)$,

$$b^{-1}(x)X\big(b(x)\big) = e^{-x\bar\omega}h^{-1}(x)X\big(h(x)\big)e^{x\bar\omega} + xX(\bar\omega), \tag{4.10}$$

where the function $b^{-1}(x)X\big(b(x)\big)$ on $\mathbf{R}$ is uniquely determined by its restriction to $[0,2\pi]$. Naturally, the derivative $X(F)$ of a function F on $\mathcal{M}_{\rm Bloch}$ is defined by using that any vector is the velocity to a smooth curve. That is, if the value of the vector field X at $b \in \mathcal{M}_{\rm Bloch}$ coincides with the velocity to the curve $\gamma(x,t)$ at $t=0$, $\gamma(x,0) = b(x)$, then for a differentiable function F we have $X(F)[b] = \frac{\rm d}{{\rm d}t}F[\gamma(x,t)]|_{t=0}$.

Arguing similarly to Section 3, it can be shown that $\Omega^{\rho_B}_{\rm Bloch}$ is weakly nondegenerate on $\mathcal{M}_{\rm Bloch}$ for any ρ_B. The *admissible* Hamiltonians that possess Hamiltonian vector fields now turn out to be those functions F on $\mathcal{M}_{\rm Bloch}$ whose derivative with respect to any vector field X exists and has the form

$$X(F) = \langle dF, X\rangle = {\rm Tr}(d_\omega F X_\omega) + \int_0^{2\pi} dx\ {\rm Tr}\big((h^{-1}d_hF)(h^{-1}X_h)\big) \tag{4.11}$$

where

$$dF = (d_hF, d_\omega F) \quad \text{with } d_hF \in T_h^*\widetilde{G},\ d_\omega F \in T^*_\omega\mathcal{A} \tag{4.12}$$

is the exterior derivative of F. We here identify $T^*_\omega\mathcal{A}$ with $\mathcal{H}$ by means of the scalar product Tr and also identify $T^*_e\widetilde{G}$ with $\widetilde{\mathcal{G}}$ by the scalar product $\int_0^{2\pi}{\rm Tr}(\cdot,\cdot)$, whereby we have $h^{-1}d_hF \in T^*_e\widetilde{G} = \widetilde{\mathcal{G}}$. It is clear that the local evaluation functions $\mathcal{M}_{\rm Bloch} \ni b \mapsto b^\Lambda_{kl}(x)$ (with the matrix elements b^Λ_{kl} taken in some representation Λ of G) are differentiable but not admissible, while e.g. the Fourier coefficients of the components of $J = \kappa b'b^{-1}$ as well as the components of ω yield admissible Hamiltonians.

Next we prove that $\kappa\Omega^{\rho_B}_{\rm Bloch}$ indeed permits to associate a unique Hamiltonian vector field, Y^F, with any Hamiltonian, F, subject to (4.11), (4.12). By definition, Y^F must satisfy

$$\langle dF, X\rangle = X(F) = \kappa\Omega^{\rho_B}_{\rm Bloch}(X, Y^F) \tag{4.13}$$

for any vector field X. To determine Y^F, we first point out that in terms of (h,ω)

$$\begin{aligned}\Omega^{\rho_B}_{\rm Bloch}(h,\omega) = -\frac{1}{2}\int_0^{2\pi} dx\ {\rm Tr}\big((h^{-1}dh)\wedge(h^{-1}dh)'\\ + 2\bar\omega(h^{-1}dh)\wedge(h^{-1}dh) - 2d\bar\omega\wedge h^{-1}dh\big) + \rho_B(\omega).\end{aligned} \tag{4.14}$$

By equating the coefficients of $h^{-1}X(h)$ and $X(\omega)$ on the two sides of (4.13), we obtain the following equations for Y^F:

$$2\pi\big(h^{-1}Y^F(h)\big)' + [h^{-1}Y^F(h),\omega] + Y^F(\omega) = -\frac{2\pi}{\kappa}h^{-1}d_hF \tag{4.15}$$

$$2\pi q_B(\omega)\big(Y^F(\omega)\big) + \int_0^{2\pi} dx\ \big(h^{-1}Y^F(h)\big)^0 = \frac{2\pi}{\kappa}d_\omega F. \tag{4.16}$$

Given d_hF and $d_\omega F$, we will determine $b^{-1}Y^F(b)$, which is equivalent to finding $h^{-1}Y^F(h)$ and $Y^F(\omega)$.

On account of (4.10), (4.15) is in fact equivalent to

$$(b^{-1}Y^F(b))'(x) = -\frac{1}{\kappa}e^{-\bar{\omega}x}(h^{-1}d_hF)(x)e^{\bar{\omega}x}, \tag{4.17}$$

whose solution is

$$b^{-1}(x)Y^F\big(b(x)\big) = b^{-1}(0)Y^F\big(b(0)\big) - \frac{1}{\kappa}\int_0^x dy\, e^{-\bar{\omega}y}(h^{-1}d_hF)(y)e^{\bar{\omega}y}. \tag{4.18}$$

Hence the only non-trivial problem is to determine the initial value

$$Q_F := b^{-1}(0)Y^F\big(b(0)\big) = h^{-1}(0)Y^F\big(h(0)\big). \tag{4.19}$$

To this end, note from (4.10) that

$$Y^F(\omega) = e^{\omega}b^{-1}(2\pi)Y^F\big(b(2\pi)\big)e^{-\omega} - b^{-1}(0)Y^F\big(b(0)\big). \tag{4.20}$$

By using (4.18), the Cartan part of (4.20) requires that

$$Y^F(\omega) = -\frac{1}{\kappa}\int_0^{2\pi} dx\,(h^{-1}d_hF)^0(x), \tag{4.21}$$

while the root part of (4.20) gives

$$e^{-\omega}Q_F^r e^{\omega} - Q_F^r = -\frac{1}{\kappa}\int_0^{2\pi} dx\, e^{-\bar{\omega}x}(h^{-1}d_hF)^r(x)e^{\bar{\omega}x}, \tag{4.22}$$

where $Q_F = Q_F^0 + Q_F^r$ according to (4.2). Then (4.22) completely determines Q_F^r as

$$Q_F^r = \frac{1}{\kappa}\sum_{\alpha\in\Phi}\frac{E_\alpha}{1-e^{-\alpha(\omega)}}\int_0^{2\pi} dx\, e^{-\alpha(\bar{\omega})x}\,\mathrm{Tr}\big((h^{-1}d_hF)(x)E^\alpha\big). \tag{4.23}$$

As for the remaining unknown, Q_F^0, (4.16) with (4.10) and (4.20) leads to the result:

$$\begin{aligned} 2\pi\kappa Q_F^0 = 2\pi d_\omega F &+ (2\pi q_B(\omega) - \pi I)\Big(\int_0^{2\pi} dx\,(h^{-1}d_hF)^0(x)\Big) \\ &+ \int_0^{2\pi} dx \int_0^x dy\,(h^{-1}d_hF)^0(y). \end{aligned} \tag{4.24}$$

In conclusion, we have found that the Hamiltonian vector field $b^{-1}Y^F(b)$ is uniquely determined and is explicitly given by (4.18) with $b^{-1}(0)Y^F\big(b(0)\big) = Q_F$ in (4.23), (4.24). In the derivation of Y^F we have crucially used that ω is restricted to the domain $\mathcal{A}\subset\mathcal{H}$. At the excluded points of $\mathcal{H}$ some denominators in (4.23) may vanish, whereby $\Omega^{\rho_B}_{\mathrm{Bloch}}$ becomes singular.

The Poisson bracket of two 'smooth enough' admissible Hamiltonians F_1 and F_2 on $\mathcal{M}_{\mathrm{Bloch}}$ is determined by the formula $\{F_1,F_2\} = \kappa\Omega^{\rho_B}_{\mathrm{Bloch}}(Y^{F_2},Y^{F_1})$ and now we extract a 'classical exchange algebra' from this formula. Analogously to the previous section, for this we consider functions of the form

$$F_\phi(h,\omega) = \int_0^{2\pi} dx\,\mathrm{Tr}\big(\phi(x)b^\Lambda(x)\big), \tag{4.25}$$

where $b^\Lambda(x)$ is taken in a representation Λ of G and $\phi(x)$ is a smooth, matrix valued, smearing-function in that representation. It is easy to check that F_ϕ is admissible if

$$\phi^{(k)}(0) = \phi^{(k)}(2\pi) = 0 \quad \forall k = 0,1,2\ldots, \tag{4.26}$$

and the exterior derivative of F_ϕ at (h,ω) is given by

$$(d_\omega F_\phi)(h,\omega) = \frac{1}{2\pi}\sum_k H^k \operatorname{Tr}\Big(H_k^\Lambda \int_0^{2\pi} dx\,\big(x\phi(x)b^\Lambda(x)\big)\Big), \tag{4.27}$$

$$\big((h^{-1}d_h F_\phi)(h,\omega)\big)(x) = \sum_a T^a \operatorname{Tr}\big(\phi(x)h^\Lambda(x)T_a^\Lambda e^{x\bar\omega^\Lambda}\big) \quad \text{for } x\in[0,2\pi]. \tag{4.28}$$

We here denote by H_k, H^k and T_a, T^a dual bases of $\mathcal{H}$ and $\mathcal{G}$, respectively. The last formula extends to a smooth 2π-periodic function on the real line precisely if (4.26) is satisfied. The Hamiltonian vector field Y^{F_ϕ} is then found to be

$$\begin{aligned}(b^{-1}Y^{F_\phi}(b))(x)& \\ &= Q_{F_\phi} - \frac{1}{\kappa}\sum_a T^a \int_0^x dy\, \operatorname{Tr}(\phi(y)b^\Lambda(y)T_a^\Lambda), \quad \text{for } x\in[0,2\pi],\end{aligned} \tag{4.29}$$

where Q_{F_ϕ} is determined as described above. By combining the preceding formulae, one can verify that

$$\begin{aligned}\{F_\chi, F_\phi\} &= \kappa\Omega^{\rho_B}_{\text{Bloch}}(Y^{F_\phi}, Y^{F_\chi}) \\ &= \int_0^{2\pi}\int_0^{2\pi} dxdy\, \operatorname{Tr}_{12}(\chi(x)\otimes\phi(y)\{b^\Lambda(x) \overset{\otimes}{,} b^\Lambda(y)\})\end{aligned} \tag{4.30}$$

holds for any ϕ, χ subject to (4.26) *provided that one has*

$$\begin{aligned}\{b^\Lambda(x) \overset{\otimes}{,} b^\Lambda(y)\}& \\ &= \frac{1}{\kappa}(b^\Lambda(x)\otimes b^\Lambda(y))\big(\widehat{\mathcal{R}}(\omega) + \frac{1}{2}\hat{I}\,\operatorname{sign}(y-x)\big)^\Lambda, \quad 0<x,y<2\pi\end{aligned} \tag{4.31}$$

with the dynamical r-matrix

$$\widehat{\mathcal{R}}(\omega) = \frac{1}{4}\sum_{\alpha\in\Phi}|\alpha|^2\coth\Big(\frac{1}{2}\alpha(\omega)\Big)E_\alpha\otimes E_{-\alpha} + \sum_{kl} q_B^{kl}(\omega)H_k\otimes H_l. \tag{4.32}$$

The local formula (4.31) completely encodes the Poisson brackets on $\mathcal{M}_{\text{Bloch}}$ since Y^{F_ϕ} can be recovered if the right hand side of (4.30) is given.

The Hamiltonian vector fields associated with $\omega_k := \operatorname{Tr}(\omega H_k)$ and with the functions $\mathcal{F}_\mu$ of J (see (3.28)) can be checked to be

$$Y^{\omega_k}\big(b(x)\big) = \frac{1}{\kappa}b(x)H_k, \quad Y^{\mathcal{F}_\mu}\big(b(x)\big) = \mu(x)b(x). \tag{4.33}$$

Thus J generates an action of the affine Kac-Moody algebra on $\mathcal{M}_{\text{Bloch}}$ centralized by the action of $\mathcal{H}$ generated by ω.

The dynamical r-matrix (4.32) is antisymmetric, and is neutral in the sense that

$$[H_k\otimes 1 + 1\otimes H_k, \widehat{\mathcal{R}}(\omega)] = 0, \tag{4.34}$$

which ensures the validity of the Jacobi identity for the three functions F_ϕ, F_χ, ω_k. Moreover, it satisfies the equation

$$[\widehat{\mathcal{R}}_{12}(\omega), \widehat{\mathcal{R}}_{23}(\omega)] + \sum_k H_1^k \frac{\partial}{\partial\omega^k}\widehat{\mathcal{R}}_{23}(\omega) + \text{cycl. perm.} = -\frac{1}{4}\hat{f} \tag{4.35}$$

that ensures the Jacobi identity for three functions of type F_ϕ. This dynamical generalization of the modified classical Yang-Baxter equation arises in other contexts as well [**25, 22, 4**] and has been much studied recently [**18, 29, 30**]. The exchange r-matrix (4.32) of the chiral WZNW Bloch waves was first obtained in [**7**], where it was also shown that it satisfies (4.35).

5. Poisson-Lie Groupoids from Chiral WZNW

It is well known [**15**] that one can associate a Poisson-Lie (PL) group with any antisymmetric solution of the modified classical Yang-Baxter equation. Remarkably, the dynamical generalizations of this equation that arise in the WZNW model permit analogous interpretations in terms of PL groupoids. In particular, this means that one can associate a PL groupoid with any chiral extension of the WZNW phase space. Below we briefly describe these groupoids, which are finite dimensional Poisson manifolds that encode the (non Kac-Moody aspects of the) infinite dimensional chiral WZNW PBs.

Roughly speaking, a groupoid is a set, say P, endowed with a 'partial multiplication' that behaves similarly to a group multiplication in the cases when it can be performed (see e.g. [**31**]). In the cases of our interest $P = S\times G\times S = \{(M^F, g, M^I)\}$, where G is a group and S is some set. The partial multiplication is defined for those triples (M^F, g, M^I) and $(\overline{M}^F, \bar{g}, \overline{M}^I)$ for which $M^I = \overline{M}^F$, and the product is

$$(M^F, g, M^I)(\overline{M}^F, \bar{g}, \overline{M}^I) := (M^F, g\bar{g}, \overline{M}^I) \quad \text{for } M^I = \overline{M}^F. \tag{5.1}$$

Thus the graph of the partial multiplication is the subset of

$$P\times P\times P = \{(M^F, g, M^I)\}\times\{(\overline{M}^F, \bar{g}, \overline{M}^I)\}\times\{(\widehat{M}^F, \hat{g}, \widehat{M}^I)\} \tag{5.2}$$

defined by the constraints

$$M^I = \overline{M}^F, \quad \widehat{M}^F = M^F, \quad \widehat{M}^I = \overline{M}^I, \quad \hat{g} = g\bar{g}, \tag{5.3}$$

where the hatted triple encodes the components of the product. A PL groupoid [**32**] is a (Lie) groupoid P, which is also a Poisson manifold in such a way that the graph of the partial multiplication is a *coisotropic* submanifold of $P\times P\times P^-$, where P^- denotes the manifold P endowed with the opposite of the PB on P. To put it differently, *the constraints that define the graph are first class.* (For $P = S\times G\times S$, this definition reduces to that of a PL group if S consists of a single point.)

Let us first recall the definition of the PL groupoids that are related to the Poisson structures on $\mathcal{M}_{\rm Bloch}$. In a more general context, these groupoids have been introduced in [**18**]. They are of the form above, where S is a domain in the dual of a Cartan subalgebra $\mathcal{H}$ of a simple Lie group G. We now identify $\mathcal{H}^*$ with $\mathcal{H}$ and take the domain to be $\mathcal{A}\subset\mathcal{H}$ considered in Section 4. For notational convenience, we further identify S with $\exp(\mathcal{A})\subset\exp(\mathcal{H})$, and denote the components of the corresponding triples as $M^I = \exp(\omega^I)$ and $M^F = \exp(\omega^F)$ for $\omega^F, \omega^I\in\mathcal{A}$. Using the standard tensorial notation and a basis H_i of $\mathcal{H}$, we then define a Poisson structure on $P_{\rm Bloch} := \mathcal{A}\times G\times\mathcal{A} = \{(\omega^F, g, \omega^I)\}$ as follows:

$$\begin{aligned}
\kappa\{g_1, g_2\}_{P_{\rm Bloch}} &= g_1 g_2 \widehat{\mathcal{R}}(\omega^I) - \widehat{\mathcal{R}}(\omega^F) g_1 g_2\\
\kappa\{g, \omega_i^I\}_{P_{\rm Bloch}} &= g H_i\\
\kappa\{g, \omega_i^F\}_{P_{\rm Bloch}} &= H_i g\\
\{\omega_i^I, \omega_j^I\}_{P_{\rm Bloch}} &= \{\omega_i^F, \omega_j^F\}_{P_{\rm Bloch}} = \{\omega_i^I, \omega_j^F\}_{P_{\rm Bloch}} = 0,
\end{aligned} \tag{5.4}$$

where κ is a constant included for comparison purposes and $\omega_i^{I(F)} = \mathrm{Tr}(H_i \omega^{I(F)})$. Equations (4.34) and (4.35) for $\widehat{\mathcal{R}}(\omega)$ are sufficient for the Jacobi identities of this PB to be satisfied [**18**]. One can also check that the graph of the partial multiplication is coisotropic.

In some sense, the PBs (5.4) on P_{Bloch} correspond to the PBs on the chiral WZNW phase space $\mathcal{M}_{\mathrm{Bloch}}$. Motivated by this, we now define a PL groupoid which is related to an arbitrary chiral extension of the WZNW phase space with generic monodromy. In this case, using an open domain $\check{G} \subset G$, we take P to be

$$P = \check{G} \times G \times \check{G}, \tag{5.5}$$

and postulate on it a PB $\{\ ,\ \}_P$ as follows:

$$\begin{aligned}
\kappa\{g_1, g_2\}_P &= g_1 g_2 \hat{r}(M^I) - \hat{r}(M^F) g_1 g_2 \\
\kappa\{g_1, M_2^I\}_P &= g_1 M_2^I \hat{\Theta}(M^I) \\
\kappa\{g_1, M_2^F\}_P &= M_2^F \hat{\Theta}(M^F) g_1 \\
\kappa\{M_1^I, M_2^I\}_P &= M_1^I M_2^I \hat{\Delta}(M^I) \\
\kappa\{M_1^F, M_2^F\}_P &= -M_1^F M_2^F \hat{\Delta}(M^F) \\
\kappa\{M_1^I, M_2^F\}_P &= 0.
\end{aligned} \tag{5.6}$$

It is easy to verify that a PB given by the ansatz (5.6) always yields a PL groupoid, since the constraints in (5.3) are first class for any choice of the $\mathcal{G} \otimes \mathcal{G}$ valued 'structure functions' $\hat{r}$, $\hat{\Theta}$, $\hat{\Delta}$ on $\check{G}$. Of course, the structure functions must satisfy a system of equations for the ansatz (5.6) to define a PB. These equations are spelled out in [**8**]. The important point is that, in fact, a *sufficient condition* for the Jacobi identity is obtained by assuming that $\hat{\Theta}(M)$ and $\hat{\Delta}(M)$ are given by (3.33) in terms of an antisymmetric solution $\hat{r}(M)$ of (3.38).

We have extracted a PL groupoid from any chiral extension of the WZNW phase space with generic monodromy by taking the triple $\hat{r}$, $\hat{\Theta}$, $\hat{\Delta}$ that arises in the WZNW model to be the structure functions of $\{\ ,\ \}_P$. It should be noticed that if $\hat{r}$ is non-dynamical, then the PL groupoid P carries the same information as the group G endowed with the corresponding Sklyanin bracket. Among our PL groupoids there are also those special cases for which $(\hat{r} - \widehat{R})$ satisfies the equivariance condition (3.44) in relation with an arbitrary constant r-matrix $\widehat{R}$ subject to (3.42). In these cases, it is possible to define commuting left and right PL actions of the group G (endowed with the PB (3.41)) on P, reflecting the corresponding PL symmetry (3.43) on the chiral WZNW phase space.

We believe that it would be interesting to study the above introduced PL groupoids further, for example to understand their quantization and relate them to the quantized (chiral) WZNW conformal field theory.

Acknowledgments

L. F. wishes to thank J. Harnad for support and hospitality in Montreal.

References

1. R. Abraham and J. Marsden, *Foundations of mechanics*, 2nd ed., Addison-Wesley, 1978.
2. A. Alekseev and S. Shatashvili, *Quantum groups and WZNW models*, Comm. Math. Phys. **133** (1990), 353–368.

3. A. Yu. Alekseev and I. T. Todorov, *Quadratic brackets from symplectic forms*, Nuclear Phys. B **421** (1994), 413–428.
4. J. Avan, O. Babelon, and E. Billey, *The Gervais-Neveu-Felder equation and the quantum Calogero-Moser systems*, Comm. Math. Phys. **178** (1996), 281–299.
5. O. Babelon, *Extended conformal algebra and the Yang-Baxter equation*, Phys. Lett. B **215** (1988), 523–529.
6. O. Babelon, F. Toppan, and L. Bonora, *Exchange algebra and the Drinfeld-Sokolov theorem*, Comm. Math. Phys. **140** (1991), 93–117.
7. J. Balog, L. Dąbrowski, and L. Fehér, *Classical r-matrix and exchange algebra in WZNW and Toda theories*, Phys. Lett. B **244** (1990), 227–234.
8. J. Balog, L. Fehér, and L. Palla, *Chiral extensions of the WZNW phase space, Poisson-Lie symmetries and groupoids*, Nuclear Phys. B **568** (2000), 501–540.
9. J. Balog, L. Fehér, and L. Palla, *Classical Wakimoto realizations of chiral WZNW Bloch waves*, J. Phys. A **33** (2000), 945–956.
10. B. Blok, *Classical exchange algebras in the Wess-Zumino-Witten model*, Phys. Lett. B **233** (1989), 359–362.
11. L. Caneschi and M. Lysiansky, *Chiral quantization of the WZW* SU(n) *model*, Nuclear Phys. B **505** (1997), 701–726.
12. M. Chu and P. Goddard, *Quantization of a particle moving on a group manifold*, Phys. Lett. B **337** (1994), 285–293; *Quantization of the* SU(N) *WZW model at level* k, Nuclear Phys. B **445** (1995), 145–168.
13. M. Chu, P. Goddard, I. Halliday, D. Olive, and A. Schwimmer, *Quantization of the Wess-Zumino-Witten model on a circle*, Phys. Lett. B **266** (1991), 71–81.
14. P. Di Francesco, P. Mathieu, and D. Sénéchal, *Conformal field theory*, Springer, New York, 1996.
15. V. G. Drinfeld, *Quantum groups*, Proc. Internat. Congress of Mathematicians (Berkeley, 1986), Amer. Math. Soc., Providence, RI, 1987, pp. 798–820.
16. P. Etingof, I. Frenkel, and A. Kirillov Jr., *Lectures on representation theory and Knizhnik-Zamolodchikov equations* (AMS, 1998).
17. P. Etingof and O. Schiffmann, *Lectures on the dynamical Yang-Baxter equations*, `math.QA/9908064`.
18. P. Etingof and A. Varchenko, *Geometry and classification of the classical dynamical Yang-Baxter equation*, Comm. Math. Phys. **192** (1998), 77–120.
19. L. Faddeev, *On the exchange matrix for WZNW model*, Comm. Math. Phys. **132** (1990), 131–138.
20. F. Falceto and K. Gawędzki, *Lattice Wess-Zumino-Witten model and quantum groups*, J. Geom. Phys. **11** (1993), 251–279.
21. V. A. Fateev and S. L. Lukyanov, *Poisson-Lie groups and classical W-algebras*, Internat. J. Modern Phys. A **7** (1992), 853–876.
22. G. Felder and C. Wieczerkowski, *Conformal blocks on elliptic curves and the Knizhnik-Zamolodchikov-Bernard equations*, Comm. Math. Phys. **176** (1996), 133–161; G. Felder, *Conformal field theory and integrable systems associated to elliptic curves*, Proc. Internat. Congress of Mathematicians (Zürich, 1994), Birkhäuser, Basel, 1995, pp. 1247–1255.
23. P. Furlan, L. K. Hadjiivanov, and I. T. Todorov, *Operator realization of the* SU(2) *WZW model*, Nuclear Phys. B **474** (1996), 497–511; *A quantum gauge group approach to the 2D* SU(n) *WZNW model*, Internat. J. Modern Phys. A **12** (1997), 23–32.
24. K. Gawędzki, *Classical origin of quantum group symmetries in Wess-Zumino-Witten conformal theory*, Comm. Math. Phys. **139** (1991), 201–213.
25. J.-L. Gervais and A. Neveu, *Novel triangle relation and absence of tachyons in Liouville string field theory*, Nuclear Phys. B **238** (1984), 125–141; E. Cremmer and J.-L. Gervais, *The quantum group structure associated with non-linearly extended Virasoro algebras*, Comm. Math. Phys. **134** (1990), 619–632.
26. C. Gómez, M. Ruiz-Altaba, and G. Sierra, *Quantum groups in two-dimensional physics*, Cambridge University Press, 1996, and references therein.
27. J. Harnad and B. A. Kupershmidt, *Sympletic geometries on* $T^*\tilde{G}$, *Hamiltonian group actions and integrable systems*, J. Geom. Phys. **16** (1995), 168–206.
28. T. Itoh and Y. Yamada, *Exchange relations in Wess-Zumino-Novikov-Witten model and quantum groups*, Prog. Theor. Phys. **85** (1991), 751–758.

29. Z. J. Liu and P. Xu, *Dirac structures and dynamical r-matrices*, math.DG/9903119.
30. J. H. Lu, *Classical dynamical r-matrices and homogeneous Poisson structures on G/H and K/T*, math/9909004.
31. K. Mackenzie, *Lie groupoids and Lie algebroids in differential geometry*, London Math. Soc. Lecture Note Ser., vol. 124, Cambridge University Press, 1987.
32. A. Weinstein, *Coisotropic calculus and Poisson groupoids*, J. Math. Soc. Japan **40** (1988), 705–727.
33. E. Witten, *Non-abelian bosonization in two dimensions*, Comm. Math. Phys. 92 (1984), 455–472.

Research Institute for Nuclear and Particle Physics, Hungarian Academy of Sciences, H-1525 Budapest 114, P.O.B. 49, Hungary
E-mail address: balog@rmki.kfki.hu

Institute for Theoretical Physics, József Attila University, H-6726 Szeged, Tisza Lajos krt 84–86, Hungary
E-mail address: lfeher@sol.cc.u-szeged.hu

Institute for Theoretical Physics, Roland Eötvös University, H-1117, Budapest, Pázmány P. sétány 1 A-ép, Hungary
E-mail address: palla@ludens.elte.hu

Centre de Recherches Mathématiques
CRM Proceedings and Lecture Notes
Volume **26**, 2000

Loop Groups, *R*-Matrices and Separation of Variables

J. Harnad

ABSTRACT. The method of linearization of integrable isospectral Hamiltonian flows generated by spectral invariants of a Lax matrix depending rationally on a spectral parameter using separation of variables in *spectral Darboux coordinates* is reviewed. A number of standard examples are given in which the flow may be linearized via abelian integrals by using the Liouville-Arnold procedure within such coordinates. The classical modified Yang-Baxter equations and associated linear and quadratic R-matrix structures are also reviewed, with emphasis on the "split" case (Adler-Kostant-Symes theorem), in the context of loop algebras and groups. The structure of the symplectic leaves for the linear case corresponding to rational coadjoint orbits of the loop algebra $\widetilde{\mathfrak{gl}}_R(r)$ is analyzed, and the linearization via abelian integrals for generic systems in $\widetilde{\mathfrak{gl}}(r)^*$ is deduced from the separation of variables in the associated spectral Darboux coordinates.

1. Isospectral Flow in Loop Algebras. I. Examples

1.1. Introduction. Lax Matrices and Spectral Curves. The general situation encountered in the isospectral approach to integrable finite dimensional systems is the following. We have a phase space $\mathcal{M}$, a Hamiltonian flow $f_t^H : p_0 \mapsto p(p_0, t)$, generated by the Hamiltonian H, and an $r \times r$ matrix $L(\lambda, p)$, depending on an auxiliary spectral parameter $\lambda \in \mathbf{C}$ and the moving point $p(t)$ in $\mathcal{M}$ in such a way that Hamilton's equations imply a Lax equation for L

$$\frac{dL}{dt} = [A, L]. \tag{1.1}$$

This in turn implies that the time dependence of L along the integral curves is given by

$$L\big(\lambda, p(t)\big) = G(\lambda, t) L_0 G(\lambda, t)^{-1}, \tag{1.2}$$

for some invertible matrix $G(\lambda, t)$, and hence the flow is isospectral for L.

The Hamiltonian H inducing the Lax equation (1.1) is usually itself a spectral invariant of L. The infinitesimal isospectral deformation matrix A also depends on λ and p, with its p dependence determined through $L(\lambda, p)$ in some particular

2000 *Mathematics Subject Classification.* 37K15, 37K20, 37K30, 37J35.

Research supported in part by the Natural Sciences and Engineering Research Council of Canada and the Fonds FCAR du Québec.

This is the final form of the paper.

way, generally governed by the R-matrix structure (which will form the subject of the second lecture). In the cases covered by the standard classical R-matrix theory [**OPRS, FT**], L is usually either a rational, trigonometric or elliptic function of λ, and the ring of spectral invariant polynomial functions of the matrix elements of L has a finite number of independent generators $\{H_1, \ldots, H_n\}$ which all Poisson commute. These may be identified as the nonconstant coefficients of the associated invariant spectral curve, defined by the characteristic equation

$$\det(L(\lambda, p) - z\mathbf{I}) = 0 \tag{1.3}$$

which, by construction, may be compactified to an algebraic curve $\mathcal{C}$ that in the rational case is an r-fold branched covering of the Riemann sphere and, in the more general elliptic case is a branched covering of the underlying elliptic curve. This also implies that all the spectral invariants, when viewed as Poisson commuting Hamiltonians, similarly generate isospectral deformations of L governed by Lax equations

$$\frac{dL}{dt_i} = [A_i, L], \tag{1.4}$$

for suitably defined A_i, where the t_i's are the flow parameters associated respectively to the H_i's and all these flows commute.

In sufficiently generic situations, given the R-matrix structure, these invariants turn out to form a maximal Poisson commuting set, and the isospectral flows they generate are completely integrable Hamiltonian systems in the Liouville sense. What also is typical in these systems is that, under a suitable complexification of the phase space, the Liouville-Arnold tori, defined by the level sets $\{H_i = c_i\}$ of the spectral invariants, are identifiable with the Jacobian variety $\mathcal{J}(\mathcal{C})$ of the associated spectral curve [**RS1, RS2, AvM1, AvM2, DKN, AHP, AHH3, AHH4, H1**].

In the following sections, we illustrate these results by four of the best known examples of integrable systems fitting into the isospectral, R-matrix framework: the Neumann-Rosochatius system of constrained harmonic oscillators on spheres; the classical Gaudin spin system (or isospectral $\mathfrak{su}(2)$ Garnier system); the periodic Toda lattice and the quasiperiodic solutions to the cubically nonlinear Schrödinger equation.

1.2. Example 1. Neumann-Rosochatius System [N, Ro, F2, AHP, AHH3, AHH4, AHH5, H1]. In this case, the phase space is the cotangent bundle of the unit $n-1$-sphere in $\mathbf{R}^n$ (identified via the metric with the tangent bundle):

$$\mathcal{M} = T^*S^{n-1} = \{(\mathbf{x}, \mathbf{y}) \in \mathbf{R}^{2n} |\ \mathbf{x}^T\mathbf{x} = 1,\ \mathbf{y}^T\mathbf{x} = 0\} \subset \mathbf{R}^n \times \mathbf{R}^n, \tag{1.5}$$

with symplectic form given by the restriction of the canonical symplectic form on $\mathbf{R}^{2n}$,

$$\omega = (d\mathbf{x}^T \wedge d\mathbf{y})|_{T^*S^{n-1}}. \tag{1.6}$$

The Rosochatius Hamiltonian is

$$\begin{aligned} H(\mathbf{x}, \mathbf{y}) &= \frac{1}{2}(\mathbf{y}^T\mathbf{y} + \mathbf{x}^T A\mathbf{x} + \boldsymbol{\xi}^T\boldsymbol{\mu}^2\boldsymbol{\xi}) \\ &= \frac{1}{2}\sum_{i=1}^{n}\left(y_i^2 + \alpha_i x_i^2 + \frac{\mu_i^2}{x_i^2}\right), \end{aligned} \tag{1.7}$$

where $\{x_i, y_i\}_{i=1\dots n}$ are the Cartesian components of $(\mathbf{x}, \mathbf{y})$, A and μ are the real, diagonal $n \times n$ matrices $A = \mathrm{diag}(\alpha_1, \dots, \alpha_n)$, $\boldsymbol{\mu} = \mathrm{diag}(\mu_1, \dots, \mu_n)$ with the α_i's distinct, and $\boldsymbol{\xi}$ is the column vector with components $\{1/x_i\}_{i=1\dots n}$. (The Neumann oscillator system is the particular case when all the μ_i's vanish.) Equivalently, we may choose the Hamiltonian as

$$\phi(\mathbf{x}, \mathbf{y}) := \frac{1}{2}\big((\mathbf{x}^T\mathbf{x})(\mathbf{y}^T\mathbf{y}) - (\mathbf{x}^T\mathbf{y})^2 + \mathbf{x}^T A\mathbf{x} + (\boldsymbol{\xi}^T\boldsymbol{\mu}^2\boldsymbol{\xi})(\mathbf{x}^T\mathbf{x})\big), \tag{1.8}$$

subject to the constraints

$$\mathbf{x}^T\mathbf{x} = 1, \qquad \mathbf{y}^T\mathbf{x} = 0. \tag{1.9}$$

The Lax matrix for this system may be chosen as

$$L(\lambda) := a(\lambda)\mathcal{N}(\lambda), \qquad a(\lambda) := \prod_{i=1}^{n}(\lambda - \alpha_i) \tag{1.10}$$

where

$$\mathcal{N}(\lambda) := \begin{pmatrix} h(\lambda) + ik(\lambda) & e(\lambda) \\ f(\lambda) & -h(\lambda) + ik(\lambda) \end{pmatrix}, \tag{1.11}$$

with

$$h(\lambda) := -\frac{1}{2}\sum_{i=1}^{n}\frac{x_i y_i}{\lambda - \alpha_i} \tag{1.12a}$$

$$e(\lambda) := -\frac{1}{2} - \frac{1}{2}\sum_{i=1}^{n}\frac{y_i^2 + \mu_i^2/x_i^2}{\lambda - \alpha_i} \tag{1.12b}$$

$$f(\lambda) := \frac{1}{2}\sum_{i=1}^{n}\frac{x_i^2}{\lambda - \alpha_i} \tag{1.12c}$$

$$k(\lambda) := \frac{1}{2}\sum_{i=1}^{n}\frac{\mu_i}{\lambda - \alpha_i}. \tag{1.12d}$$

The spectral curve in this case is hyperelliptic, generically of genus $g = n - 1$, and defined by

$$z^2 = -a(\lambda)\mathcal{P}(\lambda) - \mathcal{K}(\lambda) \tag{1.13}$$

where z is related to the eigenvalue ζ of $\mathcal{N}(\lambda)$ by

$$z = a(\lambda)\big(\zeta - ik(\lambda)\big), \tag{1.14}$$

$\mathcal{K}(\lambda)$ is the constant polynomial of degree $2n - 2$ defined by

$$\mathcal{K}(\lambda) := k^2(\lambda)a^2(\lambda), \tag{1.15}$$

and $\mathcal{P}(\lambda)$ is a polynomial of degree $n - 1$

$$\mathcal{P}(\lambda) = \sum_{i=0}^{n-1} P_i\lambda^i \tag{1.16}$$

defined by

$$\Delta(\lambda) := \det\mathcal{N}(\lambda) = -h^2(\lambda) - k^2(\lambda) - e(\lambda)f(\lambda) = \sum_{i=1}^{n}\frac{H_i}{\lambda - \alpha_i} := \frac{\mathcal{P}(\lambda)}{a(\lambda)}. \tag{1.17}$$

The leading coefficients of $\mathcal{P}(\lambda)$ are

$$P_{n-1} = \sum_{i=1}^{n} H_i = \frac{1}{4}\mathbf{x}^T\mathbf{x} = \frac{1}{4} \tag{1.18a}$$

$$P_{n-2} = \sum_{i=1}^{n} \alpha_i H_i - \sum_{i=1}^{n} \alpha_i P_{n-1} = \frac{1}{2}\phi(\mathbf{x},\mathbf{y}) - \frac{1}{4}\sum_{i=1}^{n} \alpha_i - \frac{1}{4}\sum_{\substack{i,j=1\\ j\neq i}}^{n} \mu_i\mu_j. \tag{1.18b}$$

The H_i's are defined as

$$H_i := \frac{1}{4}\sum_{\substack{j=1\\ j\neq i}}^{n} \frac{\left((x_iy_j - x_jy_i)^2 + \mu_i^2(x_j/x_i)^2 + \mu_j^2(x_i/x_j)^2 - 2\mu_i\mu_j\right)}{\alpha_i - \alpha_j} + \frac{1}{4}x_i^2, \tag{1.19}$$

and the Hamiltonian ϕ may be expressed in terms of these by

$$\phi(\mathbf{x},\mathbf{y}) = 2\sum_{i=1}^{n} \alpha_i H_i + \frac{1}{2}\sum_{\substack{i,j=1\\ i\neq j}}^{n} \mu_i\mu_j. \tag{1.20}$$

The coefficients $\{P_0, \dots, P_{n-1}\}$ form a complete set of generators for the ring of spectral invariant polynomials, and the first $n-1$ are generically functionally independent on T^*S^{n-1}. A direct calculation shows that the matrix elements of $\mathcal{N}(\lambda)$ satisfy the Poisson bracket relations

$$\{h(\lambda), e(\mu)\} = \frac{e(\lambda) - e(\mu)}{\lambda - \mu} \tag{1.21a}$$

$$\{h(\lambda), f(\mu)\} = -\frac{f(\lambda) - f(\mu)}{\lambda - \mu} \tag{1.21b}$$

$$\{e(\lambda), f(\mu)\} = 2\frac{h(\lambda) - h(\mu)}{\lambda - \mu}, \tag{1.21c}$$

from which it follows that the Poisson brackets of the matrix $\mathcal{N}(\lambda)$ with $\Delta(\mu)$ are obtained by evaluating a matrix commutator

$$\{\mathcal{N}(\lambda), \Delta(\mu)\} = -\frac{[\mathcal{N}(\lambda), \mathcal{N}(\mu)]}{\lambda - \mu}. \tag{1.22}$$

This in turn implies that the Hamiltonian equations generated by the spectral invariants all induce isospectral equations of the type (1.4), and that the Poisson brackets of $\Delta(\lambda)$ with $\Delta(\mu)$ vanish for all values of λ and μ

$$\{\Delta(\lambda), \Delta(\mu)\} = 0. \tag{1.23}$$

Therefore the spectral invariants $\{P_0, \dots, P_{n-2}\}$ all Poisson commute, and their restriction to T^*S^{n-1} forms a complete set of commuting invariants. In particular, this implies that ϕ generates a completely integrable system. The specific form of the Lax equation it generates is

$$\frac{d\mathcal{N}(\lambda)}{dt} = [A_{\mathrm{Ros}}, \mathcal{N}(\lambda)], \tag{1.24}$$

where

$$A_{\mathrm{Ros}} := \begin{pmatrix} \mathbf{x}^T\mathbf{y} & \lambda + \mathbf{y}^T\mathbf{y} + \boldsymbol{\xi}^T\boldsymbol{\mu}^2\boldsymbol{\xi} \\ -\mathbf{x}^T\mathbf{x} & -\mathbf{x}^T\mathbf{y} \end{pmatrix}. \tag{1.25}$$

To complete the Liouville-Arnold integration procedure, it is usual to introduce spheroconical coordinates $\lambda_1, \dots, \lambda_{n-1}$ on S^{n-1} which, assuming the α_i's are ordered so as to be increasing with their indices, are defined as the consecutive zeros of the rational function

$$f(\lambda) = \frac{1}{2}\sum_{i=1}^{n} \frac{x_i^2}{\lambda - \alpha_i} =: \frac{1}{2}\frac{Q(\lambda)}{a(\lambda)} = 0, \tag{1.26}$$

where the zeros $\{\lambda = \lambda_\mu\}_{\mu=1\dots n-1}$ of the degree $n-1$ polynomial

$$Q(\lambda) =: \prod_{\mu=1}^{n-1}(\lambda - \lambda_\mu) \tag{1.27}$$

are ordered so that

$$\alpha_1 < \lambda_1 < \cdots < \lambda_{n-1} < \alpha_n. \tag{1.28}$$

The corresponding canonically conjugate momenta are easily computed to be

$$\zeta_\mu = \frac{1}{2}\sum_{i=1}^{n} \frac{x_i y_i}{\lambda_\mu - \alpha_i}. \tag{1.29}$$

We thus have

$$f(\lambda_\mu) = 0, \quad \zeta_\mu = h(\lambda_\mu) \tag{1.30}$$

where $\{\pm\zeta_\mu + ik(\lambda_\mu)\}_{\mu=1\dots n-1}$ are the eigenvalues of the matrix $\mathcal{N}(\lambda)$ over the points $\{\lambda = \lambda_\mu\}_{\mu=1\dots n-1}$, where the matrix element $f(\lambda)$ vanishes, and

$$\det \mathcal{N}(\lambda_\mu) = \frac{\mathcal{P}(\lambda_\mu)}{a(\lambda_\mu)} = -\zeta_\mu^2 - k^2(\lambda_\mu). \tag{1.31}$$

It follows that the canonical 1-form on T^*S^{n-1} can be written in terms of the invariant polynomial $\mathcal{P}(\lambda)$ evaluated at the points $\lambda = \lambda_\mu$,

$$\theta = \sum_{\mu=1}^{n-1} \zeta_\mu d\lambda_\mu = \sum_{\mu=1}^{n-1} \frac{\sqrt{\mathcal{R}(\lambda_\mu)}}{a(\lambda_\mu)} d\lambda_\mu, \tag{1.32}$$

where the degree $2n-1$ polynomial $\mathcal{R}(\lambda)$ is the one defining the spectral curve

$$\mathcal{R}(\lambda) := -a(\lambda)\mathcal{P}(\lambda) - \mathcal{K}(\lambda). \tag{1.33}$$

Applying the Liouville method, we restrict θ to the leaves of the Lagrangian foliation, defined by the level sets of the invariants $\{H_i\}_{i=1\dots n}$, or equivalently, $\{P_i\}_{i=0\dots n-2}$ (recalling that P_{n-1} is already fixed by (1.18a)), and integrate to obtain the generating function

$$S(\lambda_1, \dots, \lambda_{n-1}, P_0, \dots, P_{n-2}) = \sum_{\mu=1}^{n-1} \int_{\lambda_0}^{\lambda_\mu} \frac{\sqrt{\mathcal{R}(\lambda)}}{a(\lambda)} d\lambda, \tag{1.34}$$

where λ_0 is some arbitrarily chosen base point. This is a hyperelliptic abelian integral which should be interpreted as defined on the spectral curve $\mathcal{C}_{\mathrm{Ros}}$, with the integration limits $\{\lambda_0, \dots \lambda_{n-1}\}$ understood as denoting points on this curve. Differentiation with respect to the invariants then gives the canonically conjugate coordinates

$$Q^j = \frac{\partial S}{\partial P_i} = -\frac{1}{2}\sum_{\mu=1}^{n-1} \int_{\lambda_0}^{\lambda_\mu} \frac{\lambda^j}{\sqrt{\mathcal{R}(\lambda)}} d\lambda, \quad h = 0, \dots, n-2, \tag{1.35}$$

which, in view of the relation (1.18b), evolve linearly in time

$$Q^i(t) = Q^i_0, \quad i = 0, \dots, n-3 \tag{1.36}$$
$$Q^{n-2}(t) = Q^{n-2}_0 + 2t,$$

while the conjugate P_j's remain constant. Since the integrals appearing in (1.35) form a basis for the space of holomorphic differentials on the hyperelliptic curve (1.13), this linearization is achieved by application of the Abel map to the degree $g = n - 1$ divisor

$$\mathcal{D} := \sum_{\mu=1}^{n-1} p(\lambda_\mu, z_\mu), \tag{1.37}$$

where $p(\lambda_\mu, z_\mu)$ denotes the point on the curve with coordinates $(\lambda_\mu, z_\mu = a(\lambda_\mu)\zeta_\mu)$.

The features that have appeared in this example recur, with some variations, in nearly every case of integrable systems associated with isospectral flows of Lax matrices. In the following sections we list, with fewer details, the analogous results for three of the most familiar examples: the isospectral $\mathfrak{su}(2)$ Garnier system (or classical Gaudin spin chain), the periodic Toda lattice, and the finite gap flows of the cubically nonlinear Schrödinger equation. In Lecture 2, the rational R-matrix theory which underlies all these cases will be recalled.

1.3. Example 2. The $\mathfrak{su}(2)$ Isospectral Garnier System (Classical Gaudin Spin Chain) [Gar, Gau, Sk3, AHH3, AHH4, AHH5, H2]. The phase space here is taken as the Cartesian product $\mathfrak{su}(2) \times \cdots \times \mathfrak{su}(2) = \left(\mathfrak{su}(2)\right)^n$, which may be identified with $\mathbf{R}^{3n} = (\mathbf{R}^3)^n$ by choosing a basis consisting, e.g., of the Pauli matrices

$$\sigma_1 = \begin{pmatrix} 0 & 1 \\ 1 & 0 \end{pmatrix}, \quad \sigma_2 = \begin{pmatrix} 0 & -i \\ i & 0 \end{pmatrix}, \quad \sigma_3 = \begin{pmatrix} 1 & 0 \\ 0 & -1 \end{pmatrix}. \tag{1.38}$$

Denoting the elements $\{(\mathbf{S}^1, \dots, \mathbf{S}^n), \mathbf{S}^j \in \mathfrak{su}(2) \sim \mathbf{R}^3\}$, and their components $\{S^j_a\}_{\substack{a=1\dots3 \\ j=1\dots n}}$, and making the usual identifications between $\mathfrak{su}(2)$ and $\mathfrak{su}(2)^*$, the Lie Poisson brackets are

$$\{S^j_a, S^k_b\} = \delta^{jk}\epsilon_{abc}S^j_c. \tag{1.39}$$

The symplectic leaves are coadjoint orbits $\mathcal{O} \subset \left(\mathfrak{su}^*(2)\right)^n$ which are determined by fixing the values of the Casimir invariants

$$c_j := |\mathbf{S}^j|^2 = (S^j_3)^2 + S^j_+ S^j_-, \tag{1.40}$$

where

$$S^j_\pm := S^j_1 \pm i S^j_2. \tag{1.41}$$

Choosing a set of $n+1$ real constants $\{a, \alpha_1, \dots, \alpha_n\}$ with the α_i's all distinct, we define, as before, a matrix

$$\mathcal{N}(\lambda) := \begin{pmatrix} h(\lambda) & e(\lambda) \\ f(\lambda) & -h(\lambda) \end{pmatrix} = \begin{pmatrix} ia & 0 \\ 0 & -ia \end{pmatrix} + \sum_{j=1}^{n} \frac{N_j}{\lambda - \alpha_j}, \tag{1.42}$$

where

$$h(\lambda) := ia + i\sum_{j=1}^{n} \frac{S_3^j}{\lambda - \alpha_j} = -\overline{h}(\overline{\lambda}) \tag{1.43a}$$

$$e(\lambda) := i\sum_{j=1}^{n} \frac{S_-^j}{\lambda - \alpha_j} \tag{1.43b}$$

$$f(\lambda) := i\sum_{j=1}^{n} \frac{S_+^j}{\lambda - \alpha_j} = -\overline{e(\overline{\lambda})}, \tag{1.43c}$$

and

$$N_j = i\begin{pmatrix} S_3^j & S_-^j \\ S_+^j & -S_3^j \end{pmatrix}. \tag{1.44}$$

The Poisson bracket relations (1.39) then imply the fact that $e(\lambda)$, $f(\lambda)$, $h(\lambda)$ again satisfy (1.21a)–(1.21c).

The determinant of $\mathcal{N}(\lambda)$ is again denoted

$$\Delta(\lambda) := \det \mathcal{N}(\lambda) = a^2 + \sum_{j}^{n} \frac{G_j}{\lambda - \alpha_j} + \sum_{j=1}^{n} \frac{c_j}{(\lambda - \alpha_j)^2}, \tag{1.45}$$

where

$$G_j := 2aS_3^j + 2\sum_{\substack{k=1 \\ k\neq j}}^{n} \frac{\mathbf{S}^j \cdot \mathbf{S}^k}{\alpha_j - \alpha_k} \tag{1.46}$$

are the spectral invariant Hamiltonians that generate the classcical Garnier-Gaudin system. As before, these all Poisson commute

$$\{G_i, G_j\} = 0, \quad i, j = 1, \ldots, n, \tag{1.47}$$

which may be verified either by direct computation (exercise 7.1), or as a consequence of the relations

$$\{\Delta(\lambda), \Delta(\mu)\} = 0. \tag{1.48}$$

If we denote by $\{\tau_j\}_{j=1\ldots n}$ the time parameters for the flow generated by G_j's, Hamilton's equations imply the Lax equations

$$\frac{d\mathcal{N}(\lambda)}{d\tau_j} = \{\mathcal{N}(\lambda), G_j\} = \left[\frac{N_j}{\lambda - \alpha_j}, \mathcal{N}(\lambda)\right] \tag{1.49}$$

for the matrix $\mathcal{N}(\lambda)$ (or, equivalently, for the matrix $L(\lambda)$ defined as in (1.10).)

The invariant spectral curve $\mathcal{C}_{Ga}$ defined by the characteristic equation is again hyperelliptic, generically of genus $g = n - 1$, and has the form

$$z^2 = a(\lambda)\mathcal{P}(\lambda) + C(\lambda), \tag{1.50}$$

where the polynomials $\mathcal{P}(\lambda)$, $C(\lambda)$ are defined by

$$a^2 + \sum_{j=1}^{n} \frac{G_j}{\lambda - \alpha_j} = -\frac{\mathcal{P}(\lambda)}{a(\lambda)} \tag{1.51a}$$

$$\sum_{j=1}^{n} \frac{c_j}{(\lambda - \alpha_j)^2} = -\frac{C(\lambda)}{a^2(\lambda)}. \tag{1.51b}$$

The integration of the flows is accomplished, as in the case of the previous example, by introducing a separating canonical coordinate system $\{\lambda_1, \dots, \lambda_{n-1}, q, \zeta_1, \dots, \zeta_{n-1}, P\}$ defined by

$$f(\lambda) := \frac{e^{iq}\prod_{\mu=1}^{n-1}(\lambda - \lambda_\mu)}{a(\lambda)} := e^{iq}\frac{Q(\lambda)}{a(\lambda)} \tag{1.52a}$$

$$\zeta_\mu := ia + i\sum_{i=1}^{n}\frac{S_3^j}{\lambda - \alpha_i} = h(\lambda_\mu) \tag{1.52b}$$

$$e^{iq} := i\sum_{j=1}^{n}(S_1^j + iS_2^j) = \lim_{\lambda\to\infty}(\lambda f(\lambda)), \tag{1.52c}$$

with P the total spin component given by

$$P := \sum_{j=1}^{n} S_3^j = -i\lim_{\lambda\to\infty}\big(\lambda(h(\lambda) - ia)\big). \tag{1.53}$$

From (1.51a), we see that the invariant polynomial $\mathcal{P}(\lambda)$ is of the form

$$\mathcal{P}(\lambda) = -a^2\lambda^n + \sum_{j=1}^{n-1} P_j\lambda^j, \tag{1.54}$$

with

$$P_{n-1} = -2aP + a^2\left(\sum_{j=1}^{n}\alpha_j\right). \tag{1.55}$$

We may choose the quantities $\{P_0, \dots, P_{n-2}, P\}$ as our complete set of generically independent invariants and express the invariants $\{G_1, \dots, G_n\}$ in terms of these by

$$G_k = \frac{a^2\alpha_k^n - (2aP - a^2\sum_{j=1}^{m}\alpha_j)\alpha_k^{n-1} - \sum_{j=0}^{n-2}P_j\alpha_k^j}{a'(\alpha_k)}. \tag{1.56}$$

The Liouville generating function again gives a complete solution to the Hamilton–Jacobi equation in separated form as a sum of Abelian integrals on the spectral curve $\mathcal{C}_{Ga}$

$$S(\lambda_1, \dots, \lambda_{\nu-1}, q, P_0, \dots, P_{n-2}, P) = qP + \sum_{\mu=1}^{n-1}\int_{\lambda_0}^{\lambda_\mu}\frac{\sqrt{a(\lambda)\mathcal{P}(\lambda) + C(\lambda)}}{a(\lambda)}d\lambda, \tag{1.57}$$

where again, λ_0 denotes an arbitrarily chosen base point, and the integrals are taken along a path within the curve. Introducing the coordinates conjugate to $\{P_0, \dots, P_{n-2}, P\}$,

$$Q^j := \frac{\partial S}{\partial P_j} = \frac{1}{2}\sum_{\mu=1}^{n-1}\int_{\lambda_0}^{\lambda_\mu}\frac{\lambda^j}{\sqrt{a(\lambda)\mathcal{P}(\lambda) + C(\lambda)}}d\lambda, \quad j = 0, \dots, n-2 \tag{1.58a}$$

$$Q := \frac{\partial S}{\partial P} = q + \sum_{\mu=1}^{n-1}\int_{\lambda_0}^{\lambda_\mu}\frac{a\lambda^{n-1}}{\sqrt{a(\lambda)\mathcal{P}(\lambda) + C(\lambda)}}d\lambda \tag{1.58b}$$

this again leads to a linearization of the flow in terms of abelian integrals

$$Q^j = Q^j_0 + t_j, \quad j = 0, \ldots, n-2, \tag{1.59a}$$

$$Q = Q_0 + \tau. \tag{1.59b}$$

where the flow parameters $\{t_0, \ldots, t_{n-2}, \tau\}$ correspond to the invariants $\{P_0, \ldots, P_{n-2}, P\}$, respectively. Note that the first $n-1$ of these are given by a complete set of abelian integrals of the first kind and hence, up to normalization, a linearization via the Abel map, but the last one, defining Q, involves an abelian of the third kind with simple poles at the two points over $\lambda = \infty$. When reconstructing the time dependence of the various spin components, this latter integral, however, only enters in an overall phase factor that evolves linearly with τ.

1.4. Example 3. The Periodic Toda Lattice [F1, FM, DT1, DT2, H2]. This model consists of n equal mass point particles, with exponential nearest neighbor interactions, and the particles viewed as forming a closed chain. For convenience, we extend the coordinate-momentum labeling $\{q^j, p_j\}_{j\in\mathbf{Z}}$ to all integer j's, via the periodic conditions

$$q^{j+n} = q^j, \quad p_{j+n} = p_j. \tag{1.60}$$

The Hamiltonian is then

$$h = \frac{1}{2}\sum_{i=1}^{n} p_i^2 + \sum_{j=1}^{n} e^{(q^j - q^{j+1})} \tag{1.61}$$

and Hamilton's equations are

$$\dot{q}^i = p_i, \tag{1.62a}$$

$$\dot{p}_i = e^{(q^{i-1} - q^i)} - e^{(q^i - q^{i+1})}, \quad i = 1, \ldots, n. \tag{1.62b}$$

There exist isospectral representations involving both $n \times n$ and 2×2 Lax matrices; in the following, we consider only the latter, which is defined as follows:

$$L(\lambda) := \mathcal{L}_n \mathcal{L}_{n-1} \cdots \mathcal{L}_1 := \begin{pmatrix} h(\lambda) & e(\lambda) \\ f(\lambda) & g(\lambda) \end{pmatrix}, \tag{1.63}$$

where

$$\mathcal{L}_j := \begin{pmatrix} \lambda - p_j & -e^{-q^j} \\ e^{q^j} & 0 \end{pmatrix}, \quad j = 1, \ldots, n. \tag{1.64}$$

Since

$$\det\big(\mathcal{L}(\lambda)\big) = h(\lambda)g(\lambda) - e(\lambda)f(\lambda) = 1, \tag{1.65}$$

the spectral curve $\mathcal{C}_{\mathrm{Toda}}$ is defined by

$$\det(\mathcal{L}(\lambda) - z\mathbf{I}) = z^2 - zT(\lambda) + 1 = 0 \tag{1.66}$$

where the spectral invariant

$$T(\lambda) := \mathrm{tr}\big(\mathcal{L}(\lambda)\big) = h(\lambda) + g(\lambda) \tag{1.67}$$

is a monic polynomial of degree n

$$T(\lambda) = \lambda^n + \sum_{j=0}^{n-1} P_j \lambda^j \tag{1.68}$$

whose coefficients generate the ring of spectral invariant polynomials. Again, $\mathcal{C}_{\text{Toda}}$ is hyperelliptic, generically with genus $g = n - 1$. The leading coefficients of $T(\lambda)$ are

$$P_{n-1} = -\sum_{j=1}^{n} p_j \tag{1.69a}$$

$$P_{n-2} = \frac{1}{2}\Big(\sum_{j=1}^{n} p_j\Big)^2 - \Big(\frac{1}{2}\sum_{i=1}^{n} p_i^2 + \sum_{i=1}^{n} e^{(q^i - q^{i+1})}\Big), \tag{1.69b}$$

so the Toda Hamiltonian is just the spectral invariant

$$h = \frac{1}{2}P_{n-1}^2 - P_{n-2}. \tag{1.70}$$

The matrix elements of $L(\lambda)$ may be shown inductively to satisfy the Poisson bracket relations

$$\{h(\lambda), e(\mu)\} = \frac{e(\lambda)h(\mu) - h(\lambda)e(\mu)}{\lambda - \mu} \tag{1.71a}$$

$$\{h(\lambda), f(\mu)\} = -\frac{f(\lambda)h(\mu) - h(\lambda)f(\mu)}{\lambda - \mu} \tag{1.71b}$$

$$\{g(\lambda), e(\mu)\} = -\frac{e(\lambda)g(\mu) - g(\lambda)e(\mu)}{\lambda - \mu} \tag{1.71c}$$

$$\{g(\lambda), f(\mu)\} = \frac{f(\lambda)g(\mu) - g(\lambda)f(\mu)}{\lambda - \mu} \tag{1.71d}$$

$$\{e(\lambda), f(\mu)\} = \frac{h(\lambda)g(\mu) - g(\lambda)h(\mu)}{\lambda - \mu} \tag{1.71e}$$

$$\{h(\lambda), g(\mu)\} = \frac{e(\lambda)f(\mu) - f(\lambda)e(\mu)}{\lambda - \mu}. \tag{1.71f}$$

(The proof of this involves first verifying that the same relations hold for the matrix elements of the individual terms $\mathcal{L}_j$ in the product (1.63) defining $\mathcal{L}(\lambda)$, and that if such relations hold for any pair of matrices whose relative Poisson brackets all vanish, then it also holds for the matrix product. This is the so-called *multiplicative* property of the quadratic Poisson bracket relations (1.71a)–(1.71f).) From these, it follows once again that the spectral invariants Poisson commute

$$\{T(\lambda), T(\mu)\} = 0, \tag{1.72}$$

showing that the spectral curve $\mathcal{C}_{\text{Toda}}$ is invariant under the flows generated by spectral invariants, and that the Hamiltonian equations generated by these invariants have the Lax form, since

$$\{\mathcal{L}(\lambda), T(\mu)\} = \frac{[\mathcal{L}(\lambda), \mathcal{L}(\mu)]}{\lambda - \mu}. \tag{1.73}$$

In particular, expressing the Lax matrix as

$$\mathcal{L}(\lambda) = \sum_{j=0}^{n} L_j \lambda^{n-j}, \tag{1.74}$$

and denoting the time parameters corresponding to the flows generated by spectral invariants $\{P_0, \dots, P_{n-1}\}$ as $\{t_0, \dots, t_{n-1}\}$, the resulting Lax equations are

$$\frac{d\mathcal{L}(\lambda)}{dt_j} = \left[\sum_{i=0}^{n-j-1} \lambda^{n-j-i-1} L_i, \mathcal{L}(\lambda)\right]. \tag{1.75}$$

In view of the expression (1.69b) relating the Toda Hamiltonian to the invariants P_{n-1}, P_{n-2}, the Toda flow gives rise to the Lax equation

$$\frac{d\mathcal{L}}{dt} = [\mathcal{A}, \mathcal{L}] \tag{1.76}$$

where

$$\mathcal{A} = \begin{pmatrix} -\lambda & e^{-q^1} \\ -e^{q^n} & 0 \end{pmatrix}. \tag{1.77}$$

The integration again proceeds via the Liouville method by introducing another canonical coordinate system $\{\lambda_1, \dots, \lambda_{n-1}, \zeta_1, \dots, \zeta_{n-1}, q, P)$ defined as follows. In view of the polynomial structure of $f(\lambda)$ following from the definition (1.63), we may express it in factorized form

$$f(\lambda) =: e^{q^n} \prod_{\mu=1}^{n-1} (\lambda - \lambda_\mu), \tag{1.78}$$

with the zeros $\{\lambda_1, \dots, \lambda_{n-1}\}$ viewed as functions on the phase space. These zeros can be shown to all be real, and hence so are the eigenvalues $(h(\lambda_\mu), g(\lambda_\mu) = \frac{1}{h(\lambda_\mu)})$ of $\mathcal{L}(\lambda_\mu)$, which satisfy

$$h(\lambda_\mu) + g(\lambda_\mu) = 2\cosh \zeta_\mu = T(\lambda_\mu), \tag{1.79}$$

where the functions $\{\zeta_1, \dots, \zeta_{n-1}\}$ are defined to be the logarithm of these eigenvalues

$$\zeta_\mu := \ln\big(h(\lambda_\mu)\big), \quad \mu = 1, \dots, n-1. \tag{1.80}$$

Substituting the formula (1.80) and

$$\frac{d \ln\big(f(\lambda)\big)}{d\lambda} = \sum_{\mu=1}^{n-1} \frac{1}{\lambda - \lambda_\mu} \tag{1.81}$$

into the Poisson bracket relation (1.71b) and evaluating the residues at $\lambda = \lambda_\mu$ gives

$$\{\lambda_\mu, \zeta_\nu\} = \delta_{\mu\nu}, \quad \mu, \nu = 1, \dots, n-1, \tag{1.82}$$

while the other brackets $\{\lambda_\mu, \lambda_\nu\}$, $\{\zeta_\mu, \zeta_\nu\}$ vanish because of the commutativity of $f(\lambda)$ with $f(\mu)$ and of $h(\lambda)$ with $h(\mu)$. These functions therefore form the first $n-1$ pairs of a canonical coordinate system on the Toda phase space. The remaining pair of canonical coordinates q and P are defined as

$$\begin{aligned} q &:= q^n = \ln\left(\lim_{\lambda\to\infty} \frac{f(\lambda)}{\lambda^{n-1}}\right), \\ P &:= -P_{n-1} = -\lim_{\lambda\to\infty} \left(\frac{h(\lambda) - \lambda^n}{\lambda^{n-1}}\right) = \sum_{j=1}^{n} p_j. \end{aligned} \tag{1.83}$$

These have vanishing Poisson brackets with the λ_μ's and ζ_μ's, and satisfy the canonical relation

$$\{q, P\} = 1, \tag{1.84}$$

as may be seen from their definitions and the Poisson bracket relations (1.71a)–(1.71f).

Once again, the Liouville generating function takes a completely separated form

$$S(\lambda_1, \dots, \lambda_{n-1}, q, P_0, \dots, P_{n-1}) = -qP_{n-1} + \sum_{\mu=1}^{n-1} \int_{\lambda_0}^{\lambda_\mu} \cosh^{-1}\left(\frac{T(\lambda)}{2}\right) d\lambda \tag{1.85}$$

which, although not quite an abelian integral as written here, may easily be converted into one through integration by parts.

Differentiating with respect to the P_j's , we obtain the conjugate coordinates

$$Q^j = \sum_{\mu=1}^{n-1} \int_{\lambda_0}^{\lambda_\mu} \frac{\lambda^j}{\sqrt{T^2(\lambda) - 4}} d\lambda, \quad j = 1, \dots, n-2 \tag{1.86a}$$

$$Q^{n-1} = -q_n + \sum_{\mu=1}^{n-1} \int_{\lambda_0}^{\lambda_\mu} \frac{\lambda^{n-1}}{\sqrt{T^2(\lambda) - 4}} d\lambda. \tag{1.86b}$$

Using eq. (1.70), Hamilton's equations may be trivially integrated to obtain the flow

$$Q^j(t) = Q_0^j, \quad j = 0, \dots, n-3 \tag{1.87a}$$

$$Q^{n-2}(t) = Q_0^{n-2} - t \tag{1.87b}$$

$$Q^{n-1}(t) = Q_0^{n-1} - P_{n-1}t. \tag{1.87c}$$

As in the previous example, only the first $n-1$ components $\{Q^0, \dots, Q^{n-2}\}$ involve abelian integrals of the first kind, while the last one, Q^{n-1}, is of the third kind. Completion of the integration requires, as usual, inversion formulae in which quotients of the appropriate Riemann theta functions appear. The full details may be found in the references [**F1, FM, DT1, DT2, H2**].

1.5. Example 4. Stationary NLS Flows [IK, P, AA, AHP, AHH3, AHH4, AHH5, H1]. This example concerns quasiperiodic solutions of one of the most familiar examples of integrable PDE's; the cubically nonlinear Schrödinger equation

$$iu_t + u_{xx} = 2\epsilon|u|^2 u, \tag{1.88}$$

where $u(x,t)$ is a complex valued function of the two real variables (x,t) and the constant ϵ can be normalized to have values ± 1. In the following, we only consider the case $\epsilon = -1$, but a similar approach for the case $\epsilon = +1$ is indicated in the exercises.

Prior to the imposition of invariant constraints, the phase space will just be $\mathbf{C}^n$, $n \geq 2$, with standard complex coordinates denoted $\{z_j\}_{j=1\dots n}$ and symplectic form is

$$\omega := -i d\mathbf{z}^T \wedge d\bar{\mathbf{z}} = -i \sum_{j=1}^{n} dz_j \wedge d\bar{z}_j. \tag{1.89}$$

Choosing a set of n distinct real numbers $\{\alpha_j\}_{j=1\ldots n}$, we again define a 2×2 matrix valued function

$$\mathcal{N}(\lambda) := \begin{pmatrix} h(\lambda) & e(\lambda) \\ f(\lambda) & -h(\lambda) \end{pmatrix} = \frac{1}{2}\begin{pmatrix} i & 0 \\ 0 & -i \end{pmatrix} + \sum_{j=1}^{n} \frac{N_j}{\lambda - \alpha_j}, \tag{1.90}$$

where now

$$h(\lambda) := \frac{i}{2} + \frac{i}{2}\sum_{j=1}^{n} \frac{|z_j|^2}{\lambda - \alpha_j} \tag{1.91a}$$

$$e(\lambda) := \frac{1}{2}\sum_{j=1}^{n} \frac{\bar{z}_j^2}{\lambda - \alpha_j} \tag{1.91b}$$

$$f(\lambda) := \frac{1}{2}\sum_{j=1}^{n} \frac{z_j^2}{\lambda - \alpha_j}, \tag{1.91c}$$

and

$$N_j := \frac{1}{2}\begin{pmatrix} i|z_j|^2 & \bar{z}_j^2 \\ z_j^2 & -i|z_j|^2 \end{pmatrix}. \tag{1.92}$$

The determinant is again of the form

$$\Delta(\lambda) = \det \mathcal{N}(\lambda) = -h^2(\lambda) - e(\lambda)f(\lambda) =: -\frac{\mathcal{P}(\lambda)}{a(\lambda)} \tag{1.93}$$

where $\mathcal{P}(\lambda)$ is a polynomial of degree n of the form

$$\mathcal{P}(\lambda) = -\frac{1}{4}\lambda^n + \sum_{j=0}^{n-1} P_j \lambda^j, \tag{1.94}$$

with next to leading coefficient

$$P_{n-1} = -p + \frac{1}{4}\sum_{j=1}^{n} \alpha_j, \tag{1.95a}$$

$$p := \frac{1}{2}|\mathbf{z}|^2 = -i \lim_{\lambda\to\infty}\left(\lambda\left(h(\lambda) - \frac{i}{2}\right)\right). \tag{1.95b}$$

It is easily verified that the matrix elements $e(\lambda)$, $f(\lambda)$, $h(\lambda)$ again satisfy the Poisson bracket relations (1.21a)–(1.21c), and therefore the spectral invariants of $\mathcal{N}(\lambda)$ are conserved under the Hamiltonian flows they generate, and the Poisson bracket of $\mathcal{N}(\lambda)$ with $\Delta(\lambda)$ are again given by (1.22). Defining the polynomial Lax matrix

$$\mathcal{L}(\lambda) := a(\lambda)\mathcal{N}(\lambda) = \lambda^n L_0 + \lambda^{n-1}L_1 + \cdots + L_n, \tag{1.96}$$

it follows as in the previous examples that the (generically) genus $g = n-1$ hyperelliptic curve $\mathcal{C}_{\mathrm{NLS}}$ defined by the characteristic equation

$$\det(\mathcal{L}(\lambda) - z\mathbf{I}) = z^2 - a(\lambda)\mathcal{P}(\lambda) = 0 \tag{1.97}$$

is invariant under the flows generated by the spectral invariants, and the latter all commute. In particular, denoting as x and t the flow parameters corresponding, respectively, to the Hamiltonians

$$H_x := P_{n-2}, \quad H_t := P_{n-3}, \tag{1.98}$$

the resulting Lax equations are

$$\frac{\partial \mathcal{L}}{\partial x} = [\lambda L_0 + L_1, \mathcal{L}] \tag{1.99a}$$

$$\frac{\partial \mathcal{L}}{\partial t} = [\lambda^2 L_0 + \lambda L_1 + L_2, \mathcal{L}]. \tag{1.99b}$$

Now, choosing the following invariant level sets

$$\operatorname{tr}(L_0 L_1) = P_{n-1} + \frac{1}{4}\sum_{j=1}^{n} \alpha_j = 0 \tag{1.100a}$$

$$\operatorname{tr}(L_1^2 + 2L_0 L_2) = P_{n-2} - \Big(\sum_{j=1}^{n} \alpha_j\Big) P_{n-1} + \frac{1}{4}\sum_{\substack{j,k=1\\ j<k}}^{n} \alpha_j\alpha_k = 0, \tag{1.100b}$$

it follows that L_0, L_1 and L_2 are of the form

$$L_0 = \frac{i}{2}\begin{pmatrix} 1 & 0 \\ 0 & -1 \end{pmatrix}, \quad L_1 = \begin{pmatrix} 0 & \bar{u} \\ u & 0 \end{pmatrix}, \quad L_2 = \begin{pmatrix} i|u|^2 & \bar{c} \\ c & -i|u|^2 \end{pmatrix}, \tag{1.101}$$

where

$$u := \frac{1}{2}\sum_{j=1}^{n} z_j^2. \tag{1.102}$$

Substituting in eqs. (1.99a) and equating the λ^n and λ^{n-1} terms shows that the matrix element c in L_2 is given by

$$c = iu_x \tag{1.103}$$

and

$$\frac{\partial L_2}{\partial x} = [L_1, L_2] + [L_0, L_3]. \tag{1.104}$$

Equating the λ^{n-1} terms in (1.99b) then gives

$$\frac{\partial L_1}{\partial t} = [L_0, L_3] = \frac{\partial L_2}{\partial x} - [L_1, L_0], \tag{1.105}$$

which is equivalent to the NLS equation (1.88) for $\epsilon = -1$.

The integration now proceeds just like in the previous examples. We first define the coordinate functions $\{\lambda_1, \dots, \lambda_{n-1}, q\}$ by the formulae

$$f(\lambda) = \frac{e^{iq}\prod_{\mu=1}^{n-1}(\lambda - \lambda_\mu)}{a(\lambda)}. \tag{1.106}$$

$$e^{iq} = \lim_{\lambda\to\infty} \lambda f(\lambda) = \frac{1}{2}\sum_{j=1}^{n} z_j^2 = u, \tag{1.107}$$

and the remaining functions $\{\zeta_1, \dots, \zeta_{n-1}\}$, as the values of $h(\lambda)$ at the points $\{\lambda = \lambda_\mu\}$

$$\zeta_\mu := h(\lambda_\mu) = \frac{i}{2} + \frac{i}{2}\sum_{j=1}^{n} \frac{|z_j|^2}{\lambda_\mu - \alpha_j}, \quad \mu = 1, \dots, n-1. \tag{1.108}$$

It then follows that $\{\lambda_1, \dots, \lambda_{n-1}, \zeta_1, \dots, \zeta_{n-1}, q, p\}$ define a canonical coordinate system on the (unconstrained) phase space, and so we can write the symplectic form (1.89) as

$$\omega|_{M_{\text{NLS}}} = -d\theta,$$

where

$$\theta = pdq + \sum_{j=1}^{n-2} \zeta_\mu d\lambda_\mu. \tag{1.109}$$

Fixing a level set of the invariants $\{P_0, \dots, P_{n-2}\}$, i.e., fixing the invariant spectral curve $\mathcal{C}_{\text{NLS}}$, consistently with (1.100a)–(1.100b), it follows from the definitions (1.106), (1.108), as in the previous examples, that

$$\zeta_\mu^2 = \frac{\mathcal{P}(\lambda_\mu)}{a(\lambda_\mu)}. \tag{1.110}$$

Therefore, integrating $\theta|_{\mathcal{C}_{\text{NLS}}} = dS$ along the corresponding Lagrangian leaves again gives the Liouville generating function in completely separated form as a sum of abelian integrals

$$S = qp + \sum_{\mu=1}^{n} \int_{\lambda_0}^{\lambda_\mu} \sqrt{\frac{\mathcal{P}(\lambda)}{a(\lambda)}} d\lambda. \tag{1.111}$$

Differentiating with respect to the quantities $\{P_0, \dots, P_{n-1}\}$ thus gives

$$Q^j := \frac{\partial S}{\partial P_j} = \frac{1}{2} \sum_{\mu=1}^{n-1} \int_{\lambda_0}^{\lambda_\mu} \frac{\lambda^j}{\sqrt{a(\lambda)\mathcal{P}(\lambda)}} d\lambda, \quad j = 0, \dots, n-2, \tag{1.112a}$$

$$Q^{n-1} := \frac{\partial S}{\partial P_{n-1}} = -q + \frac{1}{2} \sum_{\mu=1}^{n-1} \int_{\lambda_0}^{\lambda_\mu} \frac{\lambda^{n-2}}{\sqrt{a(\lambda)\mathcal{P}(\lambda)}} d\lambda, \tag{1.112b}$$

where eq. (1.95a) has been used in evaluating the P_{n-2} derivative in (1.112b).

The x and t flows are now computed by integrating Hamilton's equations for the Hamiltonians (1.98), giving

$$Q^j = Q_0^j + \delta_{j,n-2}x + \delta_{j,n-3}t, \quad j = 0, \dots, n-2, \tag{1.113a}$$

$$Q^{n-1} = Q_0^{n-1}. \tag{1.113b}$$

Once again, the first $n-1$ abelian integrals defining $\{Q^j\}_{j=0\dots n-2}$ are holomorphic (i.e. of the first kind) and the $\mathbf{C}^{n-1}$-valued function with components $\{Q^0, \dots, Q^{n-2}\}$ is, up to normalization, the Abel map. The quantity of interest though, in view of eq. (1.107), is q, which, by eqs. (1.112b), (1.113b) is given by

$$q = q_0 + \frac{1}{2} \sum_{\mu=1}^{n-1} \int_{\lambda_0}^{\lambda_\mu} \frac{\lambda^{n-2}}{\sqrt{a(\lambda)\mathcal{P}(\lambda)}} d\lambda, \tag{1.114}$$

where we have set $p = 1$ and absorbed the integration constant into q_0. This again involves an abelian integral of the third kind, whose integrand has simple poles at the two points over $\lambda = \infty$, with residues equal to $\pm i$. Once again, to actually compute the (x,t) dependence of the coordinate q along the flows, and hence to determine the solution $u(x,t)$ of (1.88) requires inversion formulae that involve ratios of Riemann theta functions; details of this may be found e.g., in [**AHH3, H1, H2**].

Exercises 1.

1.1 Show directly that the quantities $\{H_i\}_{i=1\ldots n}$ defined in (1.19) Poisson commute.

1.2 Prove, by computing the Jacobian $\partial(H_1,\ldots,H_n)/\partial(dy_1\ldots y_n)$, that the quantities $\{H_i\}$, $i=1,\ldots,n$, defined in (1.19) are generically independent. Show, by obtaining upper bounds on $|\mathbf{y}|^2$ and lower bounds on $|\mathbf{x}|^2$, that the intersection of the level sets of ϕ with $T^*S^{n-1}\subset\mathbf{R}^{2n}$ are compact.

1.3 Verify the Poisson bracket relations (1.21a)–(1.21c) and show that they imply (1.22) and (1.23).

1.4 Prove that the commutation relations (1.21a)–(1.21c) are satisfied by the quantities $e(\lambda)$, $f(\lambda)$, $h(\lambda)$ defined in (1.43a)–(1.43c).

1.5 Prove the multiplicative property of the Poisson brackets (1.71a)–(1.71f) indicated in their proof.

1.6 From eq. (1.22), derive the Lax equations for the flows induced by the invariants $\{P_j\}_{j=0\ldots n-2}$ defined in (1.94) for the case of the NLS phase space.

2. Isospectral Flow in Loop Algebras. II. Classical R-Matrix Theory

In this lecture, we review the theory of classical R-matrices, which serves as the structural underpinning for the examples cited above and for much of what follows. The approach used is based mainly on the formulation of Semenov-Tian-Shansky [**STS1, STS2, OPRS**], which generalizes earlier results based on the Adler-Kostant-Symes theorem [**Ad, K, Sy1, Sy2, RS1, RS2**]. A closely related approach is due to Belavin and Drin'feld [**Dr**].

2.1. Classical Yang-Baxter Equations, Commuting Invariants and Lax Equations. Let $\mathfrak{g}$ be a Lie algebra, and suppose that we have an endomorphism $R:\mathfrak{g}\to\mathfrak{g}$ such that the skew symmetric bilinear operation $[\ ,\ ]_R:\mathfrak{g}\times\mathfrak{g}\to\mathfrak{g}$ defined by

$$[\xi,\eta]_R := \frac{1}{2}[R\xi,\eta]+\frac{1}{2}[\xi,R\eta] \tag{2.1}$$

satisfies the Jacobi identity

$$[\xi,[\eta,\zeta]_R]_R+[\xi,[\eta,Z]_R]_R+[\xi,[\eta,\zeta]_R]_R=0 \quad \forall\, \xi,\eta,\zeta\in\mathfrak{g}, \tag{2.2}$$

defining a new Lie algebra structure on the space $\mathfrak{g}$, which will be denoted $\mathfrak{g}_R$. Now define a skew symmetric map $R^\wedge:\mathfrak{g}\wedge\mathfrak{g}\to\mathfrak{g}$ by

$$R^\wedge(\xi,\eta) := [R\xi,R\eta]-2R[\xi,\eta]_R \tag{2.3}$$

The following lemma gives a sufficient condition for the Jacobi identity to hold.

LEMMA 2.1. *A sufficient condition that the bracket $[\ ,\ ]_R$ satisfy the Jacobi identity is*

$$R^\wedge(\xi,\eta)=\alpha[\xi,\eta], \quad \forall\, \xi,\eta,\in\mathfrak{g}, \tag{2.4}$$

where α is any proportionality constant.

The proof follows from the fact that the Jacobi identity (2.2) is equivalent to the relation

$$[R^\wedge(\xi,\eta),\zeta]+[R^\wedge(\zeta,\xi),\eta]+[R^\wedge(\eta,\zeta),\xi]=0, \quad \xi,\eta,\zeta\in\mathfrak{g}, \tag{2.5}$$

which holds if (2.4) is satisfied because of the Jacobi identity for the original bracket $[\ ,\ ]$.

Relation (2.4) is referred to [**STS1, STS2**] as the *modified classical Yang-Baxter equation* (the unmodified case corresponding to $\alpha = 0$). There is a simple class of cases, called *split* R-matrices, that correspond to the original Adler-Kostant-Symes theorem. Here, we assume that $\mathfrak{g}$ admits a vector space decomposition

$$\mathfrak{g} = \mathfrak{h} + \mathfrak{k} \tag{2.6}$$

as the sum of two complementary subalgebras $\mathfrak{h}, \mathfrak{k} \subset \mathfrak{g}$, and define the associated *split* classical R-matrices as the difference

$$R := P_{\mathfrak{h}} - P_{\mathfrak{k}} \tag{2.7}$$

between the two projection maps

$$P_{\mathfrak{h}} : \xi_{\mathfrak{h}} + \xi_{\mathfrak{k}} \longmapsto \xi_{\mathfrak{h}} \tag{2.8a}$$

$$P_{\mathfrak{k}} : \xi_{\mathfrak{h}} + \xi_{\mathfrak{k}} \longmapsto \xi_{\mathfrak{k}} \tag{2.8b}$$

$$\xi \in \mathfrak{h}, \quad \xi_{\mathfrak{k}} \in \mathfrak{k}.$$

This is easily seen to satisfy (2.4) (with $\alpha = -1$) and the resulting modified bracket is just

$$[\xi, \eta]_R = [\xi_{\mathfrak{h}}, \eta_{\mathfrak{h}}] - [\xi_{\mathfrak{k}}, \eta_{\mathfrak{k}}]. \tag{2.9}$$

Recall now that on the dual space $\mathfrak{g}^*$ of any Lie algebra, we have the Lie Poisson bracket structure defined by

$$\{F, \widetilde{F}\}|_\mu := \langle \mu, [dF|_\mu, d\widetilde{F}|_\mu] \rangle, \quad F,\ \widetilde{F} \in C^\infty(\mathfrak{g}^*),\ \mu \in \mathfrak{g}^*, \tag{2.10}$$

where $\langle\ ,\ \rangle$ denotes the dual pairing and $(\mathfrak{g}^*)^* = \mathfrak{g}$ is used to identify the differentials $dF|_\mu$, $d\widetilde{F}|_\mu$ as elements of $\mathfrak{g}$. We define the Poisson bracket associated with any classical R-matrix as the Lie Poisson bracket associated with the modified Lie bracket $[\ ,\ ]_R$

$$\{F, \widetilde{F}\}_R|_\mu := \langle \mu, [dF|_\mu, d\widetilde{F}|_\mu]_R \rangle, \quad F,\ \widetilde{F} \in C^\infty(\mathfrak{g}^*),\ \mu \in \mathfrak{g}^*. \tag{2.11}$$

For the case of the split R-matrix (2.7), this becomes

$$\{F, \widetilde{F}\}_R = \langle \mu, [(dF_{|}\mu)_{\mathfrak{h}}, (d\widetilde{F}_{|}\mu)_{\mathfrak{h}}] - [(dF|_\mu)_{\mathfrak{k}}, (d\widetilde{F}|_\mu)_{\mathfrak{k}}] \rangle. \tag{2.12}$$

Let $\mathcal{I}(\mathfrak{g}^*)$ denote the ring of polynomial functions on $\mathfrak{g}^*$ that are invariant under the coadjoint action

$$\xi : \mu \mapsto \mathrm{ad}^*_\xi(\mu), \quad \xi \in \mathfrak{g}, \quad \mu \in \mathfrak{g}^* \tag{2.13}$$

$$\langle \mathrm{ad}^*_\xi(\mu), \eta \rangle = -\langle \mu, [\xi, \eta] \rangle, \quad \forall \eta \in \mathfrak{g},$$

i.e., which satisfy the infinitesimal invariance condition

$$\langle \mu, [\xi, df|_\mu] \rangle, = 0 \quad \forall \xi \in \mathfrak{g},\ \mu \in \mathfrak{g}^*, \quad f \in \mathcal{I}(\mathfrak{g}^*). \tag{2.14}$$

In particular, in the case of matrix algebras which are identified with their duals through the trace pairing

$$\langle \mu, \xi \rangle = \mathrm{tr}(\mu\xi), \tag{2.15}$$

$\mathcal{I}(\mathfrak{g}^*)$ is just the ring of spectral invariants.

We can now state the main result that underlies the examples given above, and very many other cases of integrable systems involving isospectral flows. (See [**OPRS, H2, STS1, STS2**].

THEOREM 2.2. *Suppose* $R \in \mathrm{End}(\mathfrak{g})$ *defines a Lie algebra structure* $\mathfrak{g}_R$ *on* $\mathfrak{g}$ *through* (2.1).

(1) *If* F, $\widetilde{F} \in \mathcal{I}(\mathfrak{g}^*)$ *is any pair of functions belonging to the center* $\mathcal{I}(\mathfrak{g}^*)$ *of the unmodified Lie Poisson algebra on* $\mathfrak{g}^*$ *(Casimir invariants), their modified Lie Poisson bracket* (2.11) *on* $\mathfrak{g}^* \sim \mathfrak{g}_R^*$ *relative to the Lie algebra structure* $\mathfrak{g}_R$ *associated to* R *vanishes*

$$\{F, \widetilde{F}\}_R = 0. \tag{2.16}$$

(2) *Hamilton's equations on* $\mathfrak{g}_R^* \sim \mathfrak{g}*$ *generated by* $F \in \mathcal{I}(\mathfrak{g}^*)$ *with respect to the modified Lie Poisson bracket structure are given by*

$$\frac{d\mu}{dt} = \frac{1}{2}\,\mathrm{ad}^*_{R(dF|_\mu)}(\mu), \tag{2.17}$$

where the ad^* *action is understood to be with respect to the unmodified Lie algebra structure* $\mathfrak{g}$. *In particular, if there exists a nonsingular* ad-*invariant quadratic form* (,) *on* $\mathfrak{g}$, *that permits an identification* $\mathfrak{g} \sim \mathfrak{g}^*$, $\mathrm{ad} \sim \mathrm{ad}^*$, *Hamilton's equations take the Lax form*

$$\frac{d\mu}{dt} = \frac{1}{2}[R(dF|_\mu), (\mu)], \quad \mu \in \mathfrak{g} \sim \mathfrak{g}^*. \tag{2.18}$$

The proof is a simple algebraic computation that may be found, e.g. in [**STS1, STS2, OPRS, H2**], and will be omitted here.

For the case of a split R-matrix, we have

$$\begin{aligned} \mathrm{ad}^*_{R(dF|_\mu)}(\mu) &= \frac{1}{2}\,\mathrm{ad}^*_{(dF|_\mu)_\mathfrak{h}}(\mu) - \frac{1}{2}\,\mathrm{ad}^*_{(dF|_\mu)_\mathfrak{k}}(\mu) \\ &= \mathrm{ad}^*_{(dF|_\mu)_\mathfrak{h}}(\mu) = -\,\mathrm{ad}^*_{(dF|_\mu)_\mathfrak{k}}(\mu). \end{aligned} \tag{2.19}$$

We therefore have the following corollary in this case, which is essentially the infinitesimal part of the Adler-Kostant-Symes theorem. (There is an additional, part, showing how the flows may be integrated using a factorization within the associated Lie group corrrsponding to the splitting (2.6) of the Lie algebra. In the general setting however, involving infinite dimensional algebras, a corrsponding Lie group need not necessarily exist, so we consider here only the infinitesimal part, which does not require the existence of a group.)

COROLLARY 2.3 (AKS theorem). *For the case of the split* R*-matrix corresponding to the decomposition* $\mathfrak{g} = \mathfrak{h} + \mathfrak{k}$, *Hamilton's equation on* $\mathfrak{g}_R^*$ *for a Hamiltonian* $F \in \mathcal{I}(\mathfrak{g}^*)$ *in the center of the unmodified Lie Poisson bracket algebra is given by*

$$\frac{d\mu}{dt} = \mathrm{ad}^*_{(dF|_\mu)_\mathfrak{h}}(\mu) = -\,\mathrm{ad}^*_{(dF|_\mu)_\mathfrak{k}}(\mu). \tag{2.20}$$

In particular, if $\mathfrak{g}$ *has an* ad-*invariant nonsingular quadratic form through which* $\mathfrak{g} \sim \mathfrak{g}^*$ *and* $\mathrm{ad} \sim \mathrm{ad}^*$ *may be identified, Hamilton's equation takes the Lax form*

$$\frac{d\mu}{dt} = [(dF|_\mu)_\mathfrak{h}), \mu] = -[(dF|_\mu)_\mathfrak{k}, \mu]. \tag{2.21}$$

2.2. Split R-Matrices and Loop Algebras. We now restrict our attention to a particular class of cases that include most of the well known examples of integrable systems associated to isospectral flows; namely, loop algebras with split R-matrix structures. Several standard works now exist dealing with this class; the monograph [**FT**] and the encyclopaedia article [**OPRS**] are particularly useful references containing many more source articles in their bibliographies.

There are many ways to define loop algebras: as formal, semi-infinite series in a loop parameter, as completions of a space of Laurent polynomials, as Fourier series, as differentiable or analytic maps, etc.—but these differ mainly in matters involving convergence and function space properties, whereas we shall primarily be concerned with invariant finite dimensional subspaces for which these differences are largely irrelevant. We therefore use an approach that is concise and convenient for present purposes, based on analytic maps. This allows us to use notions and terminology from complex function theory without having to introduce new definitions for such common things as residues, asymptotic behaviour, etc.

Let $\mathfrak{g}$ be a (finite dimensional) complex Lie algebra and let

$$\widetilde{\mathfrak{g}} = C^{\omega}(\Gamma, \mathfrak{g}) = \{\xi = \xi(\lambda), \lambda \in \Gamma\} \tag{2.22}$$

denote the space of analytic maps from a positively oriented, simple, closed curve $\Gamma \subset \mathbf{C}$ surrounding the origin in the complex λ-plane to $\mathfrak{g}$. We give $\widetilde{\mathfrak{g}}$ the structure of a Lie algebra, called the *loop algebra* associated to $\mathfrak{g}$, by defining the Lie product through pointwise evaluation

$$[\xi, \eta](\lambda) := [\xi(\lambda), \eta(\lambda)] \quad \forall\, \lambda \in \Gamma. \tag{2.23}$$

For most purposes, it is sufficient to take Γ as the unit circle, but in what follows, it is helpful to keep the choice of Γ somewhat flexible. It will also be useful to extend the complex λ-plane to the Riemann sphere $\mathbf{CP}^1$ with the understanding that Γ does not pass through the point at ∞. If $\mathfrak{G}$ is a complex Lie group whose Lie algebra is $\mathfrak{g}$, we may similarly define the associated loop group

$$\widetilde{\mathfrak{G}} = C^{\omega}(\Gamma, \mathfrak{G}) = \{g = g(\lambda), \lambda \in \Gamma\} \tag{2.24}$$

as the space of analytic maps from Γ to $\mathfrak{G}$ with group multiplication again defined by pointwise evaluation

$$gh(\lambda) := g(\lambda)h(\lambda), \quad \forall\, g, h \in \mathfrak{G}. \tag{2.25}$$

The adjoint representation $\mathrm{Ad} : \widetilde{\mathfrak{G}} \to \mathrm{End}(\widetilde{\mathfrak{g}})$ is then similarly defined by pointwise evaluation in terms of the adjoint representation of $\mathfrak{G}$. Let Γ_+ denote the interior region of Γ (containing 0) and Γ_- the exterior region (containing ∞), and let $\overline{\Gamma}_{\pm}$ denote their closures. We define two complementary subalgebras $\widetilde{\mathfrak{g}}_{\pm} \subset \widetilde{\mathfrak{g}}$, such that $\widetilde{\mathfrak{g}}_+$ consists of elements $\xi \in \widetilde{\mathfrak{g}}$ admitting an anaytic continuation into the interior region Γ_+, and $\widetilde{\mathfrak{g}}_-$ of elements admitting an analytic continuation into the exterior one Γ_-, with $\xi(\infty) = 0$. (By Liouville's theorem on bounded analytic functions, the intersection between these two subalgebras is just the zero element.)

For any $\xi \in \mathfrak{g}$, $\sigma_{\pm} \in \overline{\Gamma}_{\pm}$, we may define $\xi_{\pm} \in \widetilde{\mathfrak{g}}_{\pm}$ by the formulae

$$\xi_+(\sigma_+) := \frac{1}{2\pi i} \oint_{\Gamma(\sigma_+)} \frac{\xi}{\lambda - \sigma_+} d\lambda, \tag{2.26a}$$

$$\xi_-(\sigma_-) := -\frac{1}{2\pi i} \oint_{\Gamma(\sigma_-)} \frac{\xi}{\lambda - \sigma_-} d\lambda, \tag{2.26b}$$

where the integration contour $\Gamma(\sigma_+)$ is taken to coincide with Γ if $\sigma_+ \in \Gamma_+$, and to consist of a small deformation of Γ so as to include σ_+ in the interior if $\sigma_+ \in \Gamma$, and similarly, $\Gamma(\sigma_-)$ coincides with Γ if $\sigma_- \in \Gamma_-$, and consists of a small deformation of Γ with σ_- on the exterior if $\sigma_- \in \Gamma$. (This is well-defined since ξ is analytic within a neighborhood of the closed set Γ.)

It follows from the Cauchy integral representation that, for any $\xi \in \widetilde{\mathfrak{g}}$, we have the unique decomposition (on Γ)

$$\xi = \xi_- + \xi_+, \quad \xi_\pm \in \widetilde{\mathfrak{g}}_\pm \tag{2.27}$$

and therefore a unique decomposition of $\widetilde{\mathfrak{g}}$ as the vector space direct sum

$$\widetilde{\mathfrak{g}} = \widetilde{\mathfrak{g}}_+ + \widetilde{\mathfrak{g}}_-. \tag{2.28}$$

We denote by $P_\pm : \widetilde{\mathfrak{g}} \to \widetilde{\mathfrak{g}}_\pm$ the projection operators onto these two subalgebras. (Similar definitions may be applied to the loop group $\widetilde{\mathfrak{G}}$, to define corresponding subgroups $\widetilde{\mathfrak{G}}_\pm$, but the corresponding global factorization problem, which is related to the *matrix Riemann–Hilbert* problem, leads to many analytic subtleties and is a vast subject of its own, which we do not need to embark on here.)

Now, suppose that $\mathfrak{g}$ is endowed with a nondegenerate, Ad-invariant symmetric form $(\ ,\ )$, allowing us to identify $\mathfrak{g}$ in the usual way with its dual space $\mathfrak{g}^*$. Then we may correspondingly define a weakly nondegenerate, symmetric, Ad-invariant bilinear form $\langle\ ,\ \rangle$ on $\widetilde{\mathfrak{g}}$ by pointwise evaluation and integration around Γ:

$$\langle \xi, \eta \rangle := \frac{1}{2\pi i} \oint_\Gamma \big(\xi(\lambda), \eta(\lambda)\big) d\lambda. \tag{2.29}$$

Using this, $\widetilde{\mathfrak{g}}$ may be identified as a dense subspace of its dual space, with the coadjoint representation coinciding with the adjoint representation. This subspace, the *analytic* dual will be denoted $\widetilde{\mathfrak{g}}^*$ to maintain the conceptual difference between the algebra and its dual space. Since it is invariant under the Ad^* action, it follows that it is a Poisson subspace with respect to the Lie Poisson structure. Viewing (2.29) as defining the dual pairing between $\widetilde{\mathfrak{g}}^*$ and $\widetilde{\mathfrak{g}}$, by making suitable Taylor or Laurent expansions in a neighborhood of 0 or ∞, the annihilators $\widetilde{\mathfrak{g}}^0_+$, $\widetilde{\mathfrak{g}}^0_-$ of $\widetilde{\mathfrak{g}}_+$ and $\widetilde{\mathfrak{g}}_-$, respectively are seen to just coincide with $\widetilde{\mathfrak{g}}_+$ and $\widetilde{\mathfrak{g}}_-$ themselves. These may therefore be identified as the (analytic) dual spaces of $\widetilde{\mathfrak{g}}_-$ and $\widetilde{\mathfrak{g}}_+$, respectively

$$\widetilde{\mathfrak{g}}^0_+ \sim \widetilde{\mathfrak{g}}_+ \sim \widetilde{\mathfrak{g}}^*_-, \quad \widetilde{\mathfrak{g}}^0_+ \sim \widetilde{\mathfrak{g}}_+ \sim \widetilde{\mathfrak{g}}^*_-. \tag{2.30}$$

The coadjoint action of $\widetilde{\mathfrak{g}}_\pm$ on $\widetilde{\mathfrak{g}}^*_\pm$ is therefore defined by projection

$$\begin{gathered} \mathrm{ad}^*_{\xi_\pm} : \mu_\mp = \big(\mathrm{ad}_{\xi_\pm}(\mu_\mp)\big)_\mp, \\ \xi_\pm \in \widetilde{\mathfrak{g}}_\pm, \quad \mu_\mp \in \widetilde{\mathfrak{g}}^*_\pm \sim \widetilde{\mathfrak{g}}_\mp, \end{gathered} \tag{2.31}$$

where the $\mathrm{ad}_{\xi_\pm}$ on the right are understood to refer to the algebra $\widetilde{\mathfrak{g}}$.

We now consider the split R-matrix associated to the decomposition (2.28)

$$R := P_+ - P_-. \tag{2.32}$$

Denoting the corresponding Lie algebra structure on $\widetilde{\mathfrak{g}}$ as $\widetilde{\mathfrak{g}}_R$, we have

$$[\xi, \eta]_R = [\xi_+, \eta_+] - [\xi_-, \eta_-]. \tag{2.33}$$

Now, for each $\xi \in \mathfrak{g}$, $\sigma \in \Gamma$, $\sigma_\pm \in \Gamma_\pm$, we define the linear function, $\xi(\sigma)$ by

$$\xi(\sigma)|_{\mu \in \widetilde{\mathfrak{g}}^*} := (\mu(\sigma), \xi). \tag{2.34}$$

The following gives an explicit expression for the Lie Poisson bracket of two such linear functions corresponding to the split R-matrix (2.32).

PROPOSITION 2.4.

$$\{\xi(\sigma), \eta(\tau)\}_R = -\frac{[\xi,\eta](\sigma) - [\xi,\eta](\tau)}{\sigma - \tau} \quad \forall\ \sigma, \tau \in \Gamma, \quad \sigma \neq \tau \tag{2.35a}$$

$$\{\xi(\sigma), \eta(\sigma)\}_R = -[\xi, \eta]'(\sigma). \tag{2.35b}$$

The proof is an elementary exercise in application of the Cauchy theorem, using the fact that $\xi(\sigma)$ may be recovered as a limit

$$\xi(\sigma) = \lim_{\sigma_\pm \to \sigma} \big(\xi_+(\sigma_+) + \xi_-(\sigma_-)\big) \tag{2.36}$$

of the sum of the two associated functions $\xi_\pm$ defined by

$$\xi_\pm(\sigma_\pm)|_{\mu \in \widetilde{\mathfrak{g}}^*} = \pm\frac{1}{2\pi i}\oint_\Gamma \frac{(\mu(\lambda), \xi)}{\lambda - \sigma_\pm} d\lambda, \quad \sigma_\pm \in \Gamma_\pm. \tag{2.37}$$

As an example, we may take $\mathfrak{g} = \mathfrak{gl}(n, \mathbf{C})$ (or the real form $\mathfrak{gl}(n, \mathbf{R})$) with Ad-invariant symmetric form

$$(\xi, \eta) := \operatorname{tr}(\xi\eta), \quad \xi, \eta \in \mathfrak{gl}(n) \tag{2.38}$$

and choose for ξ and η a pair of basis elements consisting of the elementary matrices

$$\xi = e_{ji}, \quad \eta = e_{lk}, \tag{2.39}$$

which gives

$$\xi(\sigma)|_\mu = \mu_{ij}(\sigma), \quad \eta(\tau)|_\mu = \mu_{kl}(\tau). \tag{2.40}$$

It follows that the Possion brackets of these elements is given by

$$\{\mu_{ij}(\sigma), \mu_{kl}(\tau)\}_R = \frac{\mu_{il}(\sigma) - \mu_{il}(\tau)}{\sigma - \tau}\delta_{jk} - \frac{\mu_{kj}(\sigma) - \mu_{kj}(\tau)}{\sigma - \tau}\delta_{il}. \tag{2.41}$$

This often is expressed in tensorial notation as

$$\{\mu(\sigma) \overset{\otimes}{,} \mu(\tau)\} = [r(\sigma - \tau), \mu(\sigma) \otimes \mathbf{I} + \mathbf{I} \otimes \mu(\tau)] \tag{2.42}$$

where the symbol $\{\ \overset{\otimes}{,}\ \}$ signifies a simultaneous tensor product of elements in $\operatorname{End} \mathbf{C}^n \sim \mathfrak{gl}(n)$ and a Poisson bracket, and $r(\sigma - \tau) \in \operatorname{End}(\mathbf{C}^N \otimes \mathbf{C}^n)$ is defined as

$$r(\sigma - \tau) := \frac{P_{12}}{\sigma - \tau}. \tag{2.43}$$

Here $P_{12} \in \operatorname{End}(\mathbf{C}^n \otimes \mathbf{C}^n)$ signifies the permutation of the first and second factors of decomposable elements of $\mathbf{C}^n \times \mathbf{C}^n$, extended linearly to the whole space. Both $\mu(\sigma) \otimes \mathbf{I}$ and $\mathbf{I} \otimes \mu(\tau)$ are understood as elements of $\operatorname{End}(\mathbf{C}^n \otimes \mathbf{C}^n)$, and the commutator on the right signifies the operator commutator on $\operatorname{End}(\mathbf{C}^n \otimes \mathbf{C}^n)$.

Applying the R-matrix (or AKS) theorem to this case, we conclude that:

(1) The elements of the ring $\mathcal{I}(\widetilde{\mathfrak{gl}}^*(n))$ of spectral invariants Poisson commute.
(2) Hamilton's equation generated by ay element $\phi \in \mathcal{I}(\widetilde{\mathfrak{gl}}^*(n))$ induce Lax equations of the form

$$\frac{d\mu}{dt}(\lambda) = \{\mu, \phi\} = \pm[(d\phi)_\pm(\lambda), \mu(\lambda)], \tag{2.44}$$

where the differential $(d\phi)$ evaluated at $\mu \in \widetilde{\mathfrak{gl}}^*(n)$ is viewed as belonging to the algebra $\widetilde{\mathfrak{gl}}(n)$. In particular, choosing, for each $\sigma \in \Gamma$, the invariant

$$H_m(\sigma) := \frac{1}{m}\operatorname{tr}\big(\mu^m(\sigma)\big), \tag{2.45}$$

these all Poisson commute amongst themselves for the various values of $m \in \mathbf{Z}$, $\sigma \in \Gamma$

$$\{H_m(\sigma), H_n(\tau)\} = 0, \quad \forall\, m, n \in \mathbf{Z},\ \sigma, \tau \in \Gamma. \tag{2.46}$$

Denoting the corresponding flow parameter as $t_m(\sigma)$, we obtain Hamilton's equations in the form

$$\frac{d\mu(\lambda)}{dt_m(\sigma)} = \{\mu(\lambda), H_m(\sigma)\} = \frac{[\mu(\lambda), \mu^{m-1}(\sigma)]}{\lambda - \sigma}. \tag{2.47}$$

Comparing with examples 1, 2 and 4 of the previous lecture, we see that, identifying the rational Lax matrices $\mathcal{N}(\lambda)$ with the traceless part of $\mu(\lambda)$ for the case $\widetilde{\mathfrak{gl}}(2)$ (the traces, being Casimir invariants, may just be removed), we recover all the general results seen there, such as the Poisson brackets of the matrix elements (1.21a)–(1.21c), the commutator structure (1.22) of the Hamiltonian vector fields induced by the spectral invariants $\Delta(\mu)$ (which are, essentially, the quantities $H_2(\sigma)$ defined in (2.45)) and their Poisson commutativity (1.23).

To recover the quadratic Poisson bracket structure (1.71a)–(1.71f) underlying the 2×2 Lax equations governing the periodic Toda lattice, however, we must consider instead an analogous Poisson bracket structure defined on loop groups. This case, though equally important in applications, will not be developed in detail here, although this may be done in a very analogous fashion. Instead, in the next section we summarize the analogous results as far as they pertain to the loop group $\widetilde{\mathfrak{Gl}}(n)$.

2.3. Loop Groups and Quadratic Poisson Brackets. The full theory underlying the quadratic types of Poisson brackets encountered in the example of the Toda lattice involves the general notion of Poisson Lie groups and bialgebras, of which a complete discussion may be found in the references [**Dr, STS1, STS2, Sk1, Sk2, OPRS**]. Here, however, we limit ourselves to a particular subclass, associated with so-called *coboundary* bialgebras. The corresponding quadratic Poisson bracket structure on the Lie group was first found by Sklyanin [**Sk1**] and these brackets are sometimes referred to as *Sklyanin brackets.* In fact, we shall restrict ourselves immediately to the case of the loop group $\widetilde{\mathfrak{Gl}}(n)$, although the general R-matrix approach is applicable to a larger class of Lie groups, just as it is in the case of Lie algebras.

The Poisson bracket associated to the split R-matrix (2.32) may be expressed as

$$\{F, \widetilde{F}\}_r|_g = \frac{1}{2}\Big(r\big(L_g^*(dF|_g), L_g^*(\widetilde{F}|_g)\big) - r\big(R_g^*(dF|_g), R_g^*(\widetilde{F}|_g)\big)\Big), \tag{2.48}$$

where $r \in \widetilde{\mathfrak{gl}}(n) \wedge \widetilde{\mathfrak{gl}}(n)$ is defined by

$$r(\mu, \nu) = \langle \mu_+, \nu_- \rangle - \langle \mu_-, \nu_+ \rangle \tag{2.49}$$

$$\mu,\ \nu \in \widetilde{\mathfrak{gl}}^*(n) \sim \widetilde{\mathfrak{gl}}(n), \quad \mu_+,\ \nu_+ \in \widetilde{\mathfrak{gl}}_+(n), \quad \mu_-,\ \nu_- \in \widetilde{\mathfrak{gl}}_-(n).$$

Here, F and $\widetilde{F}$ are understood to be functions on the loop group space, belonging to some suitably restricted smoothness category. (It is sufficient, e.g., to again take polynomial functions of the coefficients when restricting to subspaces of rational group elements with given pole support). The differentials $dF|_g$ and $d\widetilde{F}|_g$ evaluated at the loop group element g are understood as belonging to the cotangent space to $\widetilde{\mathfrak{Gl}}(n)$ at the point g, and L_g^*, R_g^* denote the pullbacks of the left- and right translation maps under the element g, which when applied to these differentials give elements of the tangent space at the identity. The latter are, by the trace-residue loop algebra pairing defined in (2.29), (2.38), identified with elements of the Lie algebra $\widetilde{\mathfrak{gl}}(n)$, to which the bivector r defined in (2.49) is applied. The fact that the definition (2.48) actually defines a Poisson bracket on the loop group is a consequence of the general R-matrix theory.

As before, for each $\xi \in \widetilde{\mathfrak{gl}}(n)$, $\sigma \in \Gamma$, we define a function $\xi(\sigma)$ on $\widetilde{\mathfrak{Gl}}(n)$ by

$$\xi(\sigma)|_{g\in\widetilde{\mathfrak{Gl}}(n)} := \operatorname{tr}\big(\xi g(\sigma)\big). \tag{2.50}$$

(Although the group has no natural linear structure, the fact that it is realized by matrices makes this definition feasible.) It then follows from a calculation very similar to the loop algebra case that the Poisson brackets of such functions are given by the quadratic expression given in the following Proposition.

PROPOSITION 2.5.

$$\{\xi(\sigma), \eta(\tau)\}_r|_g = \frac{\operatorname{tr}\big(\xi g(\sigma)\eta g(\tau) - \xi g(\tau)\eta g(\sigma)\big)}{\sigma - \tau} \quad \textit{for } \sigma \neq \tau, \tag{2.51a}$$

$$\{\xi(\sigma), \eta(\sigma)\}_r|_g = \operatorname{tr}\big(\xi g'(\sigma)\eta g(\sigma) - \xi g(\sigma)\eta g'(\sigma)\big). \tag{2.51b}$$

In particular, choosing the elements ξ, $\eta \in \mathfrak{gl}(n)$ as same the elementary matrices as in (2.39), we obtain

$$\xi(\sigma)|_g = g_{ij}(\sigma), \quad \eta(\tau)|_g = g_{kl}(\tau). \tag{2.52}$$

In this case, eq. (2.51a) becomes

$$\{g_{ij}(\sigma), g_{kl}(\tau)\} = \frac{g_{il}(\sigma)g_{kj}(\tau) - g_{il}(\tau)g_{kj}(\sigma)}{\sigma - \tau}. \tag{2.53}$$

Equivalently, using the tensor product notation as in (2.42), this may be written

$$\{g(\sigma) \overset{\otimes}{,} g(\tau)\} = [r(\sigma - \tau), g(\sigma) \otimes g(\tau)], \tag{2.54}$$

where $r(\sigma - \tau)$ is defined in (2.43).

Finally, a theorem completely analogous to Theorem 2.2 holds in the case of Poisson Lie groups associated with coboundary bialgebras, which implies in this case that

1. If $(F, \widetilde{F})$ is any pair of functions belonging to the ring $\mathcal{I}\big(\widetilde{\mathfrak{Gl}}(n)\big)$ of Ad-invariant functions on $\widetilde{\mathfrak{Gl}}(n)$ (i.e., spectral invariants) their Poisson bracket vanishes

$$\{F, \widetilde{F}\}_r = 0. \tag{2.55}$$

2. Hamilton's equations on $\widetilde{\mathfrak{Gl}}(n)$ generated by $F \in \mathcal{I}\big(\widetilde{\mathfrak{Gl}}(n)\big)$ are of the Lax form

$$\frac{dg}{dt} = \frac{1}{2}\Big(gR\big(R_g^*(dF|_g)\big) - R\big(R_g^*(dF|_g)\big)g\Big), \quad g \in G, \tag{2.56}$$

where the left and right multiplications by g signify left and right translations to the point $g \in G$.

In particular, applying this to the class of spectral invariants defined by

$$H_m(\sigma) := \frac{1}{m}\operatorname{tr} g^m(\sigma), \quad m \in \mathbf{Z},\ \sigma \in \Gamma, \tag{2.57}$$

these Poisson commute, just as in (2.46). Denoting the corresponding flow parameter as $t_m(\sigma)$, we obtain Hamilton's equations in the form

$$\frac{dg(\lambda)}{dt_m(\sigma)} = [g(\lambda), H_m(\sigma)] = \frac{[g(\lambda), g^m(\sigma)]}{\lambda - \sigma}. \tag{2.58}$$

For the case $n = 2$, $m = 1$, identifying $g(\lambda)$ with $\mathcal{L}(\lambda)$, we recover the quadratic Poisson bracket relations (1.71a)–(1.71f) governing the Toda lattice, as well as the Poisson commutativity (1.72) of the spectral invariants and the commutator form (1.73) of the Hamiltonian vector fields they generate.

REMARK 2.6. The splitting (2.32) is not the only one possible in the case of a loop algebra. This particular splitting, and variants on it, correponds to what is usually known as the *rational R-matrix structure*, because of the form of the expression (2.43) defining the resulting Poisson bracket structure. By choosing a slightly more restrictive definition of the loop algebra, in which the curve Γ is viewed instead as the boundary of the union of a finite number of discs on an elliptic curve, with the splitting corresponding to elements that are continued to the interiors of the discs as meromorphic functions with poles of specific order at the centre; i.e., as elliptic functions, we arrive at another class of split R-matrices known as *elliptic R-matrices*, which are of at least equal importance in applications. Detailed discussions of these cases may be found, e.g., in [**OPRS, FT, Sk2**]. The algebraic geometry underlying elliptic R-matrices are dealt with in the lectures of J. Hurtubise.

In this lecture, the general phase space setting for isospectral flows in loop algebras and groups has been developed. However, to see that this does, indeed, lead to completely integrable Hamiltonian systems, as in the examples of the previous lecture, with a linearization of the flows via abelian integrals on the invariant spectral curve, we must examine more explicitly the structure of the finite dimensional Poisson subspaces that occur. It is difficult to do this in full generality, but in the next lecture the case of rational R-matrix structures on the loop algebras $\widetilde{\mathfrak{gl}}(n)$ and $\widetilde{\mathfrak{sl}}(n)$ will be dealt with in greater detail. When the underlying algebra $\mathfrak{g}$ is semisimple, it is known [**Hu**] that the finite dimensional Poisson subspaces in $\widetilde{\mathfrak{g}}^*_R$ consist of rational functions of the loop parameter, so this is the case that will be dealt with more fully in the following.

Exercises 2.

1. Deduce the relations (2.46), (2.47), directly from the Poisson bracket relations (2.41).
2. For any $\sigma \in \Gamma$ consider the invariant
$$\Delta(\sigma)|_\mu = \det\big(\mu(\sigma)\big)$$
given by evaluation of the determinant of $\mu \in \widetilde{\mathfrak{gl}}^*(n)$ at $\lambda = \sigma$. Determine the Poisson brackets $\{\mu(\lambda), \Delta(\sigma)$ analogous to those in relations (2.47).
3. Deduce the commutativity of the invariants defined in (2.57) directly from the Poisson bracket relations (2.53).
4. Prove that $\det g(\sigma)$ is a Casimir for the quadratic Poisson bracket (2.51a)–(2.51b).
5. Show that the space of $g(\sigma)$'s having a rational dependence on σ with a given pole structure is a Poisson submanifold of $\widetilde{Gl}(n)$, and that the center of the Poisson algebra on this Poisson subspace is generated by

the coefficients of the polynomial $\det k(\lambda)$, where, $k(\lambda) = a(\lambda)g(\lambda)$, and $a(\lambda)$ is the fixed monic polynomial obtained by taking the products of the denominators of the principle parts of $g(\lambda)$ at all the poles.

3. Separation of Variables in Spectral Darboux Coordinates

3.1. Rational Coadjoint Orbits in $\widetilde{\mathfrak{gl}}^*(r)_R$. We now consider finite dimensional Poisson subspaces $\mathfrak{g}_A^* \subset \widetilde{\mathfrak{gl}}^*(r)_R$ consisting of rational elements of the form

$$\mathcal{N}(\lambda) = B + \sum_{i=1}^{n}\sum_{a=1}^{m_i} \frac{N_{ia}}{(\lambda - \alpha_i)^a}, \tag{3.1}$$

where $B \in \mathfrak{gl}(r)$ is a fixed $r \times r$ matrix, $\{\alpha_i \in \mathbf{C}\}_{i=1\ldots n}$ is a set of distinct complex constants and $\{m_i \in \mathbf{N}^+\}_{i=1\ldots n}$ is a set of positive integers. Let

$$a(\lambda) := \prod_{i=1}^{n} (\lambda - \alpha_i)^{m_i}, \qquad m := \sum_{i=1}^{n} m_i,$$

and define the degree m polynomial matrix

$$\mathcal{L}(\lambda) := a(\lambda)\mathcal{N}(\lambda) =: B\lambda^m + \sum_{j=1}^{m-1} L_j \lambda^{m-j}. \tag{3.2}$$

From the results of the last lecture, the Hamiltonian flows induced by the spectral invariants of $\mathcal{N}(\lambda)$ are isospectral, and therefore the (affine) curve $\mathcal{C}_0$ defined by the characteristic equation

$$\det(\mathcal{L}(\lambda) - z\mathbf{I}) = \mathcal{P}(\lambda, z) = 0 \tag{3.3}$$

is preserved under these flows. The characteristic polynomial is of the form

$$\mathcal{P}(\lambda, z) = (-z)^r + \sum_{j=1}^{r} P_j(\lambda) z^{r-j}, \tag{3.4}$$

where $P_j(\lambda)$ is a polynomial of degree $\leq jm$ with constant leading coefficient b_j given by the spectral invariants of B

$$\det(B - z\mathbf{I}) = \sum_{j=0}^{r} b_j z^{r-j}. \tag{3.5}$$

This defines the (affine) spectral curve as an r-fold branched covering of the complex λ-plane. This may be compactified to define an algebraic curve, which we denote $\mathcal{C}_{\mathcal{N}}$, by suitably extending it over the Riemann sphere. To do so, we identify local coordinates in the ambient 2-space over a punctured neighborhood of $\lambda = \infty$ as

$$\tilde{\lambda} := \frac{1}{\lambda}, \quad \tilde{z} := \frac{z}{\lambda^m} \tag{3.6}$$

and extend the curve over a neighborhood of $\lambda = \infty$ by

$$\widetilde{\mathcal{P}}(\tilde{\lambda}, \tilde{z}) := \tilde{\lambda}^m \mathcal{P}\left(\frac{1}{\tilde{\lambda}}, \frac{\tilde{z}}{\tilde{\lambda}^m}\right) = 0. \tag{3.7}$$

This amounts to embedding $\mathcal{C}_{\mathcal{N}}$ in the total space $\mathcal{T}$ of the hyperplane section bundle $\mathcal{O}(m) \to \mathbf{CP}^1$.

Since the leading (degree jm) terms in the polyomials $\{P_j(\lambda)\}_{j=1\dots r}$ are just the coefficients $\{b_1,\dots,b_r\}$ of the characteristic polynomial (3.5), these are constants. Not all the remaining coefficients however are independent invariants on a given coadjoint orbit, since some of these are Casimir invariants in the Lie Poisson structure on $\widetilde{\mathfrak{gl}}^*(r)_R$. The number of independent spectral invarariants generating nontrivial flows will depend on the choice of coadjoint orbit $\mathcal{O}_{\mathcal{N}_0}$. The quantities $\{\alpha_i, m_i\}_{i=1\dots n}$ defining the pole structure are invariant within a given coadjoint orbit, but there may be many different orbit types sharing this same pole structure.

Denote by $\widetilde{\mathfrak{Gl}}_+(r)$ and $\widetilde{\mathfrak{Gl}}_-(r)$ the subgroups of the loop group $\widetilde{\mathfrak{Gl}}(r)$ corresponding, respectively, to the Lie algebras $\widetilde{\mathfrak{gl}}_+(r)$ and $\widetilde{\mathfrak{gl}}_-(r)$, i.e., their elements are analytic maps $g:\Gamma\rightarrow Gl(r)$ that admit holomorphic extensions to the Γ_+ and Γ_- regions, respectively, with the latter normalized so that $g(\infty)=\mathbf{I}$. The group corresponding to the modified algebra $\widetilde{\mathfrak{gl}}_R(r)$, which is essentially the direct sum of the two subalgebras $\widetilde{\mathfrak{gl}}_+(r)$ and $\widetilde{\mathfrak{gl}}_-(r)$ (with a sign reversal in the Lie bracket of the second factor), is just the product $\widetilde{\mathfrak{Gl}}_+(r)\times\widetilde{\mathfrak{Gl}}_-(r)$ (which is known sometimes as the dual group associated to the Lie Poisson group $\widetilde{\mathfrak{Gl}}(r)$). The coadjoint action of $\widetilde{\mathfrak{Gl}}_\pm(r)$ on $\widetilde{\mathfrak{gl}}_\pm(r)^*$ corresponding to the infinitesimal action (2.31) is easily computed to be conjugation composed with the $P_\pm$ projection maps

$$\mathrm{Ad}^*_{g_\pm}:\mu_\mp=\big(\mathrm{Ad}_{g_\pm}(\mu_\mp)\big)_\mp \tag{3.8}$$
$$\mu_\mp\in\widetilde{\mathfrak{gl}}(r)^*_\pm\sim\widetilde{\mathfrak{gl}}(r)_\mp,\quad g_\pm\in\widetilde{\mathfrak{Gl}}(r)_\pm,$$

where the $\mathrm{Ad}_{g_\pm}$ action on the right is understood to refer to the group $\widetilde{\mathfrak{Gl}}(r)$.

It follows that under this coadjoint action, the element $\mathcal{N}\in\widetilde{\mathfrak{gl}}_R(r)^*$ is transformed as

$$(g_+,g_-):\mathcal{N}(\lambda)\rightarrow\widetilde{\mathcal{N}}(\lambda)=B+\sum_{i=1}^{n}\sum_{a=1}^{m_i}\frac{\widetilde{N}_{ia}}{(\lambda-\alpha_i)^a}, \tag{3.9}$$

where the principal part

$$\widetilde{\mathcal{N}_i(\lambda)}=\sum_{a=1}^{m_i}\frac{\widetilde{N}_{ia}}{(\lambda-\alpha_i)^a} \tag{3.10}$$

of $\mathcal{N}(\lambda)$ at $\alpha_i\in\Gamma_\pm$ is determined in terms of the principal part

$$\widetilde{\mathcal{N}_i}(\lambda)=\sum_{a=1}^{m_i}\frac{N_{ia}}{(\lambda-\alpha_i)^a} \tag{3.11}$$

of $\mathcal{N}(\lambda)$ by evaluating the Cauchy integrals

$$\widetilde{N}_{ia}=\frac{1}{2\pi i}\oint_{\lambda=\alpha_i}g_\pm^{\pm1}\mathcal{N}_i(\lambda)g_\pm^{\mp1}(\lambda-\alpha_i)^{a-1}d\lambda, \tag{3.12}$$

and the integration path is a small, positively oriented circle centred at $\lambda=\alpha_i\in\Gamma_\pm$ and containing no other pole.

The coadjoint orbit $\mathcal{O}_{\mathcal{N}_0}$ of a given element $\mathcal{N}_0(\lambda)$ of the form (3.1) is therefore identified with the product

$$\mathcal{O}_{\mathcal{N}_{0,1}}\times\cdots\times\mathcal{O}_{\mathcal{N}_{0,n}} \tag{3.13}$$

of the orbits of the principal parts

$$\widetilde{\mathcal{N}_{0,i}(\lambda)} := \sum_{a=1}^{m_i} \frac{\widetilde{N^0}_{ia}}{(\lambda - \alpha_i)^a} \tag{3.14}$$

under the action, defined as in (3.12), of the group of $(m_i - 1)$-jets of $\mathfrak{Gl}(r)$-valued maps centred at $\lambda = \alpha_i$. The Casimir invariants, i.e., the center of the Poisson algebra on $\widetilde{\mathfrak{gl}}_R(r)^*$ restricted to the subspace $\mathfrak{g}_A^*$ of rational elements of the form (3.1) are just the invariants of all the principal parts $\{\mathcal{N}_i(\lambda)\}_{i=1\dots n}$ under this action. These may be viewed as the "generalized " spectral invariants at the points $\lambda = \alpha_i$. In the case of simple poles (where all $m_i = 1$), they are just elements of the algebra of spectral invariants of the residue matrices $\{N_{i1}\}$, generated by the invariants $\{\operatorname{tr} N_{i1}^l\}_{i=1\dots n,\ l=1\dots r-1}$.

In the following, we shall only consider "generic" classes of rational coadjoint orbits, computing their dimension as well as the number of independent spectral invariants on them. By "generic" we mean those whose stability subgroups are abelian, and have the smallest possible dimension, and hence the orbits have the largest dimension amongst those with the given pole structure. A "generic" orbit within the given pole class will be the analog of a "regular" element in $\mathfrak{gl}(r)$; that is, one whose stabilizer is of the smallest possible dimension and abelian. Since the coadjoint orbit $\mathcal{O}_{\mathcal{N}_0}$ is really a product of orbits in the dual spaces $\{\mathfrak{gl}^{(m_i-1)*}(r)\}$ of the $(m_i - 1)$th jet extensions $\mathfrak{gl}^{(m_i-1)}(r)$ of $\mathfrak{gl}(r)$, it is easy to see what the dimension of such an orbit is. We have only to consider the orbit of an element of the form

$$\mathcal{N}_{i,0}(\lambda) = \frac{N}{(\lambda - \alpha_i)^{m_i}}, \tag{3.15}$$

where $N \in \mathfrak{gl}(r)$ is regular. It follows that, under the coadjoint action of the group $\mathfrak{Gl}^{(m_i-1)}(r)$ of $m_i - 1$ jets of maps localized at $\lambda = \alpha_i$ with values in $\mathfrak{Gl}(r)$, the stability subgroup of $\mathcal{N}_{i,0}(\lambda)$ consists of $(m_i - 1)$-jets $[g(\lambda)]$ determined by their Taylor coefficients $\{g^{(k)}(\alpha_i)\}_{k=1\dots m_i-1}$, with the zeroth term $g(\alpha_i) \in \mathfrak{Gl}(r)$ in the stabilizer $\mathfrak{G}_N \subset \mathfrak{Gl}(r)$ of $N \in \mathfrak{gl}(r)$, and the quantities $\{g'g^{(-1)}(\alpha_i), \dots, g^{(m_i-1)}g^{-1}(\alpha_i)\}$in the corresponding Lie algebra $\mathfrak{g}_N$. Since N is a regular element, $\mathfrak{G}_N$ is abelian and of dimension r, and the Lie algebra $\mathfrak{g}_N$is a maximal abelian (Cartan) subalgebra. Therefore the stability subgroup of $\mathcal{N}_{i,0}$ within $\mathfrak{Gl}^{(m_i-1)}(r)$ is abelian, and of dimension $m_i r$. The dimension of the orbit of $\mathcal{N}_{i,0}$ is therefore $m_i r^2 - m_i r = m_i r(r-1)$. Taking the product over such orbits at each $\lambda = \alpha_i$ gives orbits in $\widetilde{\mathfrak{gl}}_R(r)^*$ of dimension

$$\dim \mathcal{O}_{\mathcal{N}_0} = \sum_{i=1}^{n} m_i r(r-1) = mr(r-1). \tag{3.16}$$

These are the dimensions of the maximal orbits which, in a sense, extend the notion of orbits of regular elements in $\mathfrak{gl}^*(r)$. Orbits of this type, whose stabilizers at the level of the quotient groups $\prod_{i=1}^n \mathfrak{Gl}^{(m_i-1)}(r)$ of the dual group $\widetilde{\mathfrak{Gl}}_+(r) \times \widetilde{\mathfrak{Gl}}_-(r)$ are abelian and of dimension mr, will henceforth be referred to as "generic" orbits. It follows from considerations analogous to those in $\mathfrak{gl}(r)$, that the union of such orbits forms a stratified submanifold with boundaries whose interior are an open set in the Poisson subspace $\widetilde{\mathfrak{g}}_A^* \subset \widetilde{\mathfrak{gl}}_R(r)^*$ consisting of elements of the form (3.1) (with given B, $\{a_i, m_i\}_{i=1\dots n}$).

In particular, for the case of simple poles, where all the integers m_i equal 1, the coadjoint orbits may be identified with the product of orbits of n regular elements $N_i \in \mathfrak{gl}(r)$. Since these are just the ones with simple sprectrum (but arbitrary Jordan form), such orbits are clearly "generic" in the usual sense (i.e., they form an open, dense set within any reasonable topology on the space of n-tuples of complex $r \times r$ matrices).

The next section contains the main results generalizing those that led to the complete integration via abelian integrals in the examples detailed in Lecture 1. Their proofs may be found in the reference [**AHH1, AHH2, AHH3, AHH4, AHH5**].

3.2. Spectral Darboux Coordinates and Linearization of Flows. In the following, let g denote the (geometric) genus of the spectral curve $\mathcal{C}_{\mathcal{N}}$; i.e., the dimension of the space of holomorphic abelian differentials on $\mathcal{C}_{\mathcal{N}}$. The following three theorems concern the complete (Liouville) integrability of the Hamiltonians flows on rational symplectic leaves with respect to the rational, linear Poisson bracket structures on $\widetilde{\mathfrak{gl}}_R(r)^*$ defined by (2.41). Similar results also hold for the quadratic structure (2.53) but we leave these to the references, where proofs and further computational details may be found.

THEOREM 3.1. *The dimension of a generic rational coadjoint orbit $\mathcal{O}_{\mathcal{N}_0}$ in $\widetilde{\mathfrak{gl}}_R(r)^*$ on which the spectral curves are of genus g is*

$$\dim \mathcal{O}_{\mathcal{N}_0} = 2(g + r - 1). \tag{3.17}$$

The number of independent generators $\{P_1, \ldots, P_{g+r-1}\}$ of the ring of spectral invariants $\mathcal{I}_{\mathcal{N}_0} := \mathcal{I}(\widetilde{\mathfrak{gl}}_R(r)^*)|_{\mathcal{O}_{\mathcal{N}_0}}$ restricted to $\mathcal{O}_{\mathcal{N}_0}$ is $g+r-1$. The resulting isospectral foliation, given by the level sets $\{P_1 = c_1, \ldots, P_{g+r-1} = c_{g+r-1}\}$ is therefore Lagrangian, and the resulting Hamiltonian flows generated by elements of the ring $\mathcal{I}_{\mathcal{N}_0}$ are isospectral and completely integrable.

For the following, choose a fixed vector $V_0 \in \mathbf{C}^r$, and let d_0 be the dimension of the smallest B-invariant subspace containing V_0. This is the same as the rank of the $r \times r$ matrix with columns $\{V_0, BV_0, \ldots, B^{r-1}V_0\}$. Let $\{V_1, \ldots, V_r\}$ be a basis of generalized eigenvectors of B, ordered so that $V_0 \subset \mathrm{Span}\{V_1, \ldots, V_{d_0}\}$, and let $\{\widetilde{V}_1, \ldots, \widetilde{V}_r\}$ be the dual basis of generalized eigenvectors of B^T, satisfying

$$(\widetilde{V}_j, V_k) = \delta_{jk}, \quad (\widetilde{V}_j, V_0) = 0, \quad 1 \leq j, k = r \tag{3.18}$$

where (,) denotes the standard (complex) scalar product on $\mathbf{C}^r$. Let

$$\mathcal{M}(\lambda, z) := (\widetilde{\mathcal{L}(\lambda) - z\mathbf{I}}) \tag{3.19}$$

denote the transposed matrix of cofactors of $(\widetilde{\mathcal{L}(\lambda) - z\mathbf{I}})$ (i.e., the "classical adjoint" matrix), all of whose entries are polynomials in λ and z.

THEOREM 3.2. *The system of polynomial equations*

$$\mathcal{M}(\lambda, z)V_0 = 0 \tag{3.20}$$

has $g + d_0 - 1$ generically distinct solutions $\{\lambda_\mu, z_\mu\}_{\mu=1\ldots g+d_0-1}$. Define the additional $2(r - d_0)$ functions $\{q_j, p_j\}_{j=d_0+1\ldots r}$ by

$$q_j := \ln(\widetilde{V}_j, L_1 V_0), \quad p_j := (\widetilde{V}_j, L_1 V_j), \quad j = d_0 + 1, \ldots, r, \tag{3.21}$$

and let

$$\zeta_\mu := \frac{z_\mu}{a(\lambda_\mu)}. \tag{3.22}$$

Then the functions $\{\lambda_\mu, \zeta_\mu\}_{\mu=1\dots g+d_0-1}$, $\{q_j, p_j\}_{j=d_0+1\dots r}$ *form a canonical (Darboux) coordinate system on* $\mathcal{O}_{\mathcal{N}_0}$. *That is, the orbital symplectic form* ω *resulting from restricting the rational R-matrix Poisson bracket to the orbit* $\mathcal{O}_{\mathcal{N}_0}$ *may be expressed as*

$$\omega = -d\theta, \quad \theta = \sum_{\mu=1}^{g} \zeta_\mu d\lambda_\mu + \sum_{j=d_0+1}^{r} p_j dq_j. \tag{3.23}$$

REMARK 3.3. The coordinate pairs $\{(\lambda_\mu, z_\mu)\}$ all lie on the spectral curve $\mathcal{C}_{\mathcal{N}}$, since the relations (3.20) imply that the matrix of cofactors (3.19) is not of maximal rank at these points, and hence the characteristic equation

$$\det(\mathcal{L}(\lambda_\mu) - z_\mu \mathbf{I}) = \mathcal{P}(\lambda_\mu, z_\mu, P_1, \dots, P_{r+g-1}) = 0 \tag{3.24}$$

is satisfied. (We put in the dependence on the generators $\{P_1, \dots, P_{g+r-1}\}$ of the ring of spectral invariants explicitly to recall that these equations are defined on the invariant level sets of the latter, which determine the leaves of the Lagrangian fibration.) In fact, from its definition, the matrix $\mathcal{M}(\lambda, \zeta)$ can only have either full rank (r), rank 0 (i.e., vanish) or rank 1—which occurs exactly at the solutions of the characteristic equation. This means that z_μ is an eigenvalue of $\mathcal{L}(\lambda_\mu)$ (and ζ_μ an eigenvalue of $\mathcal{N}(\lambda_\mu)$) at one of the points on the spectral curve over λ_μ, and the corresponding eigenvector is just given by the columns of the matrix $\mathcal{M}(\lambda_\mu, z_\mu)$, which are all proportional. It is also quite easy to see from the definition (3.21) that the quantities $\{p_{d_0+1}, \dots, p_r\}$ are actually elements of the algebra $\mathcal{I}_{\mathcal{N}_0}$ of spectral invariants on $\mathcal{O}_{\mathcal{N}_0}$. Therefore, in view of the spectral equation (3.24) relating the (λ_μ, z_μ) pairs on the isospectral Lagrangian leaves, there will be a complete separation of variables in the canonical coordinate system $\{\lambda_\mu, \zeta_\mu\}_{\mu=1\dots g+d_0-1}$, $\{q_j, p_j\}_{j=d_0+1\dots r}$ for all the isospectral Hamiltonian flows generated by elements of the ring $\mathcal{I}_{\mathcal{N}_0}$ of spectral invariants on $\mathcal{O}_{\mathcal{N}_0}$. In view of their definition in terms of the spectral data, these coordinates will henceforth be referred to as *spectral Darboux coordinates.* From their definition, the first set of $g + d_0 - 1$ pairs $\{\lambda_\mu, \zeta_\mu\}_{\mu=1\dots g+d_0-1}$ just determine the points at which the component of an eigenvector along V_0 vanishes.

REMARK 3.4. There is a nice invariant interpretation of the points on the curve $\mathcal{C}_{\mathcal{N}_0}$ with coordinates $\{\lambda_\mu, z_\mu\}_{\mu=1\dots g+d_0-1}$, which was in fact the original point of departure leading to the discovery that they provide a separating Darboux ccordinate system [**AHH3, AHH4, AHH5**]. A more detailed discussion of the underlying algebraic geometry may be found in [**AHH2**], but the main idea is the following. Associated to the Lax matrix $\mathcal{L}(\lambda)$ is not only the spectral curve $\mathcal{C}_{\mathcal{N}}$, but also the eigenvectors of $\mathcal{L}(\lambda)$ over the points of this curve. Generically, there is a 1-dimensional space of eigenvectors at each such point, so these eigenspaces taken together form a line bundle $E \to \mathcal{C}_{\mathcal{N}}$ over $\mathcal{C}_{\mathcal{N}}$, which is explicitly realized as a subbundle of the trivial rank-r vector bundle $\mathcal{C}_{\mathcal{N}} \times \mathbf{C}^r \to \mathcal{C}_{\mathcal{N}}$. By considering the generic degree of these eigenvector line bundles, it is possible to see that, in general, they do not admit any holomorphic sections. However, one can equivalently consider the *dual* eigenvector bundle $E^* \to \mathcal{C}_{\mathcal{N}}$, i.e., the bundle of linear forms on the (1-dimensional) eigenspaces over the points of $\mathcal{C}_{\mathcal{N}}$. Since E is just a subbundle of the trivial rank-r bundle, the dual bundle E^* is realized as a quotient of the trivial dual rank r bundle $\mathcal{C}_{\mathcal{N}} \times \mathbf{C}^{r*} \to \mathcal{C}_{\mathcal{N}}$ of linear forms on $\mathbf{C}^r$ over the points of $\mathcal{C}_{\mathcal{N}}$, modulo the image of the map $(\mathcal{L}(\lambda) - z\mathbf{I}) : \mathbf{C}^r \to \mathbf{C}^r$. The resulting line bundle $E^* \to \mathcal{C}_{\mathcal{N}}$ can be shown to generically have degree $g + r - 1$, and a space of holomorphic sections $H^0(\mathcal{C}_{\mathcal{N}}, E^*)$ of dimension $h^0(\mathcal{C}_{\mathcal{N}}, E^*) = r$. Up to linear equivalence, this line bundle is identified with the degree $g + r - 1$ divisor (i.e., unordered formal sum of points)

$$\mathcal{D}_{E^*} = \sum_{\mu=1}^{g+r-1} \mathbf{p}_\mu, \quad \mathbf{p}_\mu \in \mathcal{C}_{\mathcal{N}_0}, \tag{3.25}$$

where a holomorphic sections vanishes. In the above formulation, the choice of vector $V_0 \in \mathbf{C}^r$ determines, through the quotient, a section of the bundle $E^* \to \mathcal{C}_{\mathcal{N}_0}$ (i.e., it induces a linear form through the scalar product on $\mathbf{C}^r$). The points $\{\mathbf{p}_\mu\}$ where it vanishes either lie over $\lambda = \infty$ or are determined by the equations (3.20), so the coordinates of those points in $\mathcal{D}_{E^*}$ which lie over finite points in $\mathbf{CP}^1$ (i.e., where $\lambda \neq \infty$) are just the solutions $\{\lambda_\mu, z_\mu\}_{1\ldots g+d_0-1}$ of these equations. The number d_0 is the number of points of the given divisor that lie over ∞, which generically may be anywhere between 0 and $r-1$, depending on the nature of the spectrum of $\mathcal{L}(\lambda)$ at $\lambda = \infty$.

REMARK 3.5. The solution of the system of polynomial equations in two variables (3.20) may be reduced to the equivalent problem of finding the roots of the single polynomial in one variable

$$Q(\lambda) := \det(V_0, \mathcal{L}(\lambda)V_0, \ldots, \mathcal{L}^{r-1}(\lambda)V_0) = K \prod_{\mu=1}^{g+d_0-1} (\lambda - \lambda_\mu), \tag{3.26}$$

to determine the λ_μ's. The z_μ's are then given by the explicit rational expression

$$z_\mu := (-1)^r \left(\frac{P(\lambda_\mu)}{R(\lambda_\mu)} + \operatorname{tr} \mathcal{L}(\lambda_\mu) \right), \tag{3.27}$$

where the polynomials $P(\lambda)$ and $R(\lambda)$ are defined by

$$P(\lambda) := \det(W_0, V_0, \mathcal{L}(\lambda)V_0, \ldots, \mathcal{L}^{r-3}(\lambda)V_0, \mathcal{L}^{r-1}(\lambda)V_0), \tag{3.28a}$$

$$R(\lambda) := \det(W_0, V_0, \mathcal{L}(\lambda)V_0, \ldots, \mathcal{L}^{r-3}(\lambda)V_0, \mathcal{L}^{r-2}(\lambda)V_0) \tag{3.28b}$$

with the fixed vector $W_0 \in \mathbf{C}^r$ chosen such that the vectors:

$$\{W_0,\ V_0,\ \mathcal{L}(\lambda)V_0, \ldots, \mathcal{L}^{r-2}(\lambda)V_0\} \tag{3.29}$$

are linearly independent. Details of this simplification may be found in [**Sc, Ge, H2**].

THEOREM 3.6. *The independent spectral invariants* $\{\mathcal{P}_1, \ldots, \mathcal{P}_{g+r-1}\}$ *may be chosen such that the spectral coordinates* $\{p_{d_0+1}, \ldots, p_r\}$ *depend only the last* $r-1$ *invariants* $\{\mathcal{P}_{g+1}, \ldots, \mathcal{P}_{g+r-1}\}$ *and the abelian differentials defined on* $\mathcal{C}_{\mathcal{N}_0}$ *by*

$$\omega_\mu := \frac{\partial \mathcal{P}(\lambda, z)/\partial \mathcal{P}_\mu}{\partial \mathcal{P}(\lambda, z)/\partial z} \frac{d\lambda}{a(\lambda)}, \quad \mu = 1, \ldots, g+r-1 \tag{3.30}$$

are holomorphic for $\mu = 1, \ldots, g$, *providing a basis for the space* $H^0(\mathcal{C}_{\mathcal{N}_0}, \mathcal{K}_{\mathcal{C}_{\mathcal{N}}})$ *of sections of the canonical bundle, while the remaining ones* $\{\omega_\mu\}_{\mu=g+1\ldots g+r-1}$ *are meromorphic, with pole support at the points over* $\lambda = \infty$.

COROLLARY 3.7. *The Liouville generating function*

$$S(\lambda_\mu, z_\mu, q_j, p_j, \mathcal{P}_1, \dots, \mathcal{P}_{g+r-1})$$

determining the linearizing coordinates

$$Q_\mu := \frac{\partial S}{\partial \mathcal{P}_\mu}, \quad \mu = 1, \dots, g+r-1 \tag{3.31}$$

conjugate to the conserved spectral invariants $\{\mathcal{P}_1, \dots, \mathcal{P}_{g+r-1}\}$, *is expressible as the abelian integral*

$$S = \sum_{\nu=1}^{g+d_0-1} \int_{(\lambda_0, z_0)}^{(\lambda_\nu, z_\nu)} \frac{z d\lambda}{a(\lambda)} + \sum_{j=d_0+1}^{r} p_j q_j \tag{3.32}$$

and therefore the first g *coordinates of the linearizing map consist of a sum over holomorphic (first kind) abelian integrals*

$$Q_\mu = -\sum_{\nu=1}^{g} \int_{(\lambda_0, z_0)}^{\lambda_\nu, z_\nu)} \omega_\mu, \quad \mu = 1, \dots, g. \tag{3.33}$$

Within a choice of normalized basis, this is just given by the Abel map, whilst the remaining $r-1$ *components*

$$Q_\mu = -\sum_{\nu=1}^{g} \int_{(\lambda_0, z_0)}^{(\lambda_\nu, z_\nu)} \omega_\mu, + \sum_{j=d_0+1}^{r} q_j \frac{\partial p_j}{\partial \mathcal{P}_\mu}, \quad \mu = g+1, \dots, g+r-1, \tag{3.34}$$

involve abelian integrals of the second and third kinds, with pole support of the integrand over $\lambda = \infty$.

The linear flow equations generated by a spectral invariant Hamiltonian $H \in \mathcal{I}_\mathcal{N}$ on the level sets $\{\mathcal{P}_1 = c_1, \dots, \mathcal{P}_{g+r-1} = c_{g+r-1}\}$ is thus

$$Q_\mu = Q_\mu^0 + \frac{\partial H}{\partial \mathcal{P}_\mu}, \quad \mu = 1, \dots, g+r-1. \tag{3.35}$$

There is again a simple invariant interpretation of these linear flows—at least, of their projection to the first g components. By comparing overlapping neighborhoods $U_\pm \subset \mathcal{C}_\mathcal{N}$ that project to open discs in $\mathbf{CP}^1$ containing, respectively, the interior and exterior regions $\overline{\Gamma}_\pm$, we see that the Lax equation

$$\frac{d\mathcal{N}(\lambda)}{dt} = \pm[(d\phi)_\pm, \mathcal{N}(\lambda)] \tag{3.36}$$

implies that the eigenvector line bundle of $\mathcal{L}(\lambda)$ (or $\mathcal{N}(\lambda)$) is defined by a transition function $\rho(\lambda, z, t)$ on the overlap $U_+ \cap U_-$ that evolves linearly in t as

$$\rho(\lambda, z, t) = \rho(\lambda, z, 0) e^{Q(\lambda, z) t}, \tag{3.37}$$

where $Q(\lambda, z)$ is just the eigenvalue of $d\phi \in \widetilde{\mathfrak{gl}}_R(r)$ acting on the eigenvector of $\mathcal{L}(\lambda)$ at the point (λ, z). This determines a linear flow on the space of line bundles of given degree (Picard variety), which is a complex torus identified, by tensoring with the dual bundle at $t = 0$, as a linear flow on the Jacobian $\mathcal{J}(\mathcal{C}_\mathcal{N})$ of the curve. This torus is identified via the Abel map as a leaf of the isospectral Lagrangian foliation quotiented by the $r-1$ flows generated by the remaining invariants $\{\mathcal{P}_{g+1}, \dots, \mathcal{P}_{g+r-1}\}$.

We end this section with a brief discussion of the background and references for the material presented here. The very earliest work of course dates back to the nineteenth century studies on integration of dynamical systems using elliptic and hyperelliptic integrals [**J, N, Ro**], but these made no use of Lax equations, spectral invariants, etc.—except for the case of the Euler and Lagrange tops. R. Garnier (1919), in a remarkable paper [**Gar**], wrote down the set of compatible Lax equations

$$\frac{\partial \mathcal{A}(\lambda)}{\partial t_j} = \left[\frac{A_i}{\lambda - \alpha_j}, \mathcal{A}(\lambda)\right], \quad j = 1, \dots, n \tag{3.38a}$$

$$\mathcal{A}(\lambda) := \sum_{j=1}^{n} \frac{A_i}{\lambda - \alpha_j}, \qquad A_j \in \mathfrak{gl}(r), \tag{3.38b}$$

and integrated these by transforming them to a system of abelian integrals on the spectral curve of $\mathcal{A}(\lambda)$. He actually did this through the introduction of what we have called here the "spectral Darboux coordinates", without ever noting the Hamiltonian structure of the equations! These are, of course, just the special form of Lax equations generated by the spectral invariant Hamiltonians

$$G_j := \operatorname{res}_{\lambda=\alpha_i} \operatorname{tr}\big(\mathcal{A}^2(\lambda)\big), \quad j = 1, \dots, n \tag{3.39}$$

and may be viewed as the classical version of the $\mathfrak{gl}(r)$ Gaudin spin system, although for $r > 2$, the quadratic invariants (3.39) do not form a complete set of generators of the algebra of spectral invaraiants.

In more recent times, Flaschka and McLaughlin [**F1, F2, F3, FM**] integrated a number of finite dimensional isospectral flows via hyperelliptic integrals using what amounts to a separation of variables technique in the special case of hyperellipsoidal coordinates. Moser [**Mo1, Mo2, Mo3**] did so as well, although the Lax matrices he used did not contain an explicit spectral parameter. Previato [**P**] related the quasi-periodic solutions to the NLS equation to the Moser formulation, following the lines of previous work by Mumford [**Mu**]. Sklyanin [**Sk2, Sk3, Sk4**] applied separation of variables techniques to the solution of both classical and quantum spin systems (Gaudin, XXZ, XYZ), although his results were mainly concerned with 2×2 Lax matrices. The general case of $r \times r$ Lax matrices with rational dependence on the spectral parameter was treated first in [**AHH3, AHH4, AHH5**], where the general notion of *spectral Darboux coordinates* was introduced.

References

[AA] S. J. Al'ber and M. S. Al'ber, *Hamiltonian formalism for nonlinear Schrödinger equations and Sine-Gordon equations*, J. London Math. Soc. (2) **36** (1987), 176–192.

[Ad] M. Adler, *On a trace functional for formal pseudo-differential operators and the symplectic structure of the Korteweg-de Vries equation*, Invent. Math. **50** (1979), 219–248.

[AvM1] M. Adler and P. van Moerbeke, *Completely integrable systems, Euclidean Lie algebras, and curves*, Adv. Math. **38** (1980), 267–317.

[AvM2] M. Adler and P. van Moerbeke, *Linearization of Hamiltonian systems, Jacobi varieties and representation theory*, Adv. Math. **38** (1980), 318–379.

[AHH1] M. R. Adams, J. Harnad, and J. Hurtubise, *Dual moment maps to loop algebras*, Lett. Math. Phys. **20** (1990), 294–308.

[AHH2] M. R. Adams, J. Harnad, and J. Hurtubise, *Isospectral Hamiltonian flows in finite and infinite dimensions. II. Integration of flows*, Comm. Math. Phys. **134** (1990), 555–585.

[AHH3] M. R. Adams, J. Harnad, and J. Hurtubise, *Liouville generating function for isospectral Hamiltonian flow in loop algebras*, Integrable and Superintegrable Systems (B. A. Kupershmidt, ed.), World Scientific, Singapore, 1990, pp. 232–256.

[AHH4] M. R. Adams, J. Harnad, and J. Hurtubise, *Coadjoint orbits, spectral curves and Darboux coordinates*, The Geometry of Hamiltonian Systems (T. Ratiu, ed.), Math. Sci. Res. Inst. Publ., vol. 2, Springer, Berlin-Heidelberg-New York, 1991, pp. 9–21.

[AHH5] M. R. Adams, J. Harnad, and J. Hurtubise, *Darboux coordinates and Liouville-Arnold integration in loop algebras*, Comm. Math. Phys. **155** (1993), 385–413.

[AHP] M. R. Adams, J. Harnad, and E. Previato, *Isospectral Hamiltonian flows in finite and infinite dimensions. I. Generalised Moser systems and moment maps into loop algebras*, Comm. Math. Phys. **117** (1988), 451–500.

[DT1] E. Date and S. Tanaka, *Analogue of inverse scattering theory for the discrete Hill's equation and exact solutions for the periodic Toda lattice*, Progr. Theoret. Phys. **55** (1976), 47–465.

[DT2] E. Date and S. Tanaka, *Periodic multi-soliton solutions of Korteweg-de Vries equation and Toda lattice*, Progr. Theoret. Phys. Suppl. **59** (1976), 107–126.

[DKN] B. Dubrovin, I. Krichever, and S. Novikov, *Integrable systems*. I, Encyclopaedia Math. Sci., vol. 4, Springer, Berlin-Heidelberg-New York, 1990, pp. 173–280.

[Dr] V. G. Drinfel'd, *Hamiltonian structures on Lie groups, Lie bialgebras and the geometrical meaning of the classical Yang-Baxter equations*, Soviet Math. Dokl. **27** (1983), 69–71.

[F1] H. Flaschka, *On the Toda lattice*. I. Phys. Rev. B **9** (1974), 1924–1926; II. Progr. Theoret. Phys. **51** (1974), 703–716.

[F2] H. Flaschka, *Toward an algebro-geometrical interpretation of the Neumann system*, Tohoku Math. J. **37** (1984), 407–426.

[F3] H. Flaschka, *Relations between infinite-dimensional and finite-dimensional isospectral equations*, Nonlinear Integrable Systems—Classical Theory and Quantum Theory, World Scientific, Singapore, 1983, pp. 219–240.

[FM] H. Flaschka and D. McLaughlin, *Canonically conjugate variables for the Korteweg-de Vries equation and the Toda lattice with periodic boundary conditions*, Progr. Theoret. Phys. **55** (1976), 438–456.

[FT] L. D. Faddeev and L. A. Takhtajan, *Hamiltonian methods in the theory of solitons*, Springer, Berlin-Heidelberg-New York, 1986.

[Gar] R. Garnier, *Sur une classe de systèmes différentiels abéliens déduits de la théorie des équations linéaires*, Rend. Circ. Mat. Palermo **43** (1919), 155–191.

[Gau] M. Gaudin, *Diagonalisation d'une classe d'hamiltoniens de spin*, J. Physique I **37** (1976), 1087–1098.

[Ge] M. I. Gekhtman, *Separation of variables in the classical $Sl(N)$ magnetic chain*, Comm. Math. Phys. **167** (1995), 593–605.

[H1] J. Harnad, *Isospectral flow and Liouville-Arnold integration in loop algebras*, Geometric and Quantum Methods in Integrable Systems (G. Helminck, ed.), Lecture Notes in Phys., vol. 424, Springer, Berlin-Heidelberg-New York, 1993.

[H2] J. Harnad, *Integrable Hamiltonian systems and transformations groups*, monograph in preparation, 1999.

[Hu] J. C. Hurtubise, *The geometry of generalised Hitchin systems*, this volume.

[IK] A. R. Its and V. P. Kotljarov, *Explicit formulas for solutions of a nonlinear Schrödinger equation*, Dokl. Akad. Nauk Ukrain. SSR Ser. A no. 11, 1051 (1976), 965–968 (Russian).

[K] B. Kostant, *The solution to a generalized Toda lattice and representation theory*, Adv. Math. **34** (1979), 195–338.

[J] C. G. J. Jacobi, *Vorlesung über Dynamik*, Verlag G. Reimer, Berlin, 1884.

[Mo1] J. Moser, *Three integrable Hamiltonian systems connected with isospectral deformations*, Adv. Math. **16** (1975), 197–225.

[Mo2] J. Moser, *Various aspects of integrable systems*, Dynamical systems, CIME Lectures, Bressanone, 1978, Progr. Math., vol. 8, Birkhäuser, Boston, 1980, pp. 233–290.

[Mo3] J. Moser, *Geometry of quadrics and spectral theory*, The Chern Symposium (Berkeley, June 1979), Springer, New York, 1980, pp. 147–188.

[Mu] D. Mumford, *Tata lectures on theta*. II. *Jacobian theta functions and differential equations*, with the collaboration of C. Musili, M. Nori, E. Previato, M. Stillman, and H. Umemura, Progr. Math., vol. 43. Birkhäuser, Boston, 1984.

[N] C. Neumann, *De problemate quodam mechanico, quod ad primam integralium ultraellipticorum classem revocatur*, J. Reine Angew. Math. **56** (1859), 46–63.

[OPRS] M. M. Olshanetsky, A. M. Perelomov, A. G. Reyman, and M. A. Semenov-Tian-Shansky, *Integrable systems*. II, Encyclopaedia Math. Sci., vol. 16, Springer, Berlin-Heidelberg-New York, 1994.

[P] E. Previato, *Hyperelliptic quasi-periodic and soliton solutions of the nonlinear Schrödinger equation*, Duke Math. J. **52** (1985), 329–377.

[Ro] E. Rosochatius, Dissertation, Göttingen *Gebr. Unger*, Berlin, 1877.

[RS1] A. G. Reyman and M. A. Semenov-Tian-Shansky, *Reduction of Hamiltonian systems, affine Lie algebras and Lax equations*. I, Invent. Math. **54** (1979), 81–100.

[RS2] A. G. Reyman and M. A. Semenov-Tian-Shansky, *Reduction of Hamiltonian systems, affine Lie algebras and Lax equations*. II, Invent. Math. **63** (1981), 423–432.

[STS1] M. A. Semenov-Tian-Shansky, *What is a classical R-matrix*, Funct. Anal. Appl. **17** (1983), 259–272.

[STS2] M. A. Semenov-Tian-Shansky, *Dressing transformations and Poisson group actions*, Publ. Res. Inst. Math. Sci. **21** (1985), 1237–1260.

[Sc] D. R. D. Scott, *Classical functional Bethe ansatz for* Sl(N): *Separation of variables for the magnetic chain*, J. Math. Phys. **35** (1994), 5831–5843.

[Sk1] E. K. Sklyanin, *Algebraic structures associated with the Yang-Baxter equations*, Funktsional Anal. i Prilozhen **16** (1982), 27–34; English transl., Funct. Anal. Appl. **16** (1982), 263–270.

[Sk2] E. K. Sklyanin, *Poisson structure of a periodic classical XYZ chain*, Zap. Nauchn. Sem. Leningrad. Otdel. Math. Inst. Steklov. (LOMI) **150** (1986), 154–180; English transl., J. Soviet Math. **46** (1989), 1664–1683.

[Sk3] E. K. Sklyanin, *Separation of variables in the Gaudin model*, Zap. Nauchn. Sem. Leningrad. Otdel. Math. Inst. Steklov. (LOMI) **164** (1987), 151–169; English transl., J. Soviet Math. **47** (1989), 2473–2488.

[Sk4] E. K. Sklyanin, *Separation of variables—new trends*, Quantum Field Theory, Integrable Models and Beyond (Kyoto, 1994), Progr. Theoret. Phys. Suppl. **118** (1995), 35–60.

[Sy1] W. Symes, *Systems of Toda type, inverse spectral problems and representation theory*, Invent. Math. **59** (1980), 13–51.

[Sy2] W. Symes, *Hamiltonian group actions and integrable systems*, Physica D **1** (1980), 339–374.

Department of Mathematics and Statistics, Concordia University, 7141 Sherbrooke W., Montréal, Québec H4B 1R6, Canada and Centre de recherches mathématiques, Université de Montréal, C.P. 6128, succ. Centre-ville, Montréal, Québec H3C 3J7, Canada.

E-mail address: harnad@crm.umontreal.ca

Centre de Recherches Mathématiques
CRM Proceedings and Lecture Notes
Volume **26**, 2000

The Geometry of Generalised Hitchin Systems

J. C. Hurtubise

ABSTRACT. The generalised Hitchin systems specialise to many of the well-known integrable systems of mathematical physics. In this survey, we show how they give the systems associated to the rational, trigonometric or elliptic r-matrices, as well as the Calogero-Moser systems. We also show how they are associated in a natural way to certain symplectic surfaces, in the $\mathrm{Gl}(r, \mathbb{C})$-case, or more generally, for an arbitrary reductive group of rank r, to certain $r+1$-folds.

The Hitchin systems [**Hi1, Hi2**], and their generalisations due to Markman and Bottacin [**Ma, Bo**], are algebraically integrable systems defined on moduli spaces of "stable pairs" over a Riemann surface. These pairs are made up of a principal holomorphic G-bundle E over the Riemann surface, where G is a reductive complex group, and a holomorphic or meromorphic 1-form valued section ϕ of the associated coadjoint bundle $\mathrm{ad}^*(E)$. The Hitchin systems have become one of the most studied integrable systems in geometry, for several reasons. One is that they are the classical phase spaces for the Geometric Langlands programme of Beilinson, Drinfeld and Laumon. Another is that they naturally lead to integrable systems of Jacobian varieties, and these are of considerable interest in Seiberg-Witten theory [**Do1, DoW**]. A third reason is that they specialise to various well-studied integrable systems of mathematical physics. In our first section, we introduce the systems, and show that for the cases of rigid bundles, they become the systems associated to r-matrices on loop algebras ([**AvM, FRS, RS1, RS2**]). Specialising further, one can then obtain the various classically and currently studied integrable systems such as tops, geodesics on the ellipsoid, the Gaudin models, the Landau-Lifschitz equations, the finite gap solutions to KdV, NLS, CNLS, and so on [**AHP, AvM, Hi1, Hi2, HHu, RS1, RS2**]. The next section shows that by considering Hitchin systems for another class of groups which are not reductive but which are associated to root systems, we can obtain the Calogero-Moser systems associated to arbitrary root systems. The third section concentrates on a feature of the Hitchin systems for $G = \mathrm{Gl}(N, \mathbb{C})$. These systems are algebraically integrable systems of Jacobians, and they posses natural isomorphisms with a symmetric product of a symplectic surface; we consider systems exhibiting such a property in more generality. In the

---

1991 *Mathematics Subject Classification.* Primary 14K20; Secondary 70H05.

The author of this article would like to thank NSERC and FCAR for their support.

This is the final form of the paper.

fourth section, we look briefly at what happens in the case of arbitrary reductive groups.

Much of this material is in the literature in some form; see, in particular, [**AHH, Hu1, HuMa1, HuMa2**]. The following just attempts to present it from the particular point of view of Hitchin systems, with the aim of giving an overview.

1. Hitchin Systems and r-Matrices

Let Σ be a compact Riemann surface of genus γ. We denote the canonical bundle of Σ by K_Σ. Le D be a positive divisor on Σ; $D = \sum_i n_i p_i, p_i \in \Sigma, n_i \in \mathbb{N}$. To the divisor D, we can associate the zero-dimensional scheme $\widehat{D}$ whose components are the $(n_i - 1)$th formal neighbourhoods of the points p_i; in other words, $\widehat{D}$ consists of the points p_i with multiplicity n_i.

We fix a reductive complex group G, with Lie algebra $\mathfrak{g}$; $\mathfrak{g}$ gets identified with its dual $\mathfrak{g}^*$ by an invariant pairing. We let G_D denote the group of maps $\widehat{D} \to G$, with $\mathfrak{g}_D$ the associated Lie algebra. We also fix a degree d. Our generalised Hitchin systems will be defined on the moduli space $\mathcal{M} = \mathcal{M}_{\Sigma,G,d,D}$ of pairs (E, ϕ), where

- E is a holomorphic principal G-bundle of degree d,
- ϕ is a section in $H^0\big(\Sigma, \mathrm{ad}^*(E) \times K_\Sigma(D)\big)$, where $\mathrm{ad}^*(E)$ is the coadjoint bundle associated to E; in other terms, ϕ is a 1-form-valued section of $\mathrm{ad}^*(E)$ with poles at D. It is referred to as a Higgs field.

For these pairs to form a nice moduli space, we restrict to those satisfying an appropriate stability condition (see [**Hi1, Fa**]).

The space $\mathcal{M}$ is a Poisson manifold, and its symplectic leaves $\mathcal{M}_\mathcal{O} = \mathcal{M}_{\Sigma,G,d,D,\mathcal{O}}$ are given by asking that the polar parts of ϕ over $\widehat{D}$ lie in a fixed coadjoint orbit $\mathcal{O}$ in $\mathfrak{g}_D^*$. To obtain the Poisson structure on $\mathcal{M}$, we will reduce the cotangent bundle $T^*\mathcal{N}$ of another moduli space $\mathcal{N}$, the space of pairs (E, tr), where:

- E is a holomorphic principal G-bundle of degree d,
- tr is a trivialisation of E over $\widehat{D}$.

Concretely, we cover Σ by open sets $U_0 = \Sigma \backslash \mathrm{support}(D)$, $U_i =$ disjoint disks centred at the points p_i, $i = 1, \dots, k$. Choosing trivialisations over the open sets, the bundle is determined by the associated transition functions $T_{i0} : U_0 \cap U_i \to G$. If the trivialisations on U_i are chosen to coincide with tr at the p_i, then the functions T_{i0} encode tr also. If we consider a one-parameter family $\big(E(t), \mathrm{tr}(t)\big)$ of elements of $\mathcal{N}$, corresponding to a family of transition functions $T_{i0}(t)$, the infinitesimal variation at $t = 0$ is given by the cocycle $\dot{v}_{i0} = T_{i0}^{-1}\dot{T}_{i0}$, which represents a class in the deformation space $T\mathcal{N}|_{(E(0),\mathrm{tr}(0))} = H^1\big(\Sigma, \mathrm{ad}(E)(-D)\big)$. Dually, the cotangent space at $\big(E(0), \mathrm{tr}(0)\big)$ is given by the space $H^0\big(\Sigma, \mathrm{ad}^*(E) \otimes K_\Sigma(D)\big)$. In other terms, the cotangent bundle $T^*\mathcal{N}$ is given as the space of triples (E, tr, ϕ):

- E is a holomorphic principal G-bundle of degree d,
- tr is a trivialisation of E over the divisor D,
- ϕ is a section in $H^0\big(\Sigma, \mathrm{ad}^*(E) \otimes K_\Sigma(D)\big)$.

Concretely, the section ϕ is represented by meromorphic functions $\phi^i : U_i \to \mathfrak{g}^*$, $i = 1, \dots, r$ with poles of order at most n_i at p_i, and a holomorphic function $\phi^0 : U_0 \to \mathfrak{g}^*$, with over $U_0 \cap U_i$:

$$\phi^i = \mathrm{Ad}^*_{T_{i0}} \phi^0. \tag{1.1}$$

The group G_D of maps from $\widehat{D}$ into G acts naturally on the set of trivialisations tr, and so on $\mathcal{N}$. The action lifts to a Hamiltonian action on $T^*\mathcal{N}$.

PROPOSITION 1.1. [**Ma**] *The moment map for the action of G_D on $T^*\mathcal{N}$ is given by the expression of the polar part of ϕ in the trivialisation* tr. *The quotient $T^*\mathcal{N}/G$ can be identified (over open sets) with $\mathcal{M}$, giving a Poisson structure on $\mathcal{M}$. The symplectic leaves are then given by fixing a (co-)adjoint orbit $\mathcal{O}$ in the Lie algebra of G_D, and asking that the polar parts of ϕ lie in this orbit.*

As it is a bundle, the tangent bundle $T(T^*\mathcal{N})$ fits into the exact sequence:

$$0 \to H^0\big(\Sigma, \mathrm{ad}^*(E)(D)\big) \to T(T^*\mathcal{N}) \to H^1\big(\Sigma, \mathrm{ad}(E)(-D)\big) \to 0. \tag{1.2}$$

The first term corresponds to the tangents to the fiber, and the last to the tangents to $\mathcal{N}$, the base.

Concretely, corresponding to a family $T_{i0}(t), \phi^i(t), \phi^0(t)$ of elements of $T^*\mathcal{N}$, satisfying $\phi^i(t) = \mathrm{Ad}^*_{T_{i0}(t)}\, \phi^0(t)$, we have the tangent vectors $\dot{v}_{i0} = T_{i0}^{-1}\dot{T}_{i0}, \dot{\phi}^i, \dot{\phi}^0$; on $U_0 \cap U_0$,

$$\dot{\phi}^i = \mathrm{Ad}^*_T\big[\big(\mathrm{ad}^*_{\dot{v}_{i0}}(\phi^0)\big) + \dot{\phi}^0\big]. \tag{1.3}$$

We would like to split (1.2). Let

$$\langle\ ,\ \rangle : H^0\big(\Sigma, \mathrm{ad}^*(E)(D)\big) \times H^1\big(\Sigma, \mathrm{ad}(E)(-D)\big) \to \mathbb{C} \tag{1.4}$$

denote the Serre duality pairing; explicitly, with respect to the cover U_0, U_i, it is defined by

$$\langle a, b\rangle = \sum_i \mathrm{res}_{p_i}(a_0 \cdot b_{i0}). \tag{1.5}$$

Choose a vector space V of cocycles mapping isomorphically to $H^1\big(\Sigma, \mathrm{ad}(E)(-D)\big)$. One can split (1.2) as follows. For any section a of $\mathrm{ad}^*(E)(D)$ over U_0, let $a^{\&}$ denote the element of $H^0\big(\Sigma, \mathrm{ad}^*(E)(D)\big)$ whose pairing with elements of V is the same as that of a. One then has the isomorphism

$$\begin{aligned} T(T^*\mathcal{N}) &\to H^1\big(\Sigma, \mathrm{ad}(E)(-D)\big) \oplus H^0\big(\Sigma, \mathrm{ad}^*(E)(D)\big) \\ (\dot{v}, \dot{\phi}) &\mapsto (\dot{v}, (\dot{\phi})^{\&}). \end{aligned} \tag{1.6}$$

PROPOSITION 1.2. *Under this isomorphism, the form on $T^*\mathcal{N}$ becomes, at (E, tr, ϕ):*

$$\Omega\big((v', \phi'^{\&}), (\dot{v}, \dot{\phi}^{\&})\big) = \langle v', \dot{\phi}^{\&}\rangle - \langle \dot{v}, \phi'^{\&}\rangle + \langle [v', \dot{v}], \phi\rangle. \tag{1.7}$$

PROOF. One can parametrise $\mathcal{N}$ locally by V; indeed, if T_{i0} are the transition matrices for (E, tr), one has in a neighbourhood of the origin a map $V \to \mathcal{N}$ obtained by associating to the cocycle v the transition matrix $T_{i0} \cdot \exp(v_{i0})$; this in turn defines a map $\rho : V \times V^* = T^*V \to T^*\mathcal{N}$, which preserves the symplectic form. With respect to the splitting (1.6), the differential $d\rho$ at the origin is

$$d\rho(\dot{v}, \dot{\phi}) = \left(\dot{v}, \dot{\phi} + \frac{1}{2}(\mathrm{ad}^*_{\dot{v}}\, \phi)^{\&}\right).$$

Substituting into the standard expression for the symplectic form gives our formula. □

We have not so far mentioned any Hamiltonians on $\mathcal{M}$; we remedy this by introducing them now. Let $F_1, \ldots, F_r$ denote the ad*-invariant functions on the dual $\mathfrak{g}^*$ of the Lie algebra $\mathfrak{g}$; assume that they are homogeneous, of degrees $\delta_1, \ldots, \delta_r$. Applying F_i to ϕ gives a global section $F_i(E,\phi)$ of the bundle $K_\Sigma(D)^{\otimes\delta_i}$. If one chooses a basis for the sections, and decomposes $F_i(E,\phi)$ with respect to this basis, one obtains functions $F_{i,j}(E,\phi)$ which will be our Hamiltonians. Alternately, one can choose an element ω in $H^1\big(\Sigma, K_\Sigma^{1-\delta_i}(-\delta_i D)\big)$ and pair it with F_i using Serre duality, to obtain a function $F_{i,\omega}(E,\phi)$.

We can compute the flows on the space $T^*\mathcal{N}$ corresponding to $F_{j,\omega}$ as follows. The F_j are functions on $\mathfrak{g}^*$.The differential dF_j of F_j at a gives a map

$$\begin{aligned} DF_j(a) : \mathfrak{g}^* &\to \mathbb{C} \\ (\dot{a}) &\mapsto DF_j(a)(\dot{a}) \end{aligned} \tag{1.8}$$

which is homogeneous of degree $\delta_j - 1$ in a, linear in $\dot{a}$, and invariant.

Applying this to ϕ gives a section $DF_j(\phi)$ of the bundle $\mathrm{ad}(E) \otimes K_\Sigma(D)^{\otimes\delta_j - 1}$. (We can use the representations of ϕ in any trivialisation, by the invariance of F_j). Multiplication by the cocycle ω then gives a class $DF_j(\phi)\omega$ in $H^1\big(\Sigma, \mathrm{ad}(E)(-D)\big)$, which is the deformation space for our pairs (E, tr). Referring simply to the definitions of the Hamiltonian vector field associate to a function, as well as to the form (1.7) of the symplectic form with respect to the splitting (1.6), we obtain for the flow

$$\begin{aligned} T_{i0}^{-1}\dot{T}_{i0} &= DF_j(\phi^0)\omega_{i0}, \\ \dot{\phi}^i &= 0, \\ \dot{\phi}^0 &= 0. \end{aligned} \tag{1.9}$$

where ω_{i0} are cocycle representatives for the class ω over $U_0 \cap U_i$. We note that transition matrices for the pairs (E, tr) are not unique, and so we can modify the flows by an infinitesimal gauge transformation, that is, a coboundary $g_0 + \mathrm{Ad}_{T_{i0}^{-1}}(g_1)$

$$\begin{aligned} T_{i0}^{-1}\dot{T}_{i0} &= DF_j(\phi^0)\omega_{i0} + g_0 + \mathrm{Ad}_{T_{i0}^{-1}}(g_1), \\ \dot{\phi}^i &= \mathrm{ad}^*_{g_1}(\phi^1), \\ \dot{\phi}^0 &= -\,\mathrm{ad}^*_{g_0}(\phi^0). \end{aligned} \tag{1.10}$$

Referring to the formula for the flows and the formula for the symplectic form, we have that the Poisson bracket of $F_{k,\omega}, F_{j,\rho}$ is given by $\langle [DF_k(\phi)\omega, DF_j(\phi)\rho], \phi\rangle$, which vanishes by invariance.

These functions all descend to the quotient $\mathcal{M}$. On $\mathcal{M}$, we note that some of the functions are Casimirs: indeed, the $F_{i,\omega}$ include the pull backs under the moment map of the invariant functions on $\mathfrak{g}^*_D$, that is the invariant functions applied to the polar parts of ϕ.

There remains the question of whether or not there are enough of these quantities to give a completely integrable system. This is indeed the case. One can prove this by counting dimensions. There is, however, a more intrinsic way of seeing this, tying in to the geometry of the spectral curves of the Higgs fields. We will return to this in Sections 3 and 4.

Rigid Bundles. We will now examine three choices of base curves and degrees for which the moduli space of bundles is a point. In these cases, the moduli space

$\mathcal{N}$ becomes over an open set essentially a space of trivialisations on a fixed bundle, and the moduli space $\mathcal{M}$ is just a space of Higgs fields ϕ. The equations on the reduced space then acquire a particularly simple form: indeed, the rigidity means that the infinitesimal variations of T_{i0} can be split as cocycles:

$$DF_j(\phi^0)\omega_{i0} = (DF_j(\phi^0)\omega_{i0})_0 - T_{i0}^{-1}(DF_j(\phi^0)\omega_{i0})_i T_{i0}. \tag{1.11}$$

The equations then transform to:

$$\begin{aligned} T_{i0}^{-1}\dot{T}_{i0} &= 0, \\ \dot{\phi}^i &= \operatorname{ad}^*_{(DF_j(\phi^0)\omega_{i0})_i}(\phi^i), \\ \dot{\phi}^0 &= -\operatorname{ad}^*_{(DF_j(\phi^0)\omega_{i0})_0}(\phi^0). \end{aligned} \tag{1.12}$$

Identifying the Lie algebras with their duals, these are our Lax pairs. We will see that in these cases the symplectic leaves of $\mathcal{M}$ are simply finite dimensional symplectic leaves for certain r-matrix brackets defined on loop algebras. First our three cases:

i) *Rational curves, degree zero.* A theorem of Grothendieck [**Gro**] asserts that all G-bundles over the Riemann sphere reduce to the maximal torus H. These H-bundles are in turn classified by duals $\mathbb{C}^* \to H$ to the characters, a discrete set. There is a "norm" on the characters such that for bundles in families, the generic bundle in the family corresponds to the minimal norm, with non-generic bundles corresponding to "jumps" in the structure. For bundles of degree zero, the minimal bundle is the trivial one. This tells us that the moduli space is just a point, consisting of this trivial bundle, as the moduli cannot accommodate the discontinuities associated with these jumps. The transition functions for the only semi-stable bundle E can be taken to be the identity.

ii) *Elliptic curves, degree one.* If Σ is an elliptic curve, the stable G-bundles over Σ are classified by Atiyah [**A**] for $G = \mathrm{Gl}(n, \mathbb{C})$ and by Freedman, Morgan and Witten [**FMW**] and Donagi[**Do3**] for arbitrary G. We will restrict our attention to $\mathrm{Gl}(n, \mathbb{C})$, degree one. When the degree is coprime to n, the stable bundles are essentially rigid: they are classified by their top exterior power.

Rather than try to give a transition function for these bundles, we use an alternate approach. One of the features of an elliptic curve Σ is that its universal cover is $\mathbb{C}$: one represents Σ as $\mathbb{C}/(w_1\mathbb{Z} + w_2\mathbb{Z})$. One can lift a bundle E to $\mathbb{C}$, where it is trivial. The bundle E can then be thought of as a quotient of the trivial bundle $\mathbb{C} \times G$ over $\mathbb{C}$. If we define an equivalence relation $\simeq$ by

$$(z, g) \simeq (z + w_i, A_i(z) \cdot g), i = 1, 2$$

we can define bundles E as $\mathbb{C} \times G/\simeq$. For this to work, the *automorphy factors* $A_i(z)$ must satisfy

$$A_1(z)A_2(z + w_1) = A_2(z)A_1(z + w_2) \tag{1.13}$$

The key theorem turns out to that of Narasimhan and Seshadri [**NS, AB**], which tells us that for a stable bundle the A_i can be taken to be constant, provided one allows a puncture in the surface. More precisely, one has a flat connection over the punctured surface, whose $(0, 1)$ component defines the holomorphic structure; integrating, the A_i are essentially the holonomies of the connection around the cycles of the curve. One asks that the holonomy around the puncture be central.

If c represents this central element, the relation becomes:

$$A_1 A_2 = c A_2 A_1. \tag{1.14}$$

The choice of c is determined by the degree of the bundle.

Let us consider the case of $G = \mathrm{Gl}(r, \mathbb{C})$, and degree one. By results of [**AB**], the constant is $c = q^{-1}$, where $q = \exp(2\pi i/r)$. One can show [**HuKj**] that one can take

$$A_1 = \mathrm{diag}(1, q, q^2, \ldots, q^{r-1}), \quad A_2 = d \cdot \begin{pmatrix} 0 & 1 & 0 & \ldots & 0 \\ 0 & 0 & 1 & \ldots & 0 \\ \vdots & \vdots & \vdots & & \vdots \\ 0 & 0 & 0 & \ldots & 1 \\ 1 & 0 & 0 & \ldots & 0 \end{pmatrix}. \tag{1.15}$$

The constant d determines the top exterior power of the bundles.

In the trivialisation given by the flat connection, holomorphic sections cannot be smooth at the puncture, due to the relation (1.14); indeed, they must have a singularity of the form $z^{1/r}$·holomorphic. Lifting to $\mathbb{C}$, the sections are represented by multi-valued vector functions F defined on the inverse image in $\mathbb{C}$ of $\Sigma - \{\text{puncture}\}$, satisfying

$$F(z + \omega_1) = A_1 \cdot F(z), \; F(z + \omega_2) = A_2 \cdot F(z),$$

and of the form

$$(z)^{-1/r}(\text{holomorphic})$$

near any inverse image p of the puncture, where z is a coordinate centred at p.

Similarily, meromorphic sections of the endomorphism bundle are given by meromorphic functions on $\mathbb{C}$ satisfying

$$M(z + \omega_1) = A_1 \cdot M(z) \cdot A_1^{-1}, \quad M(z + \omega_2) = A_2 \cdot M(z) \cdot A_2^{-1}. \tag{1.16}$$

There are no singularities at the inverse images of the puncture.

iii) *Nodal curves.* One other example for which this rigidity phenomenon occurs is that of a rational degeneration of an elliptic curve. This is the Riemann sphere with two points $0, \infty$ identified. Restricting again our attention to $\mathrm{Gl}(r, \mathbb{C})$, we again consider a bundle E of degree one over the Riemann sphere with transition matrix from $U_0 = \{z \neq \infty\}$ to $U_1 = \{z \neq 0\}$ given by

$$T(z) = \begin{pmatrix} 0 & 1 & 0 & \ldots & 0 \\ 0 & 0 & 1 & \ldots & 0 \\ \vdots & \vdots & \vdots & & \vdots \\ 0 & 0 & 0 & \ldots & 1 \\ z^{-1} & 0 & 0 & \ldots & 0 \end{pmatrix}.$$

This corresponds to a sum of the trivial rank $r - 1$ bundle with the line bundle $\mathcal{O}(1)$ of degree 1. We then glue the fiber over 0 in the U_0-trivialisaation to the fiber over ∞ in the U_1 trivialisation by a multiple $d\mathbb{I}$ of the identity map, and obtain a bundle over the nodal curve. One can show [**HuKj**] that the constant d determines the top exterior power of the bundle, and that the bundle is rigid once this exterior power is fixed. We set $d = 1$.

As for the elliptic case, one can represent the vector bundle as a quotient of a trivial bundle over the universal cover $\mathbb{C}$ of $\mathbb{C}^* = \Sigma \backslash \{\text{singular point}\}$. Introduce the change of variables $z = \exp(2\pi i x)$. Sections of the bundle are represented by vector functions $F(x)$ over $\mathbb{C}$ with

$$F(x+1) = F(x)$$
$$\lim_{ix\to+\infty} F(x) = \lim_{ix\to-\infty} T\big(\exp(2\pi i x)\big)F(x).$$

Alternately, we can introduce the multi-valued change of trivialisation

$$S = \mathrm{diag}(1, z^{1/r}, z^{2/r}, \ldots, z^{n-1/r})$$

over both U_0 and U_1; the transition matrix gets conjugated to

$$STS^{-1} = z^{-1/r}A_2,$$

where A_2 is the matrix of (1.15). Lifting to $\mathbb{C}$ by changing variables $z = \exp(2\pi i x)$, sections are represented by vector functions F satisfying

$$F(x+1) = A_1F(x),$$

with

$$\lim_{ix\to+\infty} F(x) = \lim_{ix\to-\infty} \exp(-2\pi i x/r)A_2F(z).$$

Correspondingly, the sections of $\mathrm{End}(E)$ are represented by matrix valued functions $M(x)$ satisfying

$$M(x+1) = A_1M(x)A_1^{-1},$$

with

$$\lim_{ix\to+\infty} M(x) = \lim_{ix\to-\infty} A_2M(x)A_2^{-1}.$$

Rigid Bundles and the r-Matrix Systems. Let us consider $T^*\mathcal{N}$ in our three rigid cases. The formulae for the flows are given above; we want specific formulae for the symplectic form, or, dually, the Poisson structure. Let the open sets U_i be defined as above. Rigidity of the bundle is equivalent to the statement that $H^1\big(\Sigma, \mathrm{End}(E)\big) = 0$. For our first case, the bundle is indeed rigid, and the vanishing of the cohomology is tantamount to the statement that any cocycle g_{0i} : $U_0 \cap U_i \to \mathfrak{gl}(n, \mathbb{C})$ splits:

$$g_{0i} = g_0 + g_i, \quad g_0 : U_0 \to \mathbb{C},\ g_i : U_i \to \mathbb{C}. \tag{1.17}$$

If one imposes a normalisation condition of the form $g_0(\infty) = 0$, defining a subspace $\mathrm{Hol}_0\big(U_0, \mathfrak{gl}(n, \mathbb{C})\big)$ of $\mathrm{Hol}\big(U_0, \mathfrak{gl}(n, \mathbb{C})\big)$ the splitting is unique, and one has a well-known decomposition

$$\begin{aligned}\bigoplus_i \mathrm{Hol}\big(U_0 \cap U_i, \mathfrak{gl}(n, \mathbb{C})\big) &\simeq \mathrm{Hol}_0\big(U_0, \mathfrak{gl}(n, \mathbb{C})\big) \oplus \Big(\bigoplus_i \mathrm{Hol}\big(U_i, \mathfrak{gl}(n, \mathbb{C})\big)\Big)\\ &\simeq \mathrm{Hol}_+ \oplus \mathrm{Hol}_- .\end{aligned} \tag{1.18}$$

For the other two cases, the bundle is not rigid, but if one fixes the top exterior power, they are. Infinitesimally, the top exterior power corresponds to the trace, and the rigidity gets translated into the fact that $H^1\big(\Sigma, \mathfrak{sl}(E)\big) = 0$, so that the traceless component splits,

$$\begin{aligned}&\bigoplus_i \mathrm{Hol}\big(U_0 \cap U_i, \mathfrak{sl}(n, \mathbb{C})\big)\\ &\qquad\simeq \mathrm{Hol}\big(U_0, \mathfrak{sl}(n, \mathbb{C})\big) \oplus \Big(\bigoplus_i T_{i0}\,\mathrm{Hol}(U_i, \mathfrak{sl}(n, \mathbb{C}))T_{i0}^{-1}\Big)\\ &\qquad\simeq \mathrm{Hol}_+ \oplus \mathrm{Hol}_- .\end{aligned} \tag{1.19}$$

In the elliptic case, if one constructs the bundles by automorphy factors instead of transition functions, the terms Hol_+, Hol_- of (1.19) become instead:

$$\begin{aligned} \mathrm{Hol}_+ &= \{M \in \mathrm{Hol}(\mathbb{C}\backslash\{w_1\mathbb{Z}+w_2\mathbb{Z}\}, \mathfrak{sl}(n,\mathbb{C})) \mid \\ &\qquad M(z+\omega_1) = A_1 \cdot M(z) \cdot A_1^{-1}, M(z+\omega_2) = A_2 \cdot M(z) \cdot A_2^{-1}\} \\ \mathrm{Hol}_- &= \Big(\bigoplus_i \mathrm{Hol}(U_i, \mathfrak{sl}(n,\mathbb{C}))\Big). \end{aligned} \tag{1.20}$$

There are similar decompositions in the trigonometric case.

Dually to the equalities $H^1(\Sigma, \mathrm{End}(E)) = 0$, $(H^1(\Sigma, \mathfrak{sl}(E)) = 0)$, we have the equalities $H^0(\Sigma, \mathrm{End}(E) \otimes K_\Sigma) = 0$, $(H^0(\Sigma, \mathfrak{sl}(E) \otimes K_\Sigma) = 0)$, so that global $\mathrm{End}(E)$ (or $\mathfrak{sl}(E)$)-valued holomorphic 1-forms vanish. In turn, this tells us that meromorphic sections are detemined by their polar parts: the map (replacing End by $\mathfrak{sl}$ where appropriate

$$\mathrm{res}: H^0(\Sigma, \mathrm{End}(E) \otimes K_\Sigma(D)) \to H^0(\widehat{D}, \mathrm{End}(E)) \tag{1.21}$$

is injective; in the symplectic leaves of our reduction, these residues lie in a fixed coadjoint orbit of the group G_D. In our reduced space, when our bundle is rigid, we are representing tangent vectors by:

$$\begin{aligned} \dot{T}_{i0} T_{i0}^{-1} &= 0, \\ \dot{\phi}^0 &= a, \\ \mathrm{res}_{p_i}(\dot{\phi}^i) &= [c_i, \mathrm{res}(\phi^i)] \end{aligned} \tag{1.22}$$

for some a in Hol_+. Using our gauge freedom (i.e., changing trivialisations on the bundles), this can be represented in an equivalent fashion as

$$\begin{aligned} \dot{T}_{i0} T_{i0}^{-1} &= -c_i, \\ \dot{\phi}^0 &= a, \\ \mathrm{res}_{p_i}(\dot{\phi}^i) &= 0. \end{aligned} \tag{1.23}$$

Let us take a second such tangent vector:

$$\begin{aligned} \dot{T}_{i0} T_{i0}^{-1} &= -d_i, \\ \dot{\phi}^0 &= b, \\ \mathrm{res}_{p_i}(\dot{\phi}^i) &= 0. \end{aligned} \tag{1.24}$$

We evaluate the symplectic form on these two tangent vectors. Referring to our formula of the symplectic form, we obtain

$$\sum_i \langle [c_i, d_i], \mathrm{res}_{p_i}(\phi^i) \rangle.$$

Of course, this is just the pull-back of the symplectic form on the coadjoint orbits in which the polar parts live; as the moment map corresponds precisely to taking the residues, this is of course scarcely surprising.

We can of course formulate this in terms of Poisson brackets, instead of symplectic forms. Representing the differentials dF, dG of two functions F, G on $\mathcal{M}_\mathcal{O}$ by cocycles in $V \subset \mathrm{Hol}_-$, we have, dually to (1.7), the expression for the Poisson bracket of F, G at $\phi = (\phi^0, (\phi^i))$:

$$\{F, G\}(\phi) = -\langle [dF, dG], \phi^0 \rangle.$$

The same expression is true if one replaces the dF, dG by any cocycles homologous to them (with respect to $H^1(\Sigma, \mathrm{End}(E)(D))$, lying in Hol_-. Let P_+, P_- denote the projections onto the factors $\mathrm{Hol}_+, \mathrm{Hol}_-$. As $P_+(dF) = P_+(dG) = 0, P_-(dF) = dF, P_-(dG) = dG$, this can be written, setting $R = P_+ - P_-$, as

$$\{F, G\}(\phi) = \frac{1}{2}\langle [R(dF), dG)] + [dF, R(dG)], \phi^0\rangle. \tag{1.25}$$

In this representation, we have $\phi^0 \in \mathrm{Hol}_+$; we could also have chosen a set of trivialisations of our bundles with transition functions at only some of the points of D, and as the bundles are still rigid, still obtained a splitting $\mathrm{Hol}_+ \oplus \mathrm{Hol}_-$. In this setting, ϕ^0 no longer lies in Hol_-; it turns out however that the formula (1.25) still holds. The change in sign between the positive and negative projections is then essentially due to a shift in contours of integration in taking residues.

This is of course the formulation given for the r-matrix Poisson bracket, for example in [**RS2**, p. 120]. In this situation, one is on a loop algebra, and the differentials of functions are also identified with elements of the loop algebra, using an invariant pairing. The Poisson bracket is given by (1.25); the finite dimensional symplectic leaves correspond precisely to our orbits $\mathcal{M}_{\mathcal{O}}$, at least when the group is semisimple [**Hu3**]. The Hamiltonian flow induced by an invariant function F on the loop algebra is then

$$\dot{\phi} = \frac{1}{2}[\phi, R(dF)(\phi)],$$

which coincides, in our case, with the flow given above in (1.12).

2. Hitchin Systems and Calogero Systems

We next turn to another case in which the Hitchin systems turn up, albeit in modified form: the Calogero-Moser systems. These systems are associated to *root systems*, which are given by:

- A group $H = (\mathbb{C}^*)^r$, acted on by a finite group W, the Weyl group. Let $\mathfrak{h}$ denote its Lie algebra; it is equipped with a W-invariant scalar product.
- A W-invariant set $\mathcal{R}$ of homomorphisms $\tilde{\alpha} : H \to \mathbb{C}^*$. Let $\alpha :\to \mathbb{C}$ denote the corresponding Lie algebra homomorphism (a "root"), and $\mathbb{C}_\alpha$ the corresponding representation.

The root system must satisfy certain axioms, see, e.g. Humphreys [**Hum**]. The root systems are classified by Dynkin diagrams. The reduced root systems A_n, B_n, C_n, D_n, E_6, E_7, E_8, F_4, G_2 correspond to simple Lie groups; there are in addition non-reduced systems BC_n.

In canonical coordinates $(x, p) \in \mathfrak{h} \times \mathfrak{h}^*$ the system is given by the Hamiltonian

$$H_{\mathrm{CM}} = p \cdot p + \sum_{\alpha \in \mathcal{R}} m_{|\alpha|} \wp\big(\alpha(x)\big), \tag{2.1}$$

where $\wp$ is the Weierstrass $\wp$-function associated to an elliptic curve Σ, and the $m_{|\alpha|}$ are constants depending only on the norm of the root α. (This system is in fact the elliptic Calogero-Moser system; there are also trigonometric and rational versions).

There has been a considerable amount of work recently [**BCS, dP**] in finding *Lax representations* for the Calogero flows, in particular Lax pairs with a spectral parameter. This involves rewriting the equations in the form:

$$\dot{A}(\lambda) = [M(\lambda), A(\lambda)], \tag{2.2}$$

where $M(\lambda), A(\lambda)$ are elements of some loop algebra. The recent motivation comes from string theory, and has to do with the period matrices of the underlying spectral curves defined by $\det(A(\lambda) - z\mathbb{I}) = 0$, where A is now given in some representation.

Given the form (1.12) of the Hitchin flows, a natural way to look for Lax pairs is by trying to write the Calogero-Moser systems as Hitchin systems, and this for several reasons:

- Simple Lie groups are naturally associated to root systems,
- The three types of Calogero-Moser systems correspond to the three varieties of rigid Hitchin systems: rational, elliptic and trigonometric,
- The functions occurring in the Hamiltonians occur naturally when one tries to write out explicit formulae for the Higgs fields of Hitchin systems.

There is one case for which this works very well: see Krichever [**Kr**]. The Calogero-Moser system for the root system A_{n-1} corresponds to the Hitchin system for the corresponding Lie algebra $\mathfrak{sl}(n, \mathbb{C})$, where one takes the divisor D to be the origin of the elliptic curve, and the polar part of the Higgs field to lie in the coadjoint orbit of $\mathrm{diag}(1, 1, 1, \ldots, 1, -n+1)$. There are however several important problems one runs into when one tries to push this through for other root systems:

- The known Lax pairs for the Calogero-Moser systems do not in general lie in the Lie algebra corresponding to the root system. Sometimes, indeed, the Lax pairs do not even occur in a dimension corresponding to a representation of the group.
- As shown in [**Do1**], while the Calogero-Moser system is Weyl invariant, there are in general no Weyl invariant coadjoint orbits. Also, there are often no suitable coadjoint orbits of the correct dimension.
- Finally one has a Calogero-Moser system for the root systems BC_n, for which there are no corresponding groups.

The answer, following [**HuMa2**], is to consider for any root system not the associated semi-simple group associated to the system, when it exists, but the following semidirect product of the torus and the sum of the root spaces:

$$\bigoplus_{\alpha=1}^{n} \mathbb{C}_\alpha \to G \to H. \tag{2.3}$$

The Weyl group acts naturally on G. On the associated Lie algebra $\mathfrak{g}$, there is a natural pairing which is Weyl invariant and which pairs up the $-\alpha$ root space with the α root space, and pairs $\mathfrak{h}$ with itself.This identifies $\mathfrak{g}$ with $\mathfrak{g}^*$, but not equivariantly. One has:

PROPOSITION 2.1 ([**HuMa2**]). *The G-Ad^*-invariant functions on $\mathfrak{g}^*$ only depend on the root space components, and correspond to the H-invariant functions on $\bigoplus_\alpha \mathbb{C}_\alpha$. The generic coadjoint orbit is $2r$-dimensional, where $r = \dim(N')$, and is of the form*

$$\left(H\text{-orbit in } \bigoplus_{\alpha} \mathbb{C}_\alpha\right) \times \mathfrak{h}^*. \tag{2.4}$$

Moreover, $\mathfrak{g}^*$ has a $2r$-dimensional connected (W-invariant) coadjoint orbit $\mathcal{O}$.

The orbit $\mathcal{O}$ can be taken as the product of the H-orbit of $(1, 1, 1, \ldots, 1)$ in $\bigoplus_{\alpha=1}^n \mathbb{C}_\alpha$ with $\mathfrak{h}^*$.

We then consider the moduli $\mathcal{M} = \mathcal{M}_{\Sigma,G,0,\infty,\mathcal{O}}$ of pairs (E, ϕ) over the elliptic curve Σ for this group, in degree 0, with Higgs fields which possess a simple pole at the origin ∞, and whose residue lies in this coadjoint orbit.

We first discuss the bundles of degree 0 over Σ. To do this, we use the map $G \to H$, which induces a map on bundles:

$$P : \{G\text{-bundles}\} \to \{H\text{-bundles}\}. \tag{2.5}$$

An H-bundle is essentially a sum of line bundles and the space that classifies H-bundles is $Jac(\Sigma)^r = \Sigma^r$. Cover Σ by two open sets, a disk U_0 containing the origin, and $U_1 = \Sigma - \{\text{origin}\}$. Taking the natural cover $\mathfrak{h} \to \Sigma^r$, if $x = (x_1, \ldots, x_r)$ are the linear coordinates on $\mathfrak{h}$, the transition functions $T : U_0 \cap U_1 \to H$ can be taken with the components

$$(e^{x_1\zeta(z)}, e^{x_2\zeta(z)}, \ldots, e^{x_r\zeta(z)}).$$

For any root α, we can apply the corresponding character $\tilde{\alpha}$ to x to obtain the transition function of a line bundle $L_{\alpha(x)}$. with transition function $e^{\alpha(x)\zeta(z)}$. The fiber of P at T is given by

$$P^{-1}(T) = \bigoplus_{\alpha \in \mathcal{R}} H^1(\Sigma, L_{\alpha(x)}). \tag{2.6}$$

This can be seen by writing out an explicit transition function for a G-bundle, and trying to normalise. Unless $L_{\alpha(x)}$ is trivial, the group $H^1(\Sigma, L_{\alpha(x)})$ vanishes, and so over the open set $(\Sigma^r)_{\text{gen}}$ for which none of the $L_{\alpha(x)}$ are trivial, the map P is an isomorphism. For our Calogero-Moser phase space, we will restrict our attention to the set $(\Sigma^r)_{\text{gen}}$.

We next turn to the form of the Higgs fields for a bundle $x \in (\Sigma^r)_{\text{gen}}$. Under the action of H, the dual space $\mathfrak{g}^*$ decomposes as $\mathfrak{h}^* \oplus (\bigoplus_\alpha \mathbb{C}_\alpha)$, and so the expressions ϕ^i of the Higgs fields ϕ in the U_i-trivialisations do also. They are given by:

$$\begin{aligned} (\phi^0)_{\mathfrak{h}^*} &= (\phi^1)_{\mathfrak{h}^*} = p, \text{a constant in } \mathfrak{h}^*, \\ (\phi^0)_\alpha &= \rho(\alpha(x), z), \\ (\phi^1)_\alpha &= \rho(\alpha(x), z)e^{\alpha(x)\zeta(z)}. \end{aligned} \tag{2.7}$$

Here

$$\rho(x, z) = \frac{\sigma(z - x)}{\sigma(z)\sigma(x)}, \tag{2.8}$$

where σ, ζ denote the standard elliptic functions, satisfying

$$\frac{d}{dz}\log(\sigma(z)) = \zeta(z), \quad \frac{d}{dz}\zeta(z) = -\mathfrak{p}(z),$$

so that σ has a simple zero at the origin and ζ a simple pole. One finds that p, x are conjugate variables.

There are, in fact, two ways of considering the Calogero-Moser phase space; one is in terms of a full reduction of the unreduced phase space $T^*\mathcal{N}$ under the action of G, and the other is as a reduction under the action of the subgroup $\bigoplus_\alpha \mathbb{C}_\alpha$; this latter part leaves in place a trivialisation of the H-factor in the bundle over ∞. The reason is that the H-bundles possess H as a group of automorphisms; alternately, given a a pair (E, ϕ), we can normalise the pole of the Higgs field in an essentially unique way, giving back the H-trivialisation. It is this second way that we will prefer to use.

For the Hamiltonian, we note that there is a family of quadratic forms on $\mathfrak{g}$ that are both H- and W-invariant. The dimension of this space of forms is $m+1$, where m is the number of W orbits in the root system $\mathcal{R}$. Writing elements a of $\mathfrak{g}$ as $\big((a_\alpha)_{\alpha\in\mathcal{R}}, a_{\mathfrak{h}}\big)$, these are of the form

$$F_{c_{\mathfrak{h}}, c_{|\alpha|}}(a) = c_{\mathfrak{h}} a_{\mathfrak{h}} \cdot a_{\mathfrak{h}} + \sum_{\alpha\in\mathcal{R}} c_{|\alpha|} a_\alpha a_{-\alpha},$$

where $c_{\mathfrak{h}}, c_{|\alpha|}$ are constants, with $c_{|\alpha|}$ depending only on the norm $|\alpha|$.

Setting $c_{\mathfrak{h}} = 1$, and fixing the constants $c_{|\alpha|}$, we can define the Calogero-Moser Hamiltonian by applying the $F_{1,c_{|\alpha|}}$ to ϕ and taking the zero-th order term:

$$H_{\mathrm{CM}}(E,\phi) = \langle F_{1,c_{|\alpha|}}(\phi), \omega\rangle. \tag{2.9}$$

Here ω is the class in $H^1\big(\Sigma, K(-2p_0)\big)$ represented by the cocycle $\zeta(z)$. over $U_0 \cap U_1$. Computing, and using the fact that

$$\frac{\sigma\big(z-\alpha(x)\big)}{\sigma(z)\sigma\big(\alpha(x)\big)} \times \frac{\sigma\big(z-\alpha(-x)\big)}{\sigma(z)\sigma\big(\alpha(-x)\big)} = \mathfrak{p}(z) - \mathfrak{p}\big(\alpha(x)\big),$$

we have

$$H_{\mathrm{CM}}(E,\phi) = p\cdot p + \sum_{\alpha\in\mathcal{R}} -c_{|\alpha|}\mathfrak{p}\big(\alpha(x)\big). \tag{2.10}$$

We thus have an appropriate phase space (or rather a covering of it: one should quotient out the action of the Weyl group), and a natural way of obtaining the Calogero-Moser Hamiltonian; it does not, however give us the other Hamiltonians, or indicate why the Calogero-Moser system should be algebraically integrable; one also does not see in this picture the compactification of the level sets of the Jacobians.

One way to obtain these features is by embedding the system into the Hitchin systems for some $\mathrm{Gl}(V)$. We take V to be a Weyl-invariant sum of weight spaces: $V = \mathbb{C}^N = \bigoplus_w \mathbb{C}_w$. The embeddings are to be induced by embedding the group G into $\mathrm{Gl}(V)$. The torus H acts naturally on V, mapping H into the subgroup D of diagonal matrices:

$$\Xi : H \to D,$$

with on the level of Lie algebras

$$\xi : \mathfrak{h} \to \mathfrak{d}.$$

We assume that this is an embedding. It induces a natural map from H-bundles to D-bundles:

$$\widehat{\Xi} : \mathcal{M}_H \to \mathcal{M}_D.$$

We want to extend this to our picture, building a map $\widehat{\Xi}$ from our phase space of triples

$$\left(H\text{-bundle } E, H\text{-triv'n at } \infty, \text{ section } \phi_G \text{ of } E\left(\mathfrak{h}^* \oplus \Big(\bigoplus_\alpha \mathbb{C}_\alpha\Big)\right) \otimes K_\Sigma(\infty)\right),$$

to a corresponding $GL(N)$-phase space of triples:

$$\big(\text{rank } N \text{ bundle } E_{\mathrm{Gl}(N)}, \mathrm{Gl}(V)\text{-triv'n at } \infty,$$
$$\text{section } \phi_{\mathrm{Gl}(N)} \text{ of } \mathrm{End}(E_{\mathrm{Gl}(N)}) \otimes K_\Sigma(\infty)\big).$$

To do this, we extend Ξ to G, in two different ways, corresponding to the constructions of D'Hoker and Phong [**dP**], and Bordner, Corrigan and Sasaki [**BCS**]: both embeddings, unfortunately, are not homomorphisms.

i) *The d'Hoker-Phong ansatz: root spaces and shift operators.* The root spaces $\mathbb{C}_\alpha$ of a simple Lie algebra act on a representation of the group by moving weight spaces $\mathbb{C}_w$ to weight spaces $\mathbb{C}_{w+\alpha}$. It is then natural to try to think of our group acting on V via operators Sh_α on V whose matrix entries are

$$(\mathrm{Sh}_\alpha)_{w,w'} = \delta_{w-w',\alpha} C_{w,w'}.$$

We will take the constants $C_{w,w'}$ to be symmetric ($C_{w,w'} = C_{w',w}$), and Weyl invariant. We then set

$$\phi_{\mathrm{Gl}(N)} = \xi\big((\phi_G)_{\mathfrak{h}^*}\big) + \sum_{\alpha\in\mathcal{R}} (\phi_G)_\alpha \,\mathrm{Sh}_\alpha\,. \tag{2.11}$$

This embeds the Calogero-Moser system into the $\mathrm{Gl}(N)$ Hitchin phase space. Over an open dense set of the latter space, one can choose suitable trivialisations of the bundles so that the transition matrices $T = T_{10}$ can be written

$$T_{\mathrm{d}} = \mathrm{diag}\Big(\exp\big(y_i\zeta(z)\big)\Big), \quad T_{\mathrm{od}} = 0, \tag{2.12}$$

where $A = A_{\mathrm{d}} + A_{\mathrm{od}}$ denotes the splitting of a matrix into its diagonal and off-diagonal components. The y_i are constant on Σ. In turn, one represents the Higgs fields $\phi^0_{\mathrm{Gl}(N)}, \phi^1_{\mathrm{Gl}(N)} = T_{10}\phi^0_{\mathrm{Gl}(N)}$, which decompose into a sum of sections of line bundles, by

$$(\phi^0_{\mathrm{Gl}(N)})_{\mathrm{d}} = \mathrm{diag}(q_i), \tag{2.13}$$

$$\big((\phi^0_{\mathrm{Gl}(N)})_{\mathrm{od}}\big)_{w,w'} = K_{w,w'}\rho\big((w-w')(x), z\big), \tag{2.14}$$

where $q_i, K_{w,w'}$ are constants, and ρ are the functions of (2.8). Note that the $K_{w,w'}$ are the residues of the section at the origin. The Calogero-Moser locus $\mathcal{CM}$ is given by constraints

$$\mathrm{diag}(y_i) \subset \mathfrak{h}, \tag{2.15}$$

$$M_{\mathrm{od}} = 0, \tag{2.16}$$

$$\mathrm{diag}(q_i) \subset \mathfrak{h}, \tag{2.17}$$

$$\mathrm{res}_0(\phi^0_{\mathrm{Gl}(N)})_{w,w'} = C_{w,w'}\sum_\alpha \delta_{w-w',\alpha} c_{|\alpha|}. \tag{2.18}$$

PROPOSITION 2.2. *The embedding $\widehat{\Xi}$ is symplectic, and the Calogero-Moser Hamiltonian corresponds under the embedding to a multiple of* $\mathrm{res}\big(\omega\,\mathrm{tr}(\phi^2_{\mathrm{Gl}(N)})\big)$.

The Calogero-Moser flows will then be the projections, along $T\mathcal{CM}^\perp$, onto $\mathcal{CM}$ of the Hitchin flows for $\mathrm{Gl}(V)$ systems. The Hitchin flows

$$M^{-1}\dot{M} = \omega\phi^0_{\mathrm{Gl}(N)}, \quad \dot{\phi}^0_{\mathrm{Gl}(N)} = 0$$

can be rewritten, as in section 1, by modifying them by a gauge transformation. We split $\omega\phi^0_{\mathrm{Gl}(N)}$, first into its diagonal and off-diagonal components, and then write the off-diagonal term as a coboundary (function on U_0 vanishing at the origin, minus T^{-1}(function on U_1)T) plus a constant cocycle:

$$\omega\phi^0_{\mathrm{Gl}(N)} = (\omega\phi^0_{\mathrm{Gl}(N)})_{\mathrm{d}} + (\omega\phi^0_{\mathrm{Gl}(N)})^0_{\mathrm{od}} - T^{-1}(\omega\phi^0_{\mathrm{Gl}(N)})^1_{\mathrm{od}}T + (\omega\phi^0_{\mathrm{Gl}(N)})^{\mathrm{cst}}_{\mathrm{od}}.$$

This turns the flows into the equivalent ones:

$$T^{-1}\dot{T} = (\omega\phi^0_{\mathrm{Gl}(N)})_{\mathrm{d}} + (\omega\phi^0_{\mathrm{Gl}(N)})^{\mathrm{cst}}_{\mathrm{od}},$$
$$(\dot{\phi}^0_{\mathrm{Gl}(N)}) = [(\omega\phi^0_{\mathrm{Gl}(N)})^0_{\mathrm{od}}, (\phi^0_{\mathrm{Gl}(N)})],$$

The projected flows are given by

$$\begin{aligned} M^{-1}\dot{M} &= (\omega\phi^0_{\mathrm{Gl}(N)})_{\mathrm{d}} \\ (\dot{\phi}^0_{\mathrm{Gl}(N)}) &= a(x) + [(\omega\phi^0_{\mathrm{Gl}(N)})^0_{\mathrm{od}}, (\phi^0_{\mathrm{Gl}(N)})] + [\omega\phi^0_{\mathrm{Gl}(N)})^{\mathrm{cst}}_{\mathrm{od}}, (\phi^0_{\mathrm{Gl}(N)})]_d, \\ &= a(x) + [(\omega\phi^0_{\mathrm{Gl}(N)})^0_{\mathrm{od}} + (\omega\phi^0_{\mathrm{Gl}(N)})^{\mathrm{cst}}_{\mathrm{od}}, (\phi^0_{\mathrm{Gl}(N)})] \\ &- [(\omega\phi^0_{\mathrm{Gl}(N)})^{\mathrm{cst}}_{\mathrm{od}}, (\phi^0_{\mathrm{Gl}(N)})]_{\mathrm{od}}. \end{aligned} \tag{2.19}$$

Here the element $a \in \mathfrak{h}^\perp \subset \mathfrak{d}$, where $\mathfrak{d}$ is the algebra of diagonal matrices, is given by:

$$a(x) = -\pi_{\mathfrak{h}^\perp}([(\omega\phi^0_{\mathrm{Gl}(N)})^0_{\mathrm{od}} + (\omega\phi^0_{\mathrm{Gl}(N)})^{\mathrm{cst}}_{\mathrm{od}}, (\phi^0_{\mathrm{Gl}(N)})]).$$

Remembering that we are still in the unreduced phase space, (so that the trivialisations are still there) we can act by the group of automorphisms of the bundle E, which are expressed by matrices d, constant over the curve. We ask if we can rewrite (2.19) using the action of D:

$$T^{-1}\dot{T} = 0, \quad \dot{\phi}^0_{\mathrm{Gl}(N)} = [d, \phi^0_{\mathrm{Gl}(N)}],$$

so that the flows become

$$\begin{aligned} T^{-1}\dot{T} &= (\omega\phi^0_{\mathrm{Gl}(N)})_{\mathrm{d}}, \\ (\dot{\phi}^0_{\mathrm{Gl}(N)}) &= [(\omega\phi^0_{\mathrm{Gl}(N)})^0_{\mathrm{od}} + (\omega\phi^0_{\mathrm{Gl}(N)})^{\mathrm{cst}}_{\mathrm{od}} + d(x), (\phi^0_{\mathrm{Gl}(N)})], \end{aligned} \tag{2.20}$$

The choice of d can of course vary with the point (p, x) in the phase space; we just allow d to depend on the bundle: $d = d(x)$.

Let us suppose that this is the case; we then have our Lax pair. Furthermore, passing to the reduced phase space, so that the constant terms in the transition matrices are no longer meaningful, we get the flows:

$$T^{-1}\dot{T} = (\omega\phi^0_{\mathrm{Gl}(N)})_{\mathrm{d}},$$
$$(\dot{\phi}^0_{\mathrm{Gl}(N)}) = [(\omega\phi^0_{\mathrm{Gl}(N)})^0_{\mathrm{od}} + (\omega\phi^0_{\mathrm{Gl}(N)})^{\mathrm{cst}}_{\mathrm{od}}, (\phi^0_{\mathrm{Gl}(N)})],$$

which gives the Hitchin flow. In particular, one has a completely integrable system.

The constraints for the projection to the Calogero-Moser phase space are given as follows. Expand

$$\begin{aligned} (\phi^0)_{w,w+\alpha}(x,z) &= C_{w,w+\alpha}c_\alpha\big(-z^{-1} + \zeta(\alpha(x)) + O(z)\big), \\ &\overset{\mathrm{def}}{=} R_{w,w+\alpha}z^{-1} + Q_{w,w+\alpha}(x) + O(z), \\ \big((\omega\phi^0_{\mathrm{Gl}(N)})^0_{\mathrm{od}} + +\omega\phi^0_{\mathrm{Gl}(N)})^{\mathrm{cst}}_{\mathrm{od}}\big)_{w,w+\alpha} &= C_{w,w+\alpha}c_\alpha \frac{d}{d\alpha(x)}(\rho^0(\alpha(x)), z), \\ &= R_{w,w+\alpha}\mathfrak{p}\big(\alpha(x)\big) + O(z), \\ &\overset{\mathrm{def}}{=} P_{w,w+\alpha}(x) + O(z). \end{aligned}$$

We then have the necessary and sufficient constraints for $d(x)$:

$$\begin{aligned}[P(x), R]_{\rm od} &= [d'(x), R]_{\rm od},\\ 0 &= [d'(x), R]_{\rm d},\\ a(x) &= [d'(x), Q(x)]_{\rm d}.\end{aligned}$$

These are, in essence, the constraints that are given by d'Hoker and Phong [**dP**].

ii) *The Bordner-Corrigan-Sasaki ansatz: root spaces and reflection operators.* The other object associated with a root space that one can consider is the reflection operator in the root plane. A weight space $\mathbb{C}_w$ then gets mapped to the weight space $\mathbb{C}_{w'}$, where $w' = R_\alpha(w) = w - \langle\hat{\alpha}, w\rangle\alpha$. The basis vector in the α-th weight space in G then gets mapped to the permutation operator $(s_\alpha) \in \mathrm{Gl}(V)$, with entries

$$(s_\alpha)_{w.w'} = \delta_{w',R_\alpha(w)}.$$

One could then take the induced map on the level of Higgs fields; the question arises though of asking which induced map. Indeed, for the d'Hoker Phong case, all the non-zero entries in $\mathrm{Gl}(V)$ of Sh_α have weight α, and the embedding is straightforward. In this case, there are entries corresponding to multiples of α. We use the fact that the isogeny $z \to nz$ pulls back a line bundle L to L^n. For a section f of $L_{\alpha(h)}$, we define

$$\tilde{s}_\alpha(f)_{w,w'} = \sum_n (s_\alpha)_{w,w'}\delta_{w-w',n\alpha}n \cdot I_n^* f.$$

Represent the section ϕ as $\big((\phi_\alpha), \phi_{\mathfrak{h}}\big)$. We define the corresponding section $\phi_{\mathrm{Gl}(N)}$ by

$$\phi_{\mathrm{Gl}(N)} = \left(\left(\sum_\alpha \tilde{s}_\alpha(\phi_\alpha)\right) + \xi(\phi_{\mathfrak{h}})\right).$$

We have again the same procedure for the Calogero-Moser flows: we obtain an equation of the form (2.20). In this case, the constraints on d' are always solvable; we have

$$d'(x)_{w,w} = \sum_\alpha (s_\alpha)_{w,w}\frac{\partial\rho}{\partial x}(\alpha(x), 0).$$

3. Gl(N)-Hitchin Systems and Spectral Curves: Abelianisation

Let us return to the Hitchin systems for $\mathrm{Gl}(n,\mathbb{C})$, over an arbitrary curve. In this case, the Hamiltonians all occur as coefficients of the *spectral curve* S of the Higgs field. This curve lives in the total space $\mathcal{KD}$ of the line bundle $K_\Sigma(D)$, into which it is compactly embedded. If λ is a coordinate on Σ, and ζ the corresponding fiber coordinate on the canonical bundle, the spectral curve is defined by:

$$\det(\phi(\lambda) - \zeta\mathbb{I}) = 0. \tag{3.1}$$

The Lagrangian foliation is then a map $\mathcal{M} \to \{\text{family of spectral curves}\}$.

There is another datum we can associate to the pair (E, ϕ). Let $\pi : \mathcal{KD} \to \Sigma$ denote the projection. We can define a sheaf supported on the spectral curve by:

$$0 \to \pi^* E \otimes K_\Sigma^*(-D) \xrightarrow{\phi - \zeta\mathbb{I}} \pi^* E \to L \to 0. \tag{3.2}$$

When the spectral curve is smooth, reduced, the sheaf L is a line bundle supported over the spectral curve.

PROPOSITION 3.1 ([**Hu1, HuKj**]). *One can then reconstruct* (E, ϕ) *from* (S, L):

(1) $E = \pi_*(L)$,

(2) ϕ *is the map induced on* E *by multiplication by the tautological section* ζ *on* L.

In short, (E, ϕ) and (S, L) are equivalent data.

We note that the symplectic leaves of the Poisson structure on $\mathcal{M}$ correspond to fixing the intersection of the spectral curve with the divisor $\pi^{-1}(D)$. On the generic symplectic leaf, the leaf of the Lagrangian foliation at a smooth spectral curve S is a Zariski open set of the Jacobian of S. Infinitesimally, then, the tangent bundle of $\mathcal{M}$ at (S, L) fits into an exact sequence corresponding to the Lagrangian fibration:

$$0 \to H^1(S, \mathcal{O}) \to T(\mathcal{M}) \to H^0\big(S, K_S(D)\big) \to 0. \tag{3.3}$$

Here the H^1 factor corresponds to deformations of the line bundles, while the H^0-factor correspond to sections of the normal bundle, which gets identified with the bundle $K_S(D)$ by using the adjunction formula. Restricting to the symplectic leaves, which amounts to fixing the spectral curve over D, we have

$$0 \to H^1(S, \mathcal{O}) \to T(\mathcal{M}_{\mathcal{O}}) \to H^0(S, K_S) \to 0. \tag{3.4}$$

We split these two sequences at (S, L): the geometric way of doing this is to extend the line bundle to a neighbourhood of S in $\mathcal{K}_D$, giving us a way of moving the curve while keeping the line bundle fixed. One then has, at (S, L):

$$\begin{aligned} T(\mathcal{M}) &\simeq H^1(S, \mathcal{O}) \oplus H^0\big(S, K_S(D)\big), \\ T(\mathcal{M}_{\mathcal{O}}) &\simeq H^1(S, \mathcal{O}) \oplus H^0(S, K_S). \end{aligned} \tag{3.5}$$

The two factors are dual to each other: this is the geometric basis of the complete integrability. There is a natural skew form on the sum, which we will denote by ω_S.

Let us compare this with the expression (1.2) we had above for the tangent bundle of $T^*\mathcal{N}$

$$0 \to H^0\big(\Sigma, \mathrm{End}(E)(D)\big) \to T(T^*\mathcal{N}) \to H^1\big(\Sigma, \mathrm{End}(-D)\big) \to 0, \tag{3.6}$$

which we had split as

$$T(T^*\mathcal{N}) \simeq H^0\big(\Sigma, \mathrm{ad}^*(E)(D)\big) \oplus H^1\big(\Sigma, \mathrm{ad}(E)(-D)\big). \tag{3.7}$$

The natural symplectic form ω_Σ on $T^*\mathcal{N}$ is given by (1.7).

PROPOSITION 3.2. *Over* $\mathcal{M}_{\mathcal{O}}$, *we have* $\omega_S = (\omega_\Sigma)_{red}$, *where* $(\omega_\Sigma)_{red}$ *is the reduction of* ω_Σ.

Comparing (3.5) and (3.7), this can be thought of as a process of Abelianisation: at the price of passing to a cover S of the curve Σ, one reduces from rank n bundles to rank 1.

One can represent variations of the line bundle and the spectral curve fairly explicitly, if one replaces the line bundle by a corresponding divisor. The line bundles L are of fixed degree. Choosing a curve segment C intersecting the spectral curve S transversely, we can twist L by a suitable multiple nC of C so that the resulting line bundle $\tilde{L}$ has degree $g = \mathrm{genus}(S)$:

$$L \mapsto \tilde{L} = L\big(n(S \cap C)\big). \tag{3.8}$$

The line bundle $\tilde{L}$, generically, has a one dimendional space of sections. The non-zero sections have a common divisor $D(\tilde{L})$ of degree g, i.e., a sum of g points $p_1, \ldots, p_g$ of $S \subset \mathcal{KD}$, which we will suppose distinct. These points all lie in the curve S; in fact, they determine the curve, at least generically. We have thus associated to the pair (E, ϕ) an element of the g-th symmetric product $\mathrm{SP}^g(\mathcal{KD})$ of $\mathcal{KD}$:

$$\begin{aligned} \mathrm{Div} : \mathcal{M}_{\mathcal{O}} &\to \mathrm{SP}^g(\mathcal{KD}) \\ (E, \phi) &\mapsto (p_1, \ldots, p_g). \end{aligned}$$

Over an open set this map is an isomorphism.

Once again the map Div can be viewed symplectically. There is a natural map of line bundles $K_\Sigma \to K_\Sigma(D)$, which is an isomorphism away from D; it induces a map of the total spaces $\mathcal{K} \to \mathcal{KD}$. The natural cotangent symplectic form on $\mathcal{K}$ can be pushed forward to $\mathcal{KD}$ to give a meromorphic symplectic form Ω on $\mathcal{KD}$, with poles over D. Let $D_{\mathcal{O}}$ be the intersection of the divisor $\pi^{-1}(D)$ with the spectral curves of $\mathcal{M}_{\mathcal{O}}$. This intersection is independent of the spectral curve. We will assume, for simplicity, that all the points of D are of multiplicity one. Now blow up $\mathcal{KD}$ at $D_{\mathcal{O}}$, and call the result $\widetilde{\mathcal{KD}}$. The spectral curves lift naturally to $\widetilde{\mathcal{KD}}$, and one has a lift of Div:

$$\widetilde{\mathrm{Div}} : \mathcal{M}_{\mathcal{O}} \to \mathrm{SP}^g(\widetilde{\mathcal{KD}}).$$

Lifting the form Ω to $\widetilde{\mathcal{KD}}$, the resulting $\tilde{\Omega}$ is holomorphic on a neighbourhood of the lifted spectral curves. There is then a naturally induced form $\tilde{\Omega}^{(g)}$ on $\mathrm{SP}^g(\widetilde{\mathcal{KD}})$, which is holomorphic away from the (large) diagonal. We note, however, that $\mathrm{SP}^g(\widetilde{\mathcal{KD}})$ is singular along the diagonal. Fortunately, there is a natural symplectic desingularisation of $\mathrm{SP}^g(\widetilde{\mathcal{KD}})$: the Hilbert scheme $\mathrm{Hilb}^g(\widetilde{\mathcal{KD}})$ of length g zero-dimensional ideals of $\widetilde{\mathcal{KD}}$. Indeed, the variety $\mathrm{Hilb}^g(\widetilde{\mathcal{KD}})$ is smooth [**I**], and symplectic [**Be2**]. The map $\widetilde{\mathrm{Div}}$ lifts to $\mathrm{Hilb}^g(\widetilde{\mathcal{KD}})$:

$$\widehat{\mathrm{Div}} : \mathcal{M}_{\mathcal{O}} \to \mathrm{Hilb}^g(\widetilde{\mathcal{KD}}). \tag{3.9}$$

THEOREM 3.3 ([**AHH, Hu1, HuKj**]). *The map* $\widehat{\mathrm{Div}}$ *is a symplectomorphism.*

Our Lagrangian leaves in $\mathrm{Hilb}^g(\widetilde{\mathcal{KD}})$ are given by the inclusions $\mathrm{Hilb}^g(S) \subset \mathrm{Hilb}^g(\widetilde{\mathcal{KD}})$; we note that the variety $\mathrm{Hilb}^g(S) = \mathrm{SP}^g(S)$ is birational to the Jacobian of S.

Let U denote the family of spectral curves of $\mathcal{M}_{\mathcal{O}}$. Over U, there is the tautological curve $\mathbb{S} \to U$. Near a given spectral curve $S \in U$, choosing $g - 1$ curve segments C_i each intersecting nearby S' transversally in a single point, and restricting U if necessary, there is a map

$$A : \mathbb{S} \to \widetilde{\mathcal{KD}} \to \mathrm{Hilb}^g(\widetilde{\mathcal{KD}}), \tag{3.10}$$

where the map $\widetilde{\mathcal{KD}} \to \mathrm{Hilb}^g(\widetilde{\mathcal{KD}})$ adds to a point on the spectral curve S' the $g-1$ points $C_i \cap S'$. As the map factors through a two-dimensional variety, we naturally have that the pull-back $A^*(\tilde{\Omega}^{(g)})$ of the symplectic form is of rank two:

$$A^*(\tilde{\Omega}^{(g)}) \wedge A^*(\tilde{\Omega}^{(g)}) = 0.$$

We say that the Hitchin systems are of *rank two.*

The rank two property is defined more generally for an algebraically integrable system of Jacobians. Indeed, let

$$\mathbb{J} \to U$$

be such a system of Jacobians, that is a symplectic $2g$ dimensional variety fibering over a g-dimensional base U, such that the fibers $S_h, h \in U$ are isotropic and are g-dimensional Jacobians of curves. There is an associated family of curves:

$$\mathbb{S} \to U.$$

If we choose base sections, we have a fiberwise Abel map

$$I : \mathbb{S} \to \mathbb{J}$$

which commutes with the projections to U. We can pull back the symplectic form, and set

DEFINITION 3.4. The system has *rank two* if for a suitable choice of base sections the pull-back of the symplectic form satisfies

$$I^*\Omega \wedge I^*\Omega = 0.$$

THEOREM 3.5 ([**Hu1**]). (i) *Let the system be of rank 2. Under the embedding I, the variety $\mathbb{S}$ is coisotropic. Quotienting by the null foliation, one obtains, restricting U if necessary, a surface Q to which the form $I^*\Omega$ projects, defining a symplectic form ω on Q. The curves S_h all embed in Q.*

(ii) *If I, $\tilde{I}$ are two Abel maps with $I^*\Omega \wedge I^*\Omega = 0$, $\tilde{I}^*\Omega \wedge \tilde{I}^*\Omega = 0$, then $I^*\Omega = \tilde{I}^*\Omega$, when $g \geq 3$, and so Q depends only on S and not on the particular Abel map chosen. For $g = 2$, $I^*\Omega \wedge I^*\Omega$ is always zero.*

(iii) *There is a symplectic isomorphism*

$$\Phi : \mathrm{Hilb}^g(Q) \to \mathbb{J},$$

defined over a Zariski open set, between $\mathbb{J}$ and $\mathrm{Hilb}^g(Q)$. The symmetric product $\mathrm{SP}^g(S_h)$ of the curves is Lagrangian in $\mathrm{Hilb}^g(Q)$, and the restriction of Φ to $\mathrm{SP}^g(S_h)$ is the Abel map

$$\mathrm{SP}^g(S_h) \to J_h.$$

One consequence of this is that the surface encodes the integrable system. In the case of the Hitchin systems, the variety Q is $\widetilde{\mathcal{KD}}$.

One of the uses of the isomorphism Φ is that it provides natural separating coordinates for the flows on $\mathbb{J}$. Indeed, choosing Darboux coordinates λ, ζ on Q, we have natural coordinates $\lambda_\mu, \zeta_\mu, \mu = 1, \dots, g$ on $\mathrm{Hilb}^g(Q)$. Let P_α be coordinates on U. Fixing P_α fixes a curve S, and so defines ζ implictly in terms of λ: $\zeta = \zeta(\lambda, P_\alpha)$. We define a generating function by

$$F(\lambda_\mu, P_\alpha) = \sum_\mu \int^{\lambda\mu} \zeta(\lambda, P_\alpha) d\lambda.$$

As $\partial F / \partial \lambda_\mu = \zeta_\mu$, the expressions $Q_\alpha = \partial SF / \partial P_\alpha$ are coordinates which linearise the flows:

$$Q_\alpha = \sum_\mu \int^{\lambda_\mu} \frac{\partial \zeta}{\partial P_\alpha} d\lambda.$$

The integrands, in fact, are the expressions of the holomorphic differential forms on the curves, using the Poincaré residue theorem: the Q_α are Abelian integrals of the first kind. This is, of course, quite natural, as the flows should be linear on the Jacobian.

4. G-Hitchin Systems, Spectral Curves, and Prym Varieties

One can ask what the analoguous construction is for arbitrary reductive groups. It turns out that there is a beautiful construction which has been elaborated by various authors [**Fa, Sc1, Sc2, Ka, Do2**]. The idea is to generalise not the spectral curve S for $\mathrm{Gl}(r)$, but rather an $(r-1)!$-fold cover of S, consisting of all *ordered sequences of length r of "eigenvalues"*. For points over which ϕ is semi-simple, this is in natural bijection with the different diagonalisations of ϕ. It is this idea which generalises.

Indeed, for a reductive group G, fix a Cartan subgroup H and a Borel subgroup B containing H. Let $\mathfrak{h}, \mathfrak{b}$ denote the corresponding Lie algebras. For points $\lambda \in \Sigma$ over which $\phi(\lambda)$ is semi-simple, there is a Weyl group orbit of points in $K_\Sigma(D)|_\lambda \otimes \mathfrak{h}$ to which $\phi(\lambda)$ is conjugate (the different diagonalisations); more generally, there is a Weyl group orbit of points in $K_\Sigma(D)|_\lambda \otimes \mathfrak{h}$ lying in the G-orbit of $\phi(\lambda)$. Doing this over every point gives a Weyl invariant spectral curve S lying over Σ.

The line bundle L of the previous section also admits a generalisation. Lifting the bundles E to S, we find that they have natural reductions E_B to the Borel subgroup, in which ϕ lies in B; in fact, away from the branch points of the curve, one can even reduce to H. In any case, the natural projection from B to H associates to E_B an H-bundle E_H. It turns out that by choosing a square root of the canonical bundle on Σ that there is a natural way to twist E_H to obtain a W-invariant bundle $\tilde{E}_H$. This data is sufficient to reconstruct the pair (E, ϕ), at least locally. Thus, the Hitchin system can be thought of locally as a family of pairs

$$(4.1) \qquad (W\text{-invariant curves } S \subset K_\Sigma(D) \otimes \mathfrak{h}, W\text{-invariant } H\text{-bundles over } S)$$

As the polar parts of ϕ lie in fixed coadjoint orbits, the spectral curves of $\mathcal{M}_\mathcal{O}$ are fixed over D. The variety of W-invariant H-bundles on S are parametrised by an Abelian variety, the generalised Prym variety. In general, these generalised Prym varieties are defined as follows: let a finite group W acts on a curve; it then acts on its Jacobian; if the group in addition acts on an integer lattice χ (here the weight lattice of $\mathfrak{h}^*$), there is a diagonal action on the tensor product. The generalised Prym variety is the connected component of the fixed point set containing the origin:

$$Pr(S, \chi) = (J(S) \otimes_\mathbb{Z} \chi)_0^W.$$

Next, taking the infinitesimal version of (4.1) we have a corresponding exact sequence

$$(4.2) \qquad 0 \to H^1(S, \mathcal{O} \otimes \mathfrak{h})^W \to T\mathcal{M}_\mathcal{O} \to H^0(S, K_\Sigma \otimes \mathfrak{h})^W \to 0.$$

The superscript ^{W} denotes Weyl invariance. The H^1-factor corresponds to deformations of the bundles over S, while the H^0-factor corresponds to deformations of the curves. Again, we can split this sequence, by extending the bundles to a neighbourhood of the curve:

$$(4.3) \qquad T\mathcal{M}_\mathcal{O} \simeq H^0(\Sigma, \mathrm{ad}(P) \otimes K_\Sigma)^W \oplus H^1(\Sigma, \mathrm{ad}(P))^W.$$

This is a sum of two dual vector spaces, so again there is a natural skew form ω_S on the sum. Again, we have an Abelianisation of the symplectic form on $\mathcal{M}_\mathcal{O}$.

PROPOSITION 4.1 ([**Hu1, Hu2, Kj**]). *Under the correspondence* (4.1), *the symplectic form on* $\mathcal{M}_{\mathcal{O}}$ *is given by* ω_S.

One can ask what the analogue of the surface $\widetilde{\mathcal{KD}}$ is for arbitrary groups. The answer is the $r+1$ fold $\widetilde{\mathcal{KDH}}$ obtained by blowing up the total space of the vector bundle $K_\Sigma(D) \otimes \mathfrak{h}$ at the points of intersection of the spectral curves with the divisors $\pi^{-1}(D)$. This is not equipped with a symplectic form, but does have a natural $\mathfrak{h}$-valued 2-form.

As above, this corresponds to a general picture. Indeed, fix a group W and a lattice χ. Set $V = \chi \otimes_{\mathbb{Z}} \mathbb{C}$. Now let us assume we have an integrable system of Prym varieties corresponding to W and χ:

$$\mathbb{P}r \to U.$$

There are corresponding families of curves, and Jacobians, on which W acts:

$$\mathbb{S} \to U, \quad \mathbb{J} \to U.$$

We have as before an Abel map, which can be made W-equivariant:

$$I : \mathbb{S} \to \mathbb{J}.$$

PROPOSITION 4.2 ([**HuMa1**]). *There is a natural* V^**-valued closed two-form* Ω_V *on* $\mathbb{J}$ *corresponding to the symplectic form* Ω *on* $\mathbb{P}r$.

DEFINITION 4.3. We say that $\mathbb{P}r$ has rank two if $I^*(\Omega_V) \wedge I^*(\Omega_V) = 0$ as a $V^* \otimes V^*$-valued 4-form.

As for the Jacobian case, one obtains for systems of rank two a null foliation which one can quotient out to obtain a variety. One must for these theorems impose certain (reasonable) genericity conditions, which we omit here (see [**HuMa1**]). Doing this, one has:

THEOREM 4.4 ([**HuMa1**]). *Let the system* $\mathbb{P}r \to U$ *have rank two. Restricting* U *if necessary,*

There is a $r+1$ *dimensional complex manifold* X *into which the curves* S_u *all embed. It is equipped with a generically non-degenerate* V^**-valued two form* Ω_V*. The group* W *acts on* X*, preserving* Ω_V*.*

It turns out, for $r > 2$, that there is a natural codimension one foliation on X. One can quotient out this foliation and the Weyl group simultaneously, and obtain a curve Σ:

THEOREM 4.5 ([**HuMa1**]). *The manifold* X *admits a* W*-invariant fibration to a closed curve* Σ*. The quotient curves* S_u/W *are sections of* $X/W \to \Sigma$*. In particular, all the quotient curves* S_u/W *are isomorphic to* Σ*. The projection from* X *to* Σ *is such that each fiber is* Ω_V*-isotropic and the fiber over a generic point* $a \in \Sigma$ *has an affine structure modeled after the vector space* $T_a\Sigma \otimes V^*$*.*

Finally, again, the variety X encodes the integrable system. Again, imposing some genericity conditions which we will not delve into, we have:

THEOREM 4.6 ([**HuMa1**]). *Let* X *be a* $v+1$ *dimensional complex manifold, with a submersion onto a closed curve* Σ*. Let* X *be equipped with a "minimally non-degenerate"* V^**-valued two form* Ω_V*, such that the group* W *acts on* X*, preserving* Ω_V*, and preserving the fibers of the map to* Σ*. Assume that there is a smooth* W*-invariant curve* S_0 *in* X*, on which* W *acts generically freely with quotient* Σ*. Then, deforming* S_0 *in* X*, the family of smooth* W*-invariant curves* S_u *defines a rank-2 integrable system of Prym varieties.*

References

[A] M. Atiyah, *Vector bundles over an elliptic curve*, Proc. London Math. Soc. **7** (1957), 414–452.

[AB] M. Atiyah and R. Bott, *The Yang-Mills equations over Riemann surfaces*, Philos. Trans. Roy. Soc. London Ser. A **308** (1982), 523–615.

[AHH] M. R. Adams, J. Harnad, and J. Hurtubise, *Darboux coordinates and Liouville-Arnold integration in loop algebras*, Comm. Math. Phys. **155** (1993), 385–413.

[AHP] M. R. Adams, J. Harnad, and E. Previato, *Isospectral hamiltonian flows in finite and infinite dimensions. I. Generalised Moser systems and moment maps into loop algebras*, Comm. Math. Phys. **117** (1988), 451–500.

[AvM] M. Adler and P. van Moerbeke, *Completely integrable systems, Euclidean Lie algebras, and curves*, Adv. Math. **38** (1980), 267–317; *Linearization of Hamiltonian systems, Jacobi varieties and representation theory*, **38** (1980), 318–379.

[Be] A. Beauville, *Jacobiennes des courbes spectrales et systèmes hamiltoniens complètement intégrables*, Acta Math. **164** (1990), 211–235.

[Be2] A. Beauville, *Variétés Kählériennes dont la première classe de Chern est nulle*, J. Differential Geom. **18** (1983), 755–782.

[BCS] A. J. Bordner, E. Corrigan, and R. Sasaki, *Generalized Calogero-Moser models and universal Lax pair operators*, Progr. Theoret. Phys. **102** (1999), 499–529.

[Bo] F. Bottacin, *Symplectic geometry on moduli spaces of stable pairs*, Ann. Sci. École Norm. Sup. (4) **28** (1995), 391–433.

[dP] E. d'Hoker and D. H. Phong, *Calogero-Moser Lax pairs with spectral parameter for general Lie algebras*, Nuclear Phys. B **530** (1998), 537–610.

[Do1] R. Donagi, *Seiberg-Witten integrable systems*, Proc. Sympos. Pure Math., vol. 62, Part 2, Amer. Math. Soc., Providence, RI, 1997, pp. 3–43.

[Do2] R. Donagi, *Spectral covers*, Current Topics in Complex Algebraic Geometry, Math. Sci. Res. Inst. Publ., vol. 28 (Berkeley, CA 1992/93), pp. 65–86; alg-geom/9505009.

[Do3] R. Donagi, *Principal bundles on elliptic fibrations*, Asian J. Math. **1** (1997), 214–223.

[DoMa] R. Donagi and E. Markman, *Spectral covers, algebraically completely integrable Hamiltonian systems, and moduli of bundles*, Integrable Systems and Quantum Groups (R. Donagi et al., eds), Lecture Notes in Math., vol. 1620, Springer, New York, 1996, pp. 1–119.

[DoW] R. Donagi and E. Witten, *Supersymmetric Yang-Mills theory and integrable systems*, Nuclear Phys. B **460** (1996), 299; hep-th/9510101.

[Fa] G. Faltings, *Stable G-bundles and projective connections*, J. Algebraic Geom. **2** (1993), 507–568.

[FMW] R. Friedman, J. Morgan, and E. Witten, *Principal G-bundles over elliptic curves*, Math. Res. Lett. **5** (1998), 97–118.

[FRS] I. B. Frenkel, A. G. Reiman, and M. A. Semenov-Tian-Shansky, *Graded Lie algebras and completely integrable Hamiltonian systems*, Soviet Math. Dokl. **20** (1979), 811–814.

[Gro] A. Grothendieck, *Sur la classification des fibrés holomorphes sur la sphère de Riemann*, Amer. J. Math. **79** (1957), 121–138.

[HHu] J. Harnad and J. Hurtubise, *Generalised tops and moment maps into loop algebras*, J. Math. Phys. **37** (1991), 1780–1787.

[Hi1] N. J. Hitchin, *The self-duality equations on a Riemann surface*, Proc. London Math. Soc. **55** (1987), 59–126.

[Hi2] N. J. Hitchin, *Stable bundles and integrable systems*, Duke Math. J. **54** (1987), 91–114.

[Hu1] J. Hurtubise, *Integrable systems and algebraic surfaces*, Duke Math. J. **83** (1996), 19–50.

[Hu2] J. Hurtubise, *The algebraic geometry of the Kostant-Kirillov form*, J. London Math. Soc. **56** (1997), 504–518.

[Hu3] J. Hurtubise, *Finite dimensional coadjoint orbits in a loop algebra*, Lett. Math. Phys. **30** (1994), 99–104.

[HuKj] J. Hurtubise and M. Kjiri, *Separating coordinates for the generalized Hitchin systems and the classical r-matrices*, Comm. Math. Phys. (to appear).

[HuMa1] J. Hurtubise and E. Markman, *Rank 2 integrable systems of Prym varieties*, Adv. Theoret. Math. Phys. **2** (1998), 633–695.

[HuMa2] J. Hurtubise and E. Markman, *Calogero-Moser systems and Hitchin systems*, preprint.

[Hum] J. E. Humphreys, *Introduction to Lie algebras and representation theory*, Springer, New York, 1972.

[I] A. Iarrobino, *Hilbert scheme of points. Overview of the last ten years*, Algebraic Geometry—Bowdoin 1985 (S. J. Bloch, ed.), Proc. Symp. Pure Math., vol. 46 Part 2, Amer. Math. Soc., Providence, RI, 1987, pp. 297–320.

[Ka] V. Kanev, *Spectral curves, simple Lie algebras and Prym-Tjurin varieties*, Theta-Functions—Bowdoin 1987 (R. C. Gunning and L. Ehrenfencht, eds.), Proc. Symp. Pure Math., vol. 49 Part 1, Amer. Math. Soc., Providence, RI, 1989, pp. 627–645.

[Kj] M. Kjiri, *The G-generalised Hitchin systems and Prym varieties*, preprint.

[Kr] I. M. Krichever, *Elliptic solutions of the Kadomtsev-Petviashvili equation and integrable systems of particles*, Funct. Anal. Appl. **14** (1980), 282–290.

[Ma] E. Markman, *Spectral curves and integrable systems*, Compositio Math. **93** (1994), 255–290.

[OP] M. A. Olshanetsky and A. M. Perelomov, *Completely integrable Hamiltonian systems connected with semisimple Lie algebras*, Invent. Math. **37** (1976), 93–108.

[NS] M. S. Narasimhan and C. S. Seshadri *Stable and unitary bundles on a compact Riemann surface*, Ann. of Math. **82** (1965), 540–567.

[RS1] A. G. Reiman and M. A. Semenov-Tian-Shansky, *Reduction of Hamiltonian systems, affine Lie algebras and Lax equations.* I, Invent. Math. **54** (1979), 81–100; II, **63** (1981), 423–432.

[RS2] A. G. Reiman and M. A. Semenov-Tian-Shansky, *Integrable systems.* II, Dynamical Systems VII (V. I. Arnold and S. P. Novikov, eds.), Encyclopaedia Math. Sci., vol 16., Springer, Berlin, 1994, chap. 2.

[Sc1] R. Scognamillo, *Prym-Tyurin varieties and the Hitchin map*, Math. Ann. **303** (1995), 47–62.

[Sc2] R. Scognamillo, *An elementary approach to the abelianization of the Hitchin system for arbitrary reductive groups*, Compositio Math. **110** (1998), 17–37.

Department of Mathematics and Statistics, McGill University, 805 Sherbrooke W., Montréal, Québec, Canada H3A 2K6, and Centre de recherches mathématiques, Université de Montréal, C. P. 6128, succ. centre-ville, Montréal, Québec, Canada H3C 3J7.

E-mail address: hurtubis@crm.umontreal.ca

Centre de Recherches Mathématiques
CRM Proceedings and Lecture Notes
Volume **26**, 2000

Determinant Representation for Form Factors

V. E. Korepin

ABSTRACT. We study exactly solvable models of quantum statistical mechanics. Our main example is the Quantum nonlinear Schrödinger equation. This model can be solved by the algebraic Bethe Ansatz. We are interested in the expression for a form factor of the local field in the finite volume $F_N = \langle 0| \prod_{j=1}^{N} C(\mu_j)\psi(0,0) \prod_{j=1}^{N+1} B(\lambda_j)|0\rangle$. We found a determinant representation for this form factor. We think that this is a universal feature: the finite volume form factor of any exactly solvable model can be represented by a determinant.

1. Introduction

We consider exactly solvable models of statistical mechanics in one space and one time dimension. The Quantum Inverse Scattering Method and the Algebraic Bethe Ansatz are effective methods for a description of the spectrum of these models. Our aim is the evaluation of form factors of exactly solvable models in a finite volume. Our approach is based on the determinant representation. The method is described in [**5**].

In order to calculate correlation functions one has to know the form factors.

In this paper we have found the determinant representation for the form factor of the local field of the penetrable Bose gas. The scattering matrix in this model is a complicated function of momenta [**8**], this means that the model is not equivalent to free fermions. The main idea for the construction of the determinant representation is the following. We use the Algebraic Bethe Ansatz in order to reduce the evaluation of form factors to scalar products. Only special kind of the scalar products appears in the expression for form factors: one of the states is an eigenfunction of the Hamiltonian. In this case scalar products can be represented as a determinant without dual fields [**14**].

First we shall discuss our model.

The quantum nonlinear Schrödinger equation (equivalent to a Bose gas with delta-function interaction) can be described by the canonical Bose fields $\psi(x)$ and $\psi^\dagger(x)$ with the commutation relations:

$$(1.1) \qquad [\psi(x), \psi^\dagger(y)] = \delta(x-y), \quad [\psi(x), \psi(y)] = [\psi^\dagger(x), \psi^\dagger(y)] = 0,$$

1991 *Mathematics Subject Classification.* 35Q15, 35Q40, 35Q55.

Supported by the National Science Foundation (NSF) under Grant No. PHY-9605226.

This is the final form of the paper.

acting in the Fock space. The Fock vacuum $|0\rangle$ and dual vector $\langle 0|$ are important. They are defined by the relations

$$\psi(x)|0\rangle = 0, \quad \langle 0|\psi^\dagger(x) = 0, \quad \langle 0 \mid 0\rangle = 1. \tag{1.2}$$

The Hamiltonian of the model is

$$H = \int dx\, \left(\partial_x\psi^\dagger(x)\partial_x\psi(x) + c\psi^\dagger(x)\psi^\dagger(x)\psi(x)\psi(x) - h\psi^\dagger(x)\psi(x)\right). \tag{1.3}$$

Here c is the coupling constant and $h > 0$ is the chemical potential. We shall consider the repulsive case $0 < c \leq \infty$.

The spectrum of the model was first described by E. H. Lieb and W. Liniger [**11**], [**12**]. The Lax representation for the corresponding classical equation of motion

$$i\frac{\partial}{\partial t}\psi = [\psi, H] = -\frac{\partial^2}{\partial x^2}\psi + 2c\psi^\dagger\psi\psi - h\psi, \tag{1.4}$$

was found by V. E. Zakharov and A. B. Shabat [**16**]. The Quantum Inverse Scattering Method for the model was formulated by L. D. Faddeev and E. K. Sklyanin [**2**].

In this paper we shall follow the notations of [**8**]. First the model is considered in a finite periodic box of length L. Later the thermodynamic limit is considered when the length of the box L and the number of particles in the ground state go to infinity, with the ratio N/L held fixed.

The quantum nonlinear Schrödinger equation is equivalent to the Bose gas with delta-function interaction. In the sector with N particles the Hamiltonian of the Bose gas is given by

$$\mathcal{H}_N = -\sum_{j=1}^{N}\frac{\partial^2}{\partial z_j^2} + 2c\sum_{\substack{j,k=1\\ j<k}}^{N}\delta(z_k - z_j) - Nh. \tag{1.5}$$

Now a few words about the organization of the paper.

In Section 2 we shall review the Algebraic Bethe Ansatz and collect all the known facts necessary for further calculations. In Section 3 we shall calculate the form factor of the local field in a finite volume. We will represent the the form factor as a determinant. We will give arguments why we think that this formula is universal.

2. Algebraic Bethe Ansatz

Let us review some main features of the Algebraic Bethe Ansatz, which we shall use later. We consider the quantum nonlinear Schrödinger model. The starting point and central object of the Quantum Inverse Scattering Method is the R-matrix, which is a solution of the Yang-Baxter equation. For the case of the quantum nonlinear Schrödinger equation, it is of the form:

$$R(\lambda,\mu) = \begin{pmatrix} f(\mu,\lambda) & 0 & 0 & 0 \\ 0 & g(\mu,\lambda) & 1 & 0 \\ 0 & 1 & g(\mu,\lambda) & 0 \\ 0 & 0 & 0 & f(\mu,\lambda) \end{pmatrix}, \tag{2.1}$$

where

$$g(\lambda,\mu) = \frac{ic}{\lambda-\mu}, \quad f(\lambda,\mu) = \frac{\lambda-\mu+ic}{\lambda-\mu}. \tag{2.2}$$

Later we shall also use the functions

$$h(\lambda,\mu)=\frac{\lambda-\mu+ic}{ic},\quad t(\lambda,\mu)=\frac{(ic)^2}{(\lambda-\mu)(\lambda-\mu+ic)}=\frac{g(\lambda,\mu)}{h(\lambda,\mu)}. \tag{2.3}$$

Another important object is the monodromy matrix

$$T(\lambda)=\begin{pmatrix}A(\lambda) & B(\lambda)\\ C(\lambda) & D(\lambda)\end{pmatrix}. \tag{2.4}$$

The operators A, B, C, D are acting in the Fock space where the operator $\psi(x)$ was defined. Their commutation relations are given by

$$R(\lambda,\mu)\big(T(\lambda)\otimes T(\mu)\big)=\big(T(\mu)\otimes T(\lambda)\big)R(\lambda,\mu). \tag{2.5}$$

These relations are written out explicitly in Section VII.1 of [**8**].

The hermiticity properties of $T(\lambda)$ are

$$\sigma_x T^*(\bar\lambda)\sigma_x=T(\lambda), \tag{2.6}$$

so that $B^\dagger(\lambda)=C(\bar\lambda)$.

The Hamiltonian of the model can be expressed in terms of $A(\lambda)+D(\lambda)$ by means of trace identities (Section VI.3 of [**8**]). The vacuum is the eigenvector of the diagonal elements of $T(\lambda)$

$$A(\lambda)|0\rangle=a(\lambda)|0\rangle;\qquad D(\lambda)|0\rangle=d(\lambda)|0\rangle; \tag{2.7}$$

$$\langle 0|A(\lambda)=a(\lambda)\langle 0|;\qquad \langle 0|D(\lambda)=d(\lambda)\langle 0|; \tag{2.8}$$

$$a(\lambda)=\exp\left\{-\frac{iL\lambda}{2}\right\};\qquad d(\lambda)=\exp\left\{\frac{iL\lambda}{2}\right\}. \tag{2.9}$$

Later we shall also use the function

$$r(\lambda)=\frac{a(\lambda)}{d(\lambda)}=e^{-i\lambda L}. \tag{2.10}$$

The operator $C(\lambda)$ annihilates the vacuum vector and the operator $B(\lambda)$ annihilates the dual vacuum:

$$C(\lambda)|0\rangle=0,\quad \langle 0|B(\lambda)=0. \tag{2.11}$$

The Hamiltonian of the model commutes with $A(\lambda)+D(\lambda)$ and they can be diagonalized simultaneously. The eigenvectors of the Hamiltonian are

$$\prod_{j=1}^{N}B(\mu_j)|0\rangle,\quad\text{and}\quad \langle 0|\prod_{j=1}^{N}C(\mu_j), \tag{2.12}$$

if the μ_j satisfy the Bethe Equations

$$\frac{a(\mu_j)}{d(\mu_j)}\prod_{\substack{k=1\\k\neq j}}^{N}\frac{f(\mu_j,\mu_k)}{f(\mu_k,\mu_j)}=1\quad\text{or}\quad \frac{a(\mu_j)}{d(\mu_j)}\prod_{k=1}^{N}\frac{h(\mu_j,\mu_k)}{h(\mu_k,\mu_j)}=(-1)^{N-1}. \tag{2.13}$$

It is convenient to rewrite (2.13) in logarithmic form. For the ground state

$$\varphi_j+\pi\equiv L\mu_j+\sum_{k=1}^{N}i\ln\left(\frac{ic+\mu_j-\mu_k}{ic-\mu_j+\mu_k}\right)=2\pi\left(j-\frac{N+1}{2}\right). \tag{2.14}$$

It is proven in Section I.2 of [**8**] that the solutions μ_j of equation (2.14) are real.

The distribution of μ_j in the ground state in thermodynamic limit can be described by a linear integral equation. The thermodynamic limit is defined in the following way: $N \to \infty$, $L \to \infty$ and $N/L = D$ is fixed. In this limit μ_j condense ($\mu_{j+1} - \mu_j = \mathcal{O}(1/L)$) and fill the symmetric interval $[-q, q]$, where q is the value of the spectral parameter on the Fermi surface.

In the thermodynamic limit the function of local density $\rho(\mu)$ can be defined in the following way

$$\rho(\mu_j) = \lim \frac{1}{L(\mu_{j+1} - \mu_j)}. \tag{2.15}$$

The lim on the right hand side denotes the thermodynamic limit. This function satisfies the Lieb-Liniger integral equation

$$\rho(\mu) - \frac{1}{2\pi}\int_{-q}^{q} K(\nu, \mu)\rho(\nu)d\nu = \frac{1}{2\pi}. \tag{2.16}$$

Here

$$K(\nu, \mu) = \frac{2c}{c^2 + (\mu - \nu)^2}, \tag{2.17}$$

and

$$D = \frac{N}{L} = \int_{-q}^{q} d\mu \rho(\mu). \tag{2.18}$$

In this way we have described the ground state.

Now we can define the correlation function of the local fields

$$\langle \psi(0,0)\psi^\dagger(x,t)\rangle = \lim \frac{\langle 0| \prod_{j=1}^N C(\mu_j)\psi(0,0)\psi^\dagger(x,t)\prod_{j=1}^N B(\mu_j)|0\rangle}{\langle 0|\prod_{j=1}^N C(\mu_j)\prod_{j=1}^N B(\mu_j)|0\rangle}. \tag{2.19}$$

Here

$$\psi^\dagger(x,t) = e^{iHt}\psi^\dagger(x,0)e^{-iHt}. \tag{2.20}$$

We shall use the notation μ_j for the ground state only. The square of the norm of the ground state wave function (denominator of the correlation function) was found in [**6**],

$$\langle 0|\prod_{j=1}^N C(\mu_j)\prod_{j=1}^N B(\mu_j)|0\rangle = c^N\left(\prod_{\substack{j,k=1\\ j>k}}^N g(\mu_j,\mu_k)g(\mu_k,\mu_j)\right) \times\left(\prod_{j=1}^N\prod_{k=1}^N h(\mu_j,\mu_k)\right)\det{}_N\frac{\partial\varphi_j}{\partial\mu_k}. \tag{2.21}$$

Here $\partial\varphi_j/\partial\mu_k$ is the $N \times N$ matrix

$$\frac{\partial\varphi_j}{\partial\mu_k} = \delta_{jk}\left[L + \sum_{l=1}^N K(\mu_j,\mu_l)\right] - K(\mu_j,\mu_k). \tag{2.22}$$

Let us emphasize that $\det(\partial\varphi_j/\partial\mu_k) > 0$ (see Section I.2 of [**8**]). The thermodynamic limit of the square of the norm can be described by the following formula:

$$\lim\left(\frac{\det_N(\partial\varphi_j/\partial\mu_k)}{\prod_{j=1}^N 2\pi L\rho(\mu_j)}\right) = \det\left(\hat{I} - \frac{1}{2\pi}\hat{K}\right), \tag{2.23}$$

where $\hat{K}$ is an integral operator acting on some trial function $f(\lambda)$ as

$$(\hat{K}f)(\lambda) = \int_{-q}^{q} K(\lambda, \mu) f(\mu)\, d\mu. \tag{2.24}$$

The proof can be found in [**6**] (see also Section X.4 of [**8**]).

In order to calculate the correlation function we shall also need a description of excited states. We need to consider excited states which have one more particle than in the ground state:

$$\prod_{j=1}^{N+1} B(\lambda_j)|0\rangle, \quad \text{and} \quad \langle 0| \prod_{j=1}^{N+1} C(\lambda_j), \tag{2.25}$$

where the λ_j have to satisfy the Bethe Equations

$$\frac{a(\lambda_j)}{d(\lambda_j)} \prod_{\substack{k=1\\k\neq j}}^{N+1} \frac{f(\lambda_j, \lambda_k)}{f(\lambda_k, \lambda_j)} = 1 \quad \text{or} \quad \frac{a(\lambda_j)}{d(\lambda_j)} \prod_{k=1}^{N+1} \frac{h(\lambda_j, \lambda_k)}{h(\lambda_k, \lambda_j)} = (-1)^N. \tag{2.26}$$

We shall further assume that the number of particles in the ground state N is even. In order to write the logarithmic form of the Bethe Equations it is convenient to introduce

$$\tilde{\varphi}_j \equiv L\lambda_j + \sum_{\substack{k=1\\k\neq j}}^{N+1} i \ln\left(\frac{\lambda_j - \lambda_k + ic}{\lambda_j - \lambda_k - ic}\right). \tag{2.27}$$

The Bethe Equations can now be written as

$$\tilde{\varphi}_j = 2\pi n_j, \tag{2.28}$$

where n_j is an ordered set of different integers $n_{j+1} > n_j$. One can prove that all λ_j are real. In order to enumerate all the eigenstates in the sector with $N+1$ particles we have to consider all sets of ordered integers n_j. The square of the norm of the excited state is

$$\begin{aligned} \langle 0| \prod_{j=1}^{N+1} C(\lambda_j) \prod_{j=1}^{N+1} B(\lambda_j)|0\rangle \\ = c^{N+1} \left(\prod_{\substack{j,k=1\\j>k}}^{N+1} g(\lambda_j, \lambda_k) g(\lambda_k, \lambda_j)\right) \left(\prod_{j=1}^{N+1} \prod_{k=1}^{N+1} h(\lambda_j, \lambda_k)\right) \det_{N+1} \frac{\partial \tilde{\varphi}_j}{\partial \lambda_k}. \end{aligned} \tag{2.29}$$

For the excited state $\det(\partial\tilde{\varphi}_j/\partial\lambda_k)$ is also positive. We shall also mention that the scattering matrix of elementary excitations can be found in Section I.4 of [**8**]. It depends strongly on momenta; this shows that the model is not free fermionic.

Now we can define the form factor in the finite volume

$$F_N = \langle 0| \prod_{j=1}^{N} C(\mu_j) \psi(0,0) \prod_{j=1}^{N+1} B(\lambda_j)|0\rangle. \tag{2.30}$$

We shall calculate it in the next section. We shall also need the conjugated form factor

$$(2.31)\quad \langle 0|\prod_{j=1}^{N+1} C(\lambda_j)\psi^\dagger(x,t)\prod_{j=1}^{N} B(\mu_j)|0\rangle$$
$$= e^{-iht}\cdot \exp\left[it\left(\sum_{j=1}^{N+1}\lambda_j^2-\sum_{k=1}^{N}\mu_k^2\right)-ix\left(\sum_{j=1}^{N+1}\lambda_j-\sum_{k=1}^{N}\mu_k\right)\right]\cdot \overline{F}_N.$$

Here we used the fact that the energy and momentum of the eigenstate are given by the expressions

$$(2.32)\qquad E_{N+1}=\sum_{j=1}^{N+1}(\lambda_j^2-h),$$

$$(2.33)\qquad P_{N+1}=\sum_{j=1}^{N+1}\lambda_j.$$

3. Form Factor

The main purpose of the paper is to evaluate the correlation function. In the finite volume we shall use the notation

$$(3.1)\qquad \langle\psi(0,0)\psi^\dagger(x,t)\rangle_N=\frac{\langle 0|\prod_{j=1}^{N}C(\mu_j)\psi(0,0)\psi^\dagger(x,t)\prod_{j=1}^{N}B(\mu_j)|0\rangle}{\langle 0|\prod_{j=1}^{N}C(\mu_j)\prod_{j=1}^{N}B(\mu_j)|0\rangle}.$$

We shall use the standard representation of correlation function in terms of the form factors

$$(3.2)\quad \langle\psi(0,0)\psi^\dagger(x,t)\rangle_N$$
$$=\sum_{\text{all }\{\lambda\}_{N+1}}\frac{\langle 0|\prod_{j=1}^{N}C(\mu_j)\psi(0,0)\prod_{j=1}^{N+1}B(\lambda_j)|0\rangle\langle 0|\prod_{j=1}^{N+1}C(\lambda_j)\psi^\dagger(x,t)\prod_{j=1}^{N}B(\mu_j)|0\rangle}{\langle 0|\prod_{j=1}^{N+1}C(\lambda_j)\prod_{j=1}^{N+1}B(\lambda_j)|0\rangle\langle 0|\prod_{j=1}^{N}C(\mu_j)\prod_{j=1}^{N}B(\mu_j)|0\rangle}.$$

In order to calculate the form factor we need to know the action of the local field on the eigenvector. This can be found in [**3**] (see also Section XII.2 of [**8**]).

$$(3.3)\quad \psi(0,0)\prod_{j=1}^{N+1}B(\lambda_j)|0\rangle=-i\sqrt{c}\sum_{\ell=1}^{N+1}a(\lambda_\ell)\left(\prod_{\substack{m=1\\ m\neq\ell}}^{N+1}f(\lambda_\ell,\lambda_m)\right)\prod_{\substack{m=1\\ m\neq\ell}}^{N+1}B(\lambda_m)|0\rangle.$$

This permits us to represent form factor as follows

$$(3.4)\quad F_N$$
$$=-i\sqrt{c}\sum_{\ell=1}^{N+1}a(\lambda_\ell)\left(\prod_{\substack{m=1\\ m\neq\ell}}^{N+1}g(\lambda_\ell,\lambda_m)\right)\left(\prod_{\substack{m=1\\ m\neq\ell}}^{N+1}h(\lambda_\ell,\lambda_m)\right)\langle 0|\prod_{j=1}^{N}C(\mu_j)\prod_{\substack{m=1\\ m\neq\ell}}^{N+1}B(\lambda_m)|0\rangle.$$

Let us notice that the form factor is symmetric function of all the λ_j because

$$[B(\lambda_j),B(\lambda_k)]=0.$$

We now need to calculate the scalar product between the eigenvector and non-eigenvector

$$\langle 0| \prod_{j=1}^{N} C(\mu_j) \prod_{\substack{m=1\\ m\neq \ell}}^{N+1} B(\lambda_m)|0\rangle, \tag{3.5}$$

where the μ_j satisfy the Bethe Equations, but the λ_m do not. This can be done by the following theorem.

THEOREM 3.1. *The following determinant representation holds for such scalar products*:

$$\begin{aligned}\langle 0| \prod_{j=1}^{N} C(\mu_j) \prod_{j=1}^{N} B(\lambda_j)|0\rangle \\ = \left(\prod_{j=1}^{N} d(\mu_j)d(\lambda_j)\right)\left(\prod_{\substack{j,k=1\\ j>k}}^{N} g(\lambda_j,\lambda_k)g(\mu_k,\mu_j)\right)\left(\prod_{j,k=1}^{N} h(\mu_j,\lambda_k)\right)\det(M_{jk}),\end{aligned} \tag{3.6}$$

where

$$M_{jk} = \frac{g(\mu_k,\lambda_j)}{h(\mu_k,\lambda_j)} - \frac{a(\lambda_j)}{d(\lambda_j)}\frac{g(\lambda_j,\mu_k)}{h(\lambda_j,\mu_k)}\prod_{m=1}^{N}\frac{f(\lambda_j,\mu_m)}{f(\mu_m,\lambda_j)}. \tag{3.7}$$

Here the spectral parameters $\{\mu_j\}$ *satisfy the Bethe Ansatz equations* (2.13). *The spectral parameters* $\{\lambda_j\}$ *are free and do not satisfy any equations.*

This theorem was proved in [**14**].

For the scalar product, which appears in the expression for the form factor, we get

$$\begin{aligned}\langle 0| \prod_{j=1}^{N} C(\mu_j) \prod_{\substack{m=1\\ m\neq \ell}}^{N+1} B(\lambda_m)|0\rangle \\ = \prod_{\substack{j,k=1\\ j>k}}^{N} g(\mu_j,\mu_k)\cdot \prod_{\substack{j,k=1\\ j>k\\ \ell\notin\{j,k\}}}^{N+1} g(\lambda_k,\lambda_j)\cdot \prod_{j=1}^{N}\prod_{\substack{m=1\\ m\neq\ell}}^{N+1} h(\mu_j,\lambda_m) \\ \times \prod_{j=1}^{N} d(\mu_j)\cdot \prod_{\substack{m=1\\ m\neq\ell}}^{N+1} d(\lambda_m)\cdot \det{}_N M^{(\ell)}.\end{aligned} \tag{3.8}$$

Here the entries of the $N\times N$ matrix $M^{(\ell)}$ are

$$\begin{aligned}M^{(\ell)}_{jk} = t(\mu_k,\lambda_j) - r(\lambda_j)t(\lambda_j,\mu_k)\cdot \prod_{m=1}^{N}\frac{f(\lambda_j,\mu_m)}{f(\mu_m,\lambda_j)}, \\ j=1,\dots,\ell-1,\ell+1,\dots,N+1;\ k=1,\dots,N.\end{aligned} \tag{3.9}$$

Let us recall that

$$t(\lambda,\mu) = \frac{(ic)^2}{(\lambda-\mu)(\lambda-\mu+ic)} \quad\text{and}\quad r(\lambda) = \frac{a(\lambda)}{d(\lambda)}.$$

Remember that the Bethe Equations give:

$$(3.10)\qquad r(\lambda_j)=\prod_{p=1}^{N+1}\frac{h(\lambda_p,\lambda_j)}{h(\lambda_j,\lambda_p)}\quad\text{and}\quad\prod_{m=1}^{N}\frac{f(\lambda_j,\mu_m)}{f(\mu_m,\lambda_j)}=(-1)^N\prod_{m=1}^{N}\frac{h(\lambda_j,\mu_m)}{h(\mu_m,\lambda_j)},$$

or equivalently,

$$(3.11)\qquad a(\lambda_\ell)\prod_{m=1}^{N+1}h(\lambda_\ell,\lambda_m)=d(\lambda_\ell)\prod_{m=1}^{N+1}h(\lambda_m,\lambda_\ell).$$

Expression (3.9) becomes

$$(3.12)\qquad M_{jk}^{(\ell)}=t(\mu_k,\lambda_j)-t(\lambda_j,\mu_k)\Big(\prod_{p=1}^{N+1}\frac{h(\lambda_p,\lambda_j)}{h(\lambda_j,\lambda_p)}\Big)\cdot\Big(\prod_{m=1}^{N}\frac{h(\lambda_j,\mu_m)}{h(\mu_m,\lambda_j)}\Big).$$

Using the obvious equality

$$(3.13)\qquad \prod_{\substack{m=1\\m\neq\ell}}^{N+1}g(\lambda_\ell,\lambda_m)=\Big(\prod_{m=1}^{\ell-1}g(\lambda_\ell,\lambda_m)\Big)\Big(\prod_{m=\ell+1}^{N+1}g(\lambda_\ell,\lambda_m)\Big)$$

$$y=(-1)^{\ell-1}\Big(\prod_{m=1}^{\ell-1}g(\lambda_m,\lambda_\ell)\Big)\Big(\prod_{m=\ell+1}^{N+1}g(\lambda_\ell,\lambda_m)\Big),$$

and substituting (3.8) into (3.4), we have

$$(3.14)\qquad F_N=-i\sqrt{c}\sum_{\ell=1}^{N+1}(-1)^{\ell-1}\prod_{\substack{j,k=1\\j>k}}^{N}g(\mu_j,\mu_k)\prod_{\substack{j,k=1\\j>k}}^{N+1}g(\lambda_k,\lambda_j)$$

$$\times\prod_{m=1}^{N+1}h(\lambda_m,\lambda_\ell)\prod_{j=1}^{N}\prod_{\substack{m=1\\m\neq\ell}}^{N+1}h(\mu_j,\lambda_m)\prod_{j=1}^{N}d(\mu_j)\prod_{m=1}^{N+1}d(\lambda_m)\cdot\det_N M^{(\ell)}.$$

One can rewrite the determinant $\det_N M^{(\ell)}$ as

$$(3.15)\qquad \det_N M^{(\ell)}=\Big(\prod_{m=1}^{N}\prod_{\substack{j=1\\j\neq\ell}}^{N+1}\frac{1}{h(\mu_m,\lambda_j)}\Big)\cdot\Big(\prod_{p=1}^{N+1}\prod_{\substack{j=1\\j\neq\ell}}^{N+1}h(\lambda_p,\lambda_j)\Big)\cdot\det_N S^{(\ell)},$$

where

$$(3.16)\qquad \begin{gathered}S_{jk}^{(\ell)}=t(\mu_k,\lambda_j)\frac{\prod_{m=1}^{N}h(\mu_m,\lambda_j)}{\prod_{p=1}^{N+1}h(\lambda_p,\lambda_j)}-t(\lambda_j,\mu_k)\frac{\prod_{m=1}^{N}h(\lambda_j,\mu_m)}{\prod_{p=1}^{N+1}h(\lambda_j,\lambda_p)},\\ j=1,\dots,\ell-1,\ell+1,\dots,N+1;\quad k=1,\dots,N.\end{gathered}$$

Let us substitute (3.15) into (3.14),

$$(3.17)\qquad F_N=-i\sqrt{c}\prod_{\substack{j,k=1\\j>k}}^{N}g(\mu_j,\mu_k)\prod_{\substack{j,k=1\\j>k}}^{N+1}g(\lambda_k,\lambda_j)\prod_{m=1}^{N+1}\prod_{j=1}^{N+1}h(\lambda_m,\lambda_j)$$

$$\times\Big(\sum_{\ell=1}^{N+1}(-1)^{\ell+1}\det_N S^{(\ell)}\Big)\cdot\prod_{j=1}^{N}d(\mu_j)\prod_{m=1}^{N+1}d(\lambda_m).$$

In order to simplify this expression let us study

$$\mathcal{M}i = \mathcal{M}i\,\{\lambda\} \equiv \sum_{\ell=1}^{N+1} (-1)^{\ell-1} \det{}_N S^{(\ell)}. \tag{3.18}$$

Notice that $\mathcal{M}i$ is an antisymmetric function of all $\{\lambda_j\}$ because F_N is symmetric and the product of functions $g(\lambda_k, \lambda_j)$ is antisymmetric. In particular,

$$\mathcal{M}i\,\{\lambda\} = 0 \quad \text{if } \lambda_j = \lambda_k. \tag{3.19}$$

The determinant $\det S^{(\ell)}$ can be obtained from $\det S^{(N+1)}$ by interchanging λ_ℓ and λ_{N+1}. This is a special case of a permutation

$$(\lambda_1, \cdots, \lambda_\ell, \cdots, \lambda_N, \lambda_{N+1}) \longrightarrow (\lambda_1, \cdots, \lambda_{N+1}, \cdots, \lambda_N, \lambda_\ell). \tag{3.20}$$

Since $(-1)^{\ell-1}$ is the parity of this permutation,

$$\begin{aligned} \mathcal{M}i\{\lambda\} &= \sum_{\substack{\text{permutation} \\ \text{of all } \{\lambda_{N+1}\}}} (-1)^P \prod_{j=1}^{N} S_{P(j)j} \\ &= \left(1 + \frac{\partial}{\partial \alpha}\right) \det{}_N (S_{jk} - \alpha S_{N+1,k})\big|_{\alpha=0}. \end{aligned} \tag{3.21}$$

Here S_{jk} means $S_{jk}^{(N+1)}$ from (3.16) and $\det S_{jk}$ is the term $\ell = N+1$ in (3.18),

$$-\frac{\partial}{\partial \alpha} \det{}_N (S_{jk} - \alpha S_{N+1,k})\big|_{\alpha=0}$$

is the sum of N terms where each of them differs from $\det S_{jk}$ by the replacement of the ℓ-th line (corresponding to λ_ℓ) by the $(N+1)$-th line. We can use the expression (3.21) to simplify the form factor (3.17)

$$\begin{aligned} F_N = -i\sqrt{c} \prod_{\substack{j,k=1 \\ j>k}}^{N} g(\mu_j, \mu_k) \prod_{\substack{j,k=1 \\ j>k}}^{N+1} g(\lambda_k, \lambda_j) \prod_{m=1}^{N+1} \prod_{j=1}^{N+1} h(\lambda_m, \lambda_j) \\ \times \prod_{j=1}^{N} d(\mu_j) \prod_{m=1}^{N+1} d(\lambda_m) \cdot \mathcal{M}i\,\{\lambda\}. \end{aligned} \tag{3.22}$$

Summary

The main result of the paper is formula (3.22). It represents the form factor of the local field (in the finite volume) as a determinant of a matrix. The dimension of the matrix is equal to the number of particles in the ground state. It is interesting to mention that recently a similar formula was obtained in the paper [**4**] for form factors of XXZ Heisenberg spin chain.We think that determinant formulae for form factors are a universal feature of exactly solvable models.

References

1. E. Barough, B. M. McCoy, and T. T. Wu, *Zero-field susceptibility of the two-dimensional Ising model near* T_c, Phys. Rev. Lett. **31** (1973), 1409–1411.
2. L. D. Faddeev and E. K. Sklyanin, *Quantum-mechanical approach to completely integrable field theory models*, Soviet Phys. Dokl. **23** (1978), 902–904.
3. A. G. Izergin, V. E. Korepin, and N. Yu. Reshetikin, *Correlation functions in a one-dimensional Bose gas*, J. Phys. A **20** (1987), no. 14, 4799–4822.

4. N. Kitanine, J.-M. Maillet, and V. Terras, *Form factors of the XXZ Heisenberg spin-$\frac{1}{2}$ finite chain*, Nuclear Phys. B **554** (1999), 647–678; math-ph/9807020.
5. T. Kojima, V. E. Korepin, and N. A. Slavnov, *Determinant representation for dynamical correlation functions of the quantum nonlinear Schrödinger equation*, Comm. Math. Phys. **188** (1997), 657–689.
6. V. E. Korepin, *Calculation of norms of Bethe wave functions*, Comm. Math. Phys. **86** (1982), 391–418.
7. V. E. Korepin, *Dual field formulation of quantum integrable models*, Comm. Math. Phys. **113** (1987), 177–190.
8. V. E. Korepin, N. M. Bogoliubov, and A. G. Izergin, *Quantum inverse scattering method and correlation functions*, Cambridge University Press, 1993.
9. V. E. Korepin and N. A. Slavnov, *The time dependent correlation function of an impenetrable Bose gas as a Fredholm minor*, Comm. Math. Phys. **129** (1990), 103–113.
10. A. Lenard, *One-dimensional impenetrable bosons in thermal equilibrium*, J. Math. Phys. **7** (1966), no. 7, 1268–1272.
11. E. H. Lieb, *Exact analysis of an interacting Bose gas.* II. *The excitation spectrum*, Phys. Rev. **130** (1963), 1616–1624.
12. E. H. Lieb and W. Liniger, *Exact analysis of an interacting Bose gas.* I. *The general solution and the ground state*, Phys. Rev. **130** (1963), 1605–1616.
13. E. W. Montroll, R. B. Potts, and J. C. Ward, *Correlations and spontaneous magnetization of the two-dimensional Ising model*, J. Math. Phys. **4** (1963), no. 2, 308–322.
14. N. A. Slavnov, *Calculation of scalar products of wave functions and form-factors in the framework of the algebraic Bethe ansatz*, Theoret. and Math. Phys. **79** (1989), 502–508.
15. C. N. Yang and C. P. Yang, *Thermodynamics of a one-dimensional system of bosons with repulsive delta-function interaction*, J. Math. Phys. **10** (1969), no. 7, 1115–1122.
16. V. E. Zakharov and A. B. Shabat, *Exact theory of two-dimensional self-focusing and one-dimensional self-modulation of waves in nonlinear media*, Soviet Physics JETP **34** (1972), 62–69.

INSTITUTE FOR THEORETICAL PHYSICS, STATE UNIVERSITY OF NEW YORK AT STONY BROOK, STONY BROOK, NY 11794-3840, USA

E-mail address: `korepin@insti.physics.sunysb.edu`

Centre de Recherches Mathématiques
CRM Proceedings and Lecture Notes
Volume **26**, 2000

Isomonodromic Deformations in Genus Zero and One: Algebro-Geometric Solutions and Schlesinger Transformations

D. A. Korotkin

ABSTRACT. We review some recent developments in the theory of isomonodromic deformations on the Riemann sphere and the elliptic curve. For both cases we show how to derive Schlesinger transformations together with their action on the τ-function, and construct classes of solutions in terms of multidimensional theta-functions.

1. Introduction

The theory of isomonodromic deformations of ordinary matrix differential equations of the type

$$\frac{d\Psi}{d\lambda} = A(\lambda)\Psi, \tag{1.1}$$

where $A(\lambda)$ is a matrix-valued meromorphic function on $\overline{\mathbb{C}}$, is a classical area intimately related to the matrix Riemann-Hilbert problem on the Riemann sphere. Over the last 20 years this has become a powerful tool in areas like soliton theory, statistical mechanics, theory of random matrices, quantum field theory, etc. The main object associated with the isomonodromic deformation equations is the so-called τ-function.

After the classical work of Schlesinger [**28**] the important contributions to the development of the subject were made in the papers of Jimbo, Miwa and their collaborators in the early 80's [**14, 15, 12, 13**].

There are only a few cases where the matrix Riemann-Hilbert problem may be solved explicitly in terms of known special functions.

However, as was already discovered by Schlesinger himself, there exists a large class of transformations which allow to get an infinite chain of new solutions starting from known ones. They share the characteristic feature that they shift the eigenvalues of the residues of the connection $A(\lambda)$ in (1.1) by integer or half-integer values, thus changing the associated monodromies by sign only. These transformations—nowadays called Schlesinger transformations—were systematically studied in [**12,**

1991 *Mathematics Subject Classification.* 35Q15.
This is the final form of the paper.

13]. In particular, it turns out that when written in terms of the τ-functions the superposition laws of these transformations provide a big supply of discrete integrable systems.

Recently, in the papers [**17, 5**] a class of 2×2 Riemann-Hilbert problems with arbitrary off-diagonal monodromy matrices was solved in terms of multidimensional theta-functions. The equations for the τ-function were integrated in the paper [**17**] to give the following result:

$$\tau(\{\lambda_j\}) = [\det \mathcal{A}]^{-1/2} \prod_{j<k} (\lambda_j - \lambda_k)^{-1/8} \Theta \begin{bmatrix} \mathbf{p} \\ \mathbf{q} \end{bmatrix} (0 \mid \mathbf{B}),$$

where all the objects associated to the auxiliary hyperelliptic curve are defined below in Section 2.3.

The natural question of generalizing the theory of isomonodromic deformations on the sphere to higher genus surfaces was addressed by several authors. Here, we mention the contributions of Okamoto [**26, 27**] and Iwasaki [**11**].

For the case of the torus, recently two different explicit forms of equations of isomonodromic deformations were proposed. In [**23**] the present author and Samtleben studied isomonodromic deformations of non-singlevalued meromorphic connections on the torus whose "twists" (which determine the transformation of the connection $A(\lambda)$ with respect to tracing along basic cycles of the torus) vary with respect to the deformation parameters. The isomonodromic deformation equations for these connections therefore contain transcendental dependence on the dynamical variables, which makes it difficult to analyse this system in a way analogous to the Schlesinger system on the sphere. On the other hand, Takasaki [**29**] considered connections on the torus whose twists remain invariant with respect to the parameters of deformation. In Takasaki's form, the equations of isomonodromic deformations have already the same degree of non-linearity as the ordinary Schlesinger system.

In the paper [**20**], transformations of Schlesinger type were constructed for elliptic isomonodromic deformations in Takasaki form, and the action of these transformations on the elliptic version of the τ-function was derived. Here we review these results, and, in addition, present the generalization of results of the paper [**17**] to the elliptic case. We show how to solve a certain class of Riemann-Hilbert problems on the torus in terms of Prym theta-functions. In turn, this allows us to construct a class of algebro-geometric solutions of the elliptic Schlesinger system.

In Section 2 we introduce the Schlesinger system on the Riemann sphere. For the 2×2 case we discuss the elementary Schlesinger transformations together with their action on the τ-function and, following [**17**], derive a class of algebro-geometric solutions of the Schlesinger system in terms of theta-functions of the auxiliary hyperelliptic curve. In Section 3 we describe equations of elliptic isomonodromic deformations with constant twists [**29**], and, following [**20**], construct an elliptic version of elementary Schlesinger transformations. The new result of this paper—the construction of algebro-geometric solutions of the elliptic Schlesinger system in terms of Prym theta-functions—is presented in Section 3.

2. Schlesinger System on the Riemann Sphere: Schlesinger Transformations and Algebro-Geometric Solutions

2.1. Schlesinger System. Consider the following ordinary linear differential equation (1.1) for a matrix-valued function $\Psi(\lambda) \in \mathrm{SL}(2, \mathbb{C})$ and

(2.1) $$A(\lambda) = \sum_{j=1}^{N} \frac{A_j}{\lambda - \lambda_j},$$

where the residues $A_j \in \mathfrak{sl}(2, \mathbb{C})$ are independent of λ. Regularity at $\lambda = \infty$ requires

(2.2) $$\sum_{j=1}^{N} A_j = 0,$$

and allows to further impose the initial condition $\Psi(\lambda = \infty) = I$. The matrix $\Psi(\lambda)$ defined in this way lives on the universal covering X of $\mathbb{C}P^1 \setminus \{\lambda_1, \dots, \lambda_N\}$. Its asymptotic expansion near the singularities λ_j is given by

(2.3) $$\Psi(\lambda) = G_j \Psi_j \cdot (\lambda - \lambda_j)^{T_j} C_j,$$

with $G_j, C_j \in \mathrm{SL}(2, \mathbb{C})$ constant, $\Psi_j = I + \mathcal{O}(\lambda - \lambda_j) \in \mathrm{SL}(2, \mathbb{C})$ holomorphic around $\lambda = \lambda_j$, and where T_j is a traceless diagonal matrix with eigenvalues $\pm t_j$. The residues A_j of (2.1) are encoded in the local expansion as

(2.4) $$A_j = G_j T_j G_j^{-1}.$$

Upon analytic continuation around $\lambda = \lambda_j$, the function $\Psi(\lambda)$ in $\mathbb{C}P^1 \setminus \{\lambda_1, \dots, \lambda_N\}$ changes by right multiplication with some monodromy matrices M_j

(2.5) $$\Psi(\lambda) \to \Psi(\lambda) M_j, \qquad M_j = C_j^{-1} e^{2\pi i T_j} C_j.$$

In the sequel we shall consider the generic case where no t_j is an integer or half-integer.

The assumption of independence of all monodromy matrices M_i of the positions of the singularities λ_j: $\partial M_i / \partial \lambda_j = 0$ is called the isomonodromy condition; it implies the following dependence of $\Psi(\lambda)$ on λ_j:

(2.6) $$\frac{\partial \Psi}{\partial \lambda_j} = -\frac{A_j}{\lambda - \lambda_j} \Psi,$$

as follows from (2.3) and the normalization of $\Psi(\lambda)$ at ∞. Compatibility of (1.1) and (2.6) then is equivalent to the classical Schlesinger system [**28**]:

(2.7) $$\frac{\partial A_j}{\partial \lambda_i} = \frac{[A_j, A_i]}{\lambda_j - \lambda_i}, \quad i \neq j, \quad \frac{\partial A_j}{\partial \lambda_j} = -\sum_{i \neq j} \frac{[A_j, A_i]}{\lambda_j - \lambda_i},$$

describing the dependence of the residues A_j on the λ_i. Obviously, the eigenvalues t_j of the A_j are integrals of motion of the Schlesinger system. The functions G_j have the following dependence on λ_j [**15**]:

(2.8) $$\frac{\partial G_j}{\partial \lambda_i} = \frac{A_i G_j}{\lambda_i - \lambda_j}, \; i \neq j, \qquad \frac{\partial G_j}{\partial \lambda_j} = -\sum_{i \neq j} \frac{A_i G_j}{\lambda_i - \lambda_j},$$

which obviously implies (2.7).

To introduce the notion of the τ-function for the Schlesinger system, one notes that (2.7) is a multi-time Hamiltonian system [**14**] with respect to the Poisson structure on the residues A_j

(2.9) $$\{A_i^\alpha, A_j^\beta\} = \delta_{ij} \varepsilon^{\alpha\beta\gamma} A_j^\alpha c,$$

(α, β, γ denoting $\mathfrak{sl}(2)$ algebra indices with the completely antisymmetric structure constants $\varepsilon^{\alpha\beta\gamma}$) and Hamiltonians

$$H_i = \frac{1}{4\pi i}\oint_{\lambda_i} \operatorname{tr} A^2(\lambda)\, d\lambda = \frac{1}{2}\sum_{j\neq i} \frac{\operatorname{tr} A_i A_j}{\lambda_j - \lambda_i}. \tag{2.10}$$

Explicitly, (2.7) takes the form

$$\frac{\partial A_j}{\partial \lambda_i} = \{H_i, A_j\}, \tag{2.11}$$

and all the Hamiltonians H_j Poisson-commute.

The τ-function $\tau(\{\lambda_j\})$ of the Schlesinger system then is defined as the generating functions of the Hamiltonians

$$\frac{\partial \ln \tau}{\partial \lambda_j} = H_j, \tag{2.12}$$

where compatibility of these equations follows from (2.7). This τ-function is closely related to the Fredholm determinant of a certain integral operator associated to the Riemann-Hilbert problem (see [**8**] for details).

2.2. Schlesinger Transformations on the Riemann Sphere. Schlesinger transformations are symmetry transformations of the Schlesinger system (2.7) which map a given solution $\{A_j(\{\lambda_i\})\}$ to another solution $\{\hat{A}_j(\{\lambda_i\})\}$ with the same number and positions of poles λ_j such that the related eigenvalues t_j are shifted by integer or half-integer values $t_j \to t_j + n_j/2$, $n_j \in \mathbb{Z}$. The monodromy matrices M_j hence remain invariant or change sign under this transformation. We shall restrict ourselves to elementary Schlesinger transformations, which change only two t_j's, say, t_k and t_l for $k \neq l$ by $\pm 1/2$. The transformed variables will be denoted by $\widehat{\Psi}$, $\hat{A}_j$, $\hat{t}_j$, etc. Without loss of generality we consider the case

$$\hat{t}_j = \begin{cases} t_j + \frac{1}{2} & \text{for } j = k, l \\ t_j & \text{otherwise.} \end{cases} \tag{2.13}$$

Our presentation here mainly follows [**16**]. For the transformed function $\widehat{\Psi}$ we make the ansatz

$$\widehat{\Psi}(\lambda) = F(\lambda)\Psi(\lambda), \tag{2.14}$$

with

$$F(\lambda) = \sqrt{\frac{\lambda - \lambda_k}{\lambda - \lambda_l}}\, S_+ + \sqrt{\frac{\lambda - \lambda_l}{\lambda - \lambda_k}}\, S_-, \tag{2.15}$$

where the matrices $S_\pm$ do not depend on λ and are uniquely determined by [**16**]:

$$S_\pm^2 = S_\pm, \quad S_+ + S_- = I, \quad S_+ G_l^1 = S_- G_k^1 = 0. \tag{2.16}$$

By G_j^α we denote here the α-th column of the matrix G_j ($\alpha = 1, 2$). Combining the columns G_k^1 and G_l^1 into a 2×2 matrix

$$G = (G_k^1, G_l^1), \tag{2.17}$$

we can deduce from (2.16) the following simple formula for $S_\pm$:

$$S_\pm = G P_\pm G^{-1}, \tag{2.18}$$

with projection matrices

$$P_+ = \begin{pmatrix} 1 & 0 \\ 0 & 0 \end{pmatrix}, \quad P_- = \begin{pmatrix} 0 & 0 \\ 0 & 1 \end{pmatrix}.$$

It is easy to check, using the local expansion of Ψ at the singularities λ_j (2.3) and the defining relations for $S_\pm$ (2.18), that the transformed function $\widehat{\Psi}$ has a local expansion at λ_j of the form (2.3) with the same matrices C_j and the desired transformation (2.13) of the t_j. The matrices G_j change to new matrices $\widehat{G}_j$. Thus, $\widehat{\Psi}$ satisfies the system

$$\frac{\partial \widehat{\Psi}}{\partial \lambda} = \sum_{j=1}^{N} \frac{\hat{A}_j}{\lambda - \lambda_j} \widehat{\Psi}, \quad \frac{\partial \widehat{\Psi}}{\partial \lambda_j} = -\frac{\hat{A}_j}{\lambda - \lambda_j} \widehat{\Psi}, \tag{2.19}$$

where the functions $\hat{A}_j(\{\lambda_i\})$ build a new solution of the Schlesinger system (2.7).

On the level of the residues A_j, the form of the Schlesinger transformation is not very transparent; however, it turns out that the associated τ-function transforms in a rather simple way. Namely, for $\widehat{\Psi}$ we find

$$\operatorname{tr} \hat{A}^2 = \operatorname{tr} A^2 + 2 \operatorname{tr}\left[F^{-1} \frac{dF}{d\lambda} A\right] + \operatorname{tr}\left[F^{-1} \frac{dF}{d\lambda}\right]^2. \tag{2.20}$$

For example, the Hamiltonians H_j for $j \neq k, l$ transform as follows:

$$\begin{aligned} \widehat{H}_j - H_j &= \left(\frac{1}{\lambda_j - \lambda_k} - \frac{1}{\lambda_j - \lambda_l}\right) \operatorname{tr}[A_j S_+] \\ &= \frac{\operatorname{tr}[A_j G P_+ G^{-1}]}{\lambda_j - \lambda_k} + \frac{\operatorname{tr}[A_j G P_- G^{-1}]}{\lambda_j - \lambda_l} \\ &= \operatorname{tr}\left[\frac{\partial G}{\partial \lambda_j} G^{-1}\right] \end{aligned}$$

according to (2.8). Hence the transformed τ-function $\hat{\tau}$ is given by

$$\hat{\tau} = f(\lambda_k, \lambda_l) \det G \cdot \tau$$

with some function $f(\lambda_k, \lambda_l)$ to be determined from the transformation of H_k, H_l. Taking into account the transformation of the Hamiltonians H_k and H_l following from (2.20) we find the following formula describing the action of the elementary Schlesinger transformation (2.13) on the τ-function:

$$\hat{\tau}(\{\lambda_j\}) = \{(\lambda_k - \lambda_l)^{-1/2} \det G\} \cdot \tau(\{\lambda_j\}). \tag{2.21}$$

Other elementary Schlesinger transformations may be obtained in a similar way by building the matrix G from G_k^1 and G_l^2 instead of (2.17), etc. Moreover, all such transformations with different k and l may be superposed to get the general Schlesinger transformation which simultaneously shifts an arbitrary number of the t_j by some integer or half-integer constants. These general transformations were studied in detail in [**15, 12, 13**].

2.3. Algebro-Geometric Solutions of the Schlesinger System. Let us take $N = 2g+2$ and introduce the hyperelliptic curve $\mathcal{L}$ of genus g by the equation

$$w^2 = \prod_{j=1}^{2g+2} (\lambda - \lambda_j) \tag{2.22}$$

with branch cuts $[\lambda_{2j+1}, \lambda_{2j+2}]$. Let us choose the canonical basis of cycles (a_j, b_j), $j = 1, \dots, g$, such that the cycle a_j encircles the branch cut $[\lambda_{2j+1}, \lambda_{2j+2}]$. Cycle b_j starts from one bank of the branch cut $[\lambda_1, \lambda_2]$, goes to the second sheet through the branch cut $[\lambda_{2j+1}, \lambda_{2j+2}]$, and comes back to another bank of the branch cut $[\lambda_1, \lambda_2]$.

The dual basis of holomorphic 1-forms on $\mathcal{L}$ is given by $(\lambda^{k-1} d\lambda)/w$, $k = 1, \dots, g$.

Let us introduce two $g \times g$ matrices of a- and b-periods of these 1-forms:

$$\mathcal{A}_{kj} = \oint_{a_j} \frac{\lambda^{k-1}\, d\lambda}{w}, \quad \mathcal{B}_{kj} = \oint_{b_j} \frac{\lambda^{k-1}\, d\lambda}{w}. \tag{2.23}$$

The holomorphic 1-forms

$$dU_k = \frac{1}{w} \sum_{j=1}^{g} (\mathcal{A}^{-1})_{kj} \lambda^{j-1}\, d\lambda \tag{2.24}$$

satisfy the normalization conditions $\oint_{a_j} dU_k = \delta_{jk}$.

The matrices $\mathcal{A}$ and $\mathcal{B}$ define the symmetric $g \times g$ matrix of b-periods of the curve $\mathcal{L}$: $\mathbf{B} = \mathcal{A}^{-1}\mathcal{B}$.

Let us cut the curve $\mathcal{L}$ along all basic cycles to get the fundamental polygon $\hat{\mathcal{L}}$. For any meromorphic 1-form dW on $\mathcal{L}$ we can define the integral $\int_Q^P dW$, where the integration contour lies inside of $\hat{\mathcal{L}}$ (if dW is meromorphic, the value of this integral might also depend on the choice of the integration contour inside $\hat{\mathcal{L}}$). The vector of Riemann constants corresponding to our choice of the initial point of this map is given by the formula $K_j = j/2 + \frac{1}{2}\sum_{k=1}^{g} \mathbf{B}_{jk}$ (see [**7**]).

The characteristic with components $\mathbf{p} \in \mathbb{C}^g/2\mathbb{C}^g$, $\mathbf{q} \in \mathbb{C}^g/2\mathbb{C}^g$ is called the half-integer characteristic: the half-integer characteristics are in one-to-one correspondence with the half-periods $\mathbf{Bp} + \mathbf{q}$. To any half-integer characteristic we can assign parity which by definition coincides with the parity of the scalar product $4\langle \mathbf{p}, \mathbf{q} \rangle$.

The odd characteristics which will be of importance for us in the sequel correspond to any given subset $S = \{\lambda_{i_1}, \dots, \lambda_{i_{g-1}}\}$ of $g - 1$ arbitrary non-coinciding branch points. The odd half-period associated to the subset S is given by

$$\mathbf{Bp}^S + \mathbf{q}^S = \sum_{j=1}^{g-1} \int_{\lambda_1}^{\lambda_{i_j}} dU - K, \tag{2.25}$$

where $dU = (dU_1, \dots, dU_g)^t$. Denote by $\Omega \subset \mathbb{C}$ the neighbourhood of the infinite point $\lambda = \infty$, such that Ω does not overlap with the projections of any of the basic cycles onto the λ-plane. Let the 2×2 matrix-valued function $\Phi(\lambda)$ be defined in the domain Ω of the first sheet of $\mathcal{L}$ by the following formula,

$$\Phi(\lambda \in \Omega_\lambda) = \begin{pmatrix} \varphi(\lambda) & \varphi(\lambda^*) \\ \psi(\lambda) & \psi(\lambda^*) \end{pmatrix}, \tag{2.26}$$

where the functions φ and ψ are defined in the fundamental polygon $\hat{\mathcal{L}}$ by the formulas:

$$\varphi(\lambda) = \Theta\begin{bmatrix}\mathbf{p}\\ \mathbf{q}\end{bmatrix}\left(\int_{\lambda_1}^{\lambda} dU + \int_{\lambda_1}^{\lambda_\varphi} dU \,\Big|\, \mathbf{B}\right)\Theta\begin{bmatrix}\mathbf{p}^S\\ \mathbf{q}^S\end{bmatrix}\left(\int_{\lambda_\varphi}^{\lambda} dU \,\Big|\, \mathbf{B}\right), \tag{2.27}$$

$$\psi(\lambda) = \Theta\begin{bmatrix}\mathbf{p}\\\mathbf{q}\end{bmatrix}\left(\int_{\lambda_1}^{\lambda} dU + \int_{\lambda_1}^{\lambda_\psi} dU\,\Big|\,\mathbf{B}\right)\Theta\begin{bmatrix}\mathbf{p}^S\\\mathbf{q}^S\end{bmatrix}\left(\int_{\lambda_\psi}^{\lambda} dU\,\Big|\,\mathbf{B}\right), \tag{2.28}$$

with two arbitrary (possibly $\{\lambda_j\}$-dependent) points λ_φ, $\lambda_\psi \in \mathcal{L}$ and arbitrary constant complex characteristic $\left[\begin{smallmatrix}\mathbf{p}\\\mathbf{q}\end{smallmatrix}\right]$; $*$ is the involution on $\mathcal{L}$ interchanging the sheets. An odd theta characteristic $\left[\begin{smallmatrix}\mathbf{p}^S\\\mathbf{q}^S\end{smallmatrix}\right]$ corresponds to an arbitrary subset S of $g-1$ branch points via eq. (2.25).

Since the domain Ω does not overlap with the projections of the basic cycles of $\mathcal{L}$ onto the λ-plane, the domain Ω^* does not overlap with the boundary of $\hat{\mathcal{L}}$, and the functions $\varphi(\lambda^*)$ and $\psi(\lambda^*)$ in (2.26) are uniquely defined by (2.3), (2.3) for $\lambda \in \Omega$.

Now choose some sheet of the universal covering X, define a new function $\Psi(\lambda)$ in the subset Ω of this sheet by the formula

$$\Psi(\lambda \in \Omega) = \sqrt{\frac{\det \Phi(\infty^1)}{\det \Phi(\lambda)}}\,\Phi^{-1}(\infty^1)\Phi(\lambda) \tag{2.29}$$

and extend it to the rest of X by analytic continuation.

The function $\Psi(\lambda)$ (2.29) transforms as follows with respect to the tracing around the basic cycles of $\mathcal{L}$ (by T_{a_j} and T_{b_j} we denote the corresponding operators of analytic continuation):

$$T_{a_j}[\Psi(\lambda)] = \Psi(\lambda)e^{2\pi i p_j \sigma_3}; \quad T_{b_j}[\Psi(\lambda)] = \Psi(\lambda)e^{-2\pi i q_j \sigma_3}.$$

The following statement proved in the paper [**17**] claims that the function Ψ satisfies the condition of isomonodromy, and therefore provides a class of solutions of the Schlesinger system:

THEOREM 2.1. *Let* $\mathbf{p}, \mathbf{q} \in \mathbb{C}^g$ *be an arbitrary set of* $2g$ *constants such that the characteristic* $\left[\begin{smallmatrix}\mathbf{p}\\\mathbf{q}\end{smallmatrix}\right]$ *is not a half-integer. Then:*

(1) *The function* $\Psi(Q \in X)$ *defined by* (2.29) *is independent of* λ_φ *and* λ_ψ, *and satisfies the linear system* (2.1) *with*

$$A_j \equiv \operatorname{res}|_{\lambda=\lambda_j}\{\Psi_\lambda \Psi^{-1}\}, \tag{2.30}$$

which in turn solves the Schlesinger system (2.7).

(2) *The monodromies* (2.5) *of* $\Psi(\lambda)$ *around the points* λ_j *are given by*

$$M_j = \begin{pmatrix} 0 & -m_j \\ m_j^{-1} & 0 \end{pmatrix}, \tag{2.31}$$

where the constants m_j *may be expressed in terms of* $\mathbf{p}$ *and* $\mathbf{q}$ *as follows:*

$$m_1 = i \qquad m_2 = i\exp\left\{-2\pi i \sum_{k=1}^{g} p_k\right\}$$

$$m_{2j+1} = -i\exp\left\{2\pi i q_j - 2\pi i \sum_{k=j}^{g} p_k\right\}$$

$$m_{2j+2} = i\exp\left\{2\pi i q_j - 2\pi i \sum_{k=j+1}^{g} p_k\right\}$$

for $j = 1, \ldots, g$.

(3) *The τ-function, corresponding to the solution (2.4) of the Schlesinger system, has the following form:*

$$\tau(\{\lambda_j\}) = [\det \mathcal{A}]^{-1/2} \prod_{j<k} (\lambda_j - \lambda_k)^{-1/8} \Theta \begin{bmatrix} \mathbf{p} \\ \mathbf{q} \end{bmatrix} (0 \mid \mathbf{B}). \tag{2.32}$$

3. Elliptic Isomonodromic Deformations: Schlesinger Transformations and Algebro-Geometric Solutions

3.1. Isomonodromic Deformations on the Torus. Consider the elliptic curve E with periods 1 and μ together with the canonical basis of cycles (a, b). A (naive) straightforward generalization of the idea of isomonodromic deformations from the complex plane to the torus E runs into difficulties related to the absence of meromorphic functions on the torus with just one simple pole. An independent variation of the simple poles of a meromorphic connection A on the torus preserving the monodromies around the singularities and basic cycles is impossible for the following simple reason. The existence of such a deformation would imply a version of (2.6) with the function $A_j/(\lambda - \lambda_j)$ on the right hand side being replaced by a meromorphic function with only one simple pole on the torus, which gives rise to a contradiction. Therefore, one of the underlying assumptions has to be relaxed.

For example, one may consider the case where not all the poles of the connection A are varied independently. Another possibility is the assumption that some of the poles of A are of order higher than one [**27**]. A third alternative which we shall consider here, is to relax the condition of single-valuedness of the connection A on E and to assume that A has "twists" with respect to analytic continuation along the basic cycles a and b, i.e.,

$$A(\lambda + 1) = QA(\lambda)Q^{-1}, \quad A(\lambda + \mu) = RA(\lambda)R^{-1},$$

where the matrices Q, R do not depend on λ. By a gauge transformation of the form $A \to SAS^{-1} + dSS^{-1}$ with S holomorphic but possibly multi-valued, one may bring the connection into a form where $Q = I$ and $R = e^{\kappa\sigma_3}$, where σ_α denotes the Pauli matrices:

$$\sigma_1 = \begin{pmatrix} 0 & 1 \\ 1 & 0 \end{pmatrix}, \quad \sigma_2 = \begin{pmatrix} 0 & i \\ -i & 0 \end{pmatrix}, \quad \sigma_3 = \begin{pmatrix} 1 & 0 \\ 0 & -1 \end{pmatrix}.$$

The equations of isomonodromic deformations with this choice of the twist were considered in [**23**], where the multi-valuedness of A had a natural origin in the holomorphic gauge fixing of the Chern-Simons theory on the punctured torus. The resulting equations however are rather complicated in comparison with the Schlesinger system on the sphere. This is due to the fact that the twist κ itself becomes a dynamical variable—i.e., changes under isomonodromic deformations—and in the generic situation has a highly non-trivial λ_j-dependence. Therefore, instead of being bilinear with respect to the dynamical variables, this Schlesinger system on the torus becomes highly transcendental.

An alternative form of the elliptic Schlesinger system was proposed by Takasaki [**29**] who considered the restriction $Q = \sigma_3$, $R = \sigma_1$, related to the classical limit of Etingof's elliptic version of the Knizhnik-Zamolodchikov-Bernard system on the torus [**6**]. This choice of fixing the twists turns out to be compatible with the isomonodromic deformation equations, therefore essentially simplifying the dynamics as compared to [**23**]. It leads to the study of isomonodromic deformations of the system

$$\begin{aligned}\frac{d\Psi}{d\lambda} &= A(\lambda)\Psi,\\ A(\lambda) &\equiv \sum_{j=1}^{N}\sum_{\alpha=1}^{3} A_j^\alpha w_\alpha(\lambda-\lambda_j)\sigma_\alpha,\end{aligned} \tag{3.1}$$

with $\lambda \in \mathbb{C}$. The functions w_α on the torus are defined in the appendix (see (A.3)). The connection $A(\lambda)$ obviously has only simple poles on E and the following twist properties, cf. (A.4):

$$A(\lambda+1) = \sigma_3 A(\lambda)\sigma_3, \quad A(\lambda+\mu) = \sigma_1 A(\lambda)\sigma_1. \tag{3.2}$$

Since the residues of all w_α at $\lambda = 0$ coincide, the residue of $A(\lambda)$ at λ_j is

$$A_j \equiv \sum_\alpha A_j^\alpha \sigma_\alpha.$$

As in the case of the Riemann sphere, the function Ψ has regular singularities at $\lambda = \lambda_j$ with the same local properties (2.3)–(2.5). The twist properties of Ψ take the form

$$\Psi(\lambda+1) = \sigma_3\Psi(\lambda)M_a \quad \Psi(\lambda+\mu) = \sigma_1\Psi(\lambda)M_b, \tag{3.3}$$

with monodromy matrices M_a, M_b along the basic cycles of the torus. Moreover, as in the case of the Riemann sphere, $\Psi(\lambda)$ has monodromies M_j around the singularities λ_j.

The isomonodromy condition on the torus requires that all monodromies M_j, M_a and M_b are independent of the positions of the singularities λ_j and the module μ of the torus. As on the Riemann sphere this implies that the function $\partial\Psi/\partial\lambda_j\Psi^{-1}$ has the only simple pole at $\lambda = \lambda_j$ with residue $-A_j$. In addition, it has the following twist properties:

$$\begin{aligned}\frac{\partial\Psi}{\partial\lambda_j}\Psi^{-1}(\lambda+1) &= \sigma_3\frac{\partial\Psi}{\partial\lambda_j}\Psi^{-1}(\lambda)\sigma_3,\\ \frac{\partial\Psi}{\partial\lambda_j}\Psi^{-1}(\lambda+\mu) &= \sigma_1\frac{\partial\Psi}{\partial\lambda_j}\Psi^{-1}(\lambda)\sigma_1.\end{aligned}$$

Therefore,

$$\frac{\partial\Psi}{\partial\lambda_j} = -\sum_{\alpha=1}^{3} A_j^\alpha w_\alpha(\lambda-\lambda_j)\sigma_\alpha\Psi. \tag{3.4}$$

To derive the equation with respect to the module μ we observe that $\partial\Psi/\partial\mu\,\Psi^{-1}$ is holomorphic at $\lambda = \lambda_j$ (but not at $\lambda = \lambda_j + \mu$) and has twist properties

$$\begin{aligned}\frac{\partial\Psi}{\partial\mu}\Psi^{-1}(\lambda+1) &= \sigma_3\frac{\partial\Psi}{\partial\mu}\Psi^{-1}(\lambda)\sigma_3,\\ \frac{\partial\Psi}{\partial\mu}\Psi^{-1}(\lambda+\mu) &= \sigma_1\left(\frac{\partial\Psi}{\partial\mu}\Psi^{-1}(\lambda) - \frac{\partial\Psi}{\partial\lambda}\Psi^{-1}(\lambda)\right)\sigma_1.\end{aligned}$$

Taking into account the periodicity properties of the functions $\mathcal{Z}_\alpha$ (A.6), this therefore implies

$$\frac{\partial\Psi}{\partial\mu} = \sum_{j=1}^{N}\sum_{\alpha=1}^{3} A_j^\alpha \mathcal{Z}_\alpha(\lambda-\lambda_j)\sigma_\alpha\Psi. \tag{3.5}$$

The compatibility conditions of the equations (3.1), (3.4) and (3.5) then yield the λ_i and μ dependence of the residues A_j. The result is summarized in the following

THEOREM 3.1 ([**29**]). *The isomonodromic deformations of the system* (3.1) *are described by the following elliptic version of the Schlesinger system:*

$$\begin{cases} \frac{dA_j}{d\lambda_i} = \left[A_j, \sum_{\alpha=1}^{3} A_i^\alpha w_\alpha(\lambda_j - \lambda_i)\sigma_\alpha\right], \quad i \neq j, \\ \frac{dA_j}{d\lambda_j} = -\sum_{i\neq j}\left[A_j, \sum_{\alpha=1}^{3} A_i^\alpha w_\alpha(\lambda_j - \lambda_i)\sigma_\alpha\right], \\ \frac{dA_j}{d\mu} = -\sum_{i=1}^{N}\left[A_j, \sum_{\alpha=1}^{3} A_i^\alpha Z_\alpha(\lambda_j - \lambda_i)\sigma_\alpha\right]. \end{cases} \tag{3.6}$$

The corresponding equations for the matrices G_j from (2.3) take a form analogous to the equations (2.8) on the Riemann sphere:

$$\begin{aligned} \frac{\partial G_j}{\partial \lambda_i} &= \sum_\alpha A_i^\alpha w_\alpha(\lambda_i - \lambda_j)\sigma_\alpha G_j, \\ \frac{\partial G_j}{\partial \lambda_j} &= -\sum_{i=1}^{N}\sum_\alpha A_i^\alpha w_\alpha(\lambda_i - \lambda_j)\sigma_\alpha G_j. \end{aligned} \tag{3.7}$$

The system (3.6) admits a multi-time Hamiltonian formulation with respect to the Poisson structure (2.9) on the residues

$$\{A_i^\alpha, A_j^\beta\} = \delta_{ij}\varepsilon^{\alpha\beta\gamma}A_j^\alpha c. \tag{3.8}$$

The Hamiltonians describing deformation with respect to the variables λ_i and to the module μ of the torus are respectively given by

$$H_i = \frac{1}{4\pi i}\oint_{\lambda_i} \operatorname{tr} A^2(\lambda)\, d\lambda = \sum_{j\neq i}\sum_\alpha A_j^\alpha A_i^\alpha w_\alpha(\lambda_j - \lambda_i), \tag{3.9}$$

$$H_\mu = -\frac{1}{2\pi i}\oint_{a} \operatorname{tr} A^2(\lambda)\, d\lambda = -\sum_{i,j}\sum_\alpha A_i^\alpha A_j^\alpha Z_\alpha(\lambda_i - \lambda_j). \tag{3.10}$$

The representation of H_μ as a contour integral along the basic a-cycle in (3.1) was derived in [**20**]. All Hamiltonians Poisson-commute as a direct consequence of (3.8).

The τ-function of the elliptic Schlesinger system (3.6) is defined as the generating function $\tau(\{\lambda_j\}, \mu)$ of the Hamiltonians

$$\frac{\partial \ln \tau}{\partial \lambda_j} = H_j, \qquad \frac{\partial \ln \tau}{\partial \mu} = H_\mu; \tag{3.11}$$

it is uniquely determined up to an arbitrary $(\mu, \{\lambda_j\})$-independent multiplicative constant. Compatibility of the equations (3.11) is a corollary of the elliptic Schlesinger system.

3.2. Schlesinger Transformations for Elliptic Isomonodromic Deformations. The natural generalization of the notion of Schlesinger transformations on the Riemann sphere to the elliptic case was given in the paper [**20**]. Starting from any solution of the elliptic Schlesinger system (3.6) with associated function

Ψ satisfying (3.1) and (3.3), we construct a new solution $\hat{A}_j$, $\widehat{\Psi}$ with eigenvalues $\hat{t}_j$ which differ from the t_j by integer or half-integer values. In particular, we will consider the elliptic analog of the elementary Schlesinger transformation (2.13) on the Riemann sphere. The following construction was inspired by the papers [**3, 4**].

As an elliptic analog of the function $F(\lambda)$ from (2.15) we shall choose the following ansatz

$$F(\lambda) = \frac{f(\lambda)}{\sqrt{\det f(\lambda)}},$$
$$f(\lambda) = \frac{1}{2} + \sum_{\alpha=1}^{3} J_\alpha w_\alpha \Big(\lambda - \frac{1}{2}(\lambda_k + \lambda_l)\Big)\sigma_\alpha, \tag{3.12}$$

where the functions $J_\alpha(\lambda_j, \mu)$ depend on G_k and G_l and will be defined below. The elementary elliptic Schlesinger transformation is described by the following

THEOREM 3.2 ([**20**]). *Let the functions $\{A_j(\{\lambda_i\})\}$ satisfy the elliptic Schlesinger system* (3.6) *with twist properties* (3.2), *and let the function Ψ satisfy the associated linear system* (3.1). *For two arbitrary non-coinciding poles λ_k and λ_l, define the new function*

$$\widehat{\Psi}(\lambda) \equiv F(\lambda)\Psi(\lambda), \tag{3.13}$$

where $F(\lambda)$ is given by formula (3.12), *and define the λ-independent coefficients J_α by*

$$\sum_\alpha J_\alpha w_\alpha \Big(\frac{1}{2}(\lambda_k - \lambda_l)\Big)\sigma_\alpha \equiv -\frac{1}{2} G \sigma_3 G^{-1}; \tag{3.14}$$

as above, we denote by G the matrix (2.17) *containing the first columns of the matrices G_k and G_l.*

Then the function $\widehat{\Psi}(\lambda)$ satisfies the equations (3.1), (3.4), (3.5) *and the twist conditions* (3.3) *with the transformed functions*

$$\hat{A}_j(\{\lambda_i\}) \equiv \mathrm{res}_{\lambda=\lambda_j}\left\{\frac{d\widehat{\Psi}}{d\lambda}\widehat{\Psi}^{-1}\right\}. \tag{3.15}$$

In turn, the functions $\hat{A}_j$ satisfy the elliptic Schlesinger system (3.6). *For the eigenvalues t_j we have*

$$\hat{t}_j = \begin{cases} t_j + \frac{1}{2} & \text{for } j = k, l \\ t_j & \text{otherwise.} \end{cases}$$

The monodromy matrices $\widehat{M_j}$, $\widehat{M_a}$ and $\widehat{M_b}$ of the function $\widehat{\Psi}$ coincide with the monodromies of Ψ, except for $\widehat{M_k} = -M_k$ and $\widehat{M_l} = -M_l$.

PROOF. The proper local behaviour of function $\widehat{\Psi}$ at the singularities λ_j is ensured by the relations

$$S_\pm^2 = S_\pm, \quad S_+ + S_- = I, \quad S_+ G_l^1 = S_- G_k^1 = 0; \tag{3.16}$$

for

$$S_\pm \equiv \frac{1}{2} \mp \sum_\alpha J_\alpha w_\alpha \Big(\frac{1}{2}(\lambda_k - \lambda_l)\Big)\sigma_\alpha = G P_\pm G^{-1},$$

which in complete analogy to (2.16) describe the annihilation of the vectors G_k^1 and G_l^1 by the matrices $f(\lambda_k)$ and $f(\lambda_l)$, respectively. Obviously, equations (3.16) are a consequence of (3.14). Similarly to the case of the sphere, it is then easy to verify that (3.16) provide the required asymptotical expansions (2.3) for the function $\widehat{\Psi}$ with parameters $\widehat{G}_j$, C_j and $\hat{t}_j$.

Concerning the global behavior of $\widehat{\Psi}$ we note that the prefactor $(\det f(\lambda))^{-1/2}$ in (3.12) provides the condition $\det \widehat{\Psi} = 1$ and kills the simple pole of $f(\lambda)$ at $\lambda = (\lambda_k + \lambda_l)/2$. Therefore, the only singularities of $F(\lambda)$ on E are the zeros of $\det f(\lambda)$. Since $\det f(\lambda)$ has only one pole—this is the second order pole at $\lambda = (\lambda_k + \lambda_l)/2$—it must have also two zeros on E whose sum according to Abel's theorem equals $\lambda_k + \lambda_l$. According to (3.16) these are precisely λ_k and λ_l. It remains to check that $\widehat{\Psi}$ satisfies conditions (3.3) with the same matrices M_a and M_b. This follows from the twist properties

$$f(\lambda + 1) = \sigma_3 f(\lambda)\sigma_3, \quad f(\lambda + \mu) = \sigma_1 f(\lambda)\sigma_1,$$

which in turn follow from (3.12) and the periodicity properties (A.4) of the functions $w_j(\lambda)$. □

As a result of rather long calculations one can prove the elliptic analog of formula (2.21) describing the transformation of the τ-function under the action of the elliptic Schlesinger transformations.

THEOREM 3.3 ([**20**]). *The τ-function $\hat{\tau}$ corresponding to the Schlesinger-transformed solution $\hat{A}_j$ (3.15) of the elliptic Schlesinger system is related to the τ-function corresponding to the solution A_j as follows*

$$\hat{\tau}(\{\lambda_j\}, \mu) = \left\{ \left[w_1 w_2 w_3 \left(\frac{\lambda_k - \lambda_l}{2} \right) \right]^{1/2} \det[G J^{1/2}] \right\} \cdot \tau(\{\lambda_j\}, \mu), \tag{3.17}$$

where G is the matrix (2.17) containing the first columns of the matrices G_k, G_l,

$$J \equiv \sum_{\alpha=1}^{3} J_A \sigma_A,$$

and the functions J_α are defined in terms of G via (3.14).

The natural open problem arising here is to construct elliptic generalizations of integrable chains associated to ordinary Schlesinger system [**12, 13**].

In the next section we shall present the extension of the construction of algebro-geometric solutions of the Schlesinger system to the case of elliptic isomonodromic deformations.

3.3. Algebro-Geometric Solutions of the Elliptic Schlesinger System. To construct theta-functional solutions of the elliptic Schlesinger system (3.6) let us assume that $N = 2g$ and introduce the two-sheet covering $\mathcal{L}$ of the torus E with branch points $\lambda_1, \ldots, \lambda_{2g}$. The genus of $\mathcal{L}$ equals $g + 1$. Denote by $*$ the involution of $\mathcal{L}$ interchanging the sheets of the covering. Let us choose the canonical basis of cycles on $\mathcal{L}$ in such a way (see figure 6.2 on p.215 of [**2**]) that

$$a_1^* = -a_{g+1}, \quad b_1^* = -b_{g+1}$$

$$a_j^* = -a_j, \quad b_j^* = -b_j, \quad j = 2, \ldots, g.$$

The basic holomorphic differentials $dU_1, \ldots, dU_{g+1}$ on $\mathcal{L}$ normalized by

$$\oint_{a_j} dU_k = \delta_{jk}, \quad j, k = 1, \ldots, g+1,$$

transform as follows under the action of involution $*$:

$$dU_1(P^*) = -dU_{g+1}(P), \quad dU_j(P^*) = -dU_j(P), \quad j = 2, \dots, g. \tag{3.18}$$

Let us introduce the following Prym differentials dV_j, $j = 1, \dots, g$:

$$dV_1 = \frac{1}{2}(dU_1 + dU_{g+1}), \quad dV_j = dU_j, \quad j = 2, \dots, g, \tag{3.19}$$

and symmetric $g \times g$ matrix of their b-periods:

$$\Pi_{jk} = \oint_{b_j} dV_k \tag{3.20}$$

which has positively-defined imaginary part.

REMARK 3.4. The differentials dV_j and the matrix Π were first introduced by Bobenko [**2**] in the studies of classical tops admitting an elliptic Lax representation. These objects are related to the standard Prym differentials dW_j and the standard Prym matrix Π^{Prym} as follows: $dW_1 = 2dV_1$, $dW_j = dV_j$, $j = 2, \dots, g$; $\Pi = 2S\Pi^{Prym}S$, where S is the diagonal matrix $S = \text{diag}(\frac{1}{2}, 1, \dots, 1)$.

Denote by $\hat{\mathcal{L}}$ the universal covering of the curve $\mathcal{L}$.

THEOREM 3.5. *Define the* 2×2 *matrix-valued function* $\Phi(P)$ *on* $\hat{\mathcal{L}}$ *by the formulas*

$$\Phi(P) = \Phi(P) = \begin{pmatrix} \varphi(P) & \varphi(P) \\ \psi(P) & \psi(P) \end{pmatrix}, \tag{3.21}$$

where

$$\varphi(P) = \Theta\begin{bmatrix}\mathbf{p} \\ \mathbf{q}\end{bmatrix}\left(\int_{\lambda_1}^P dV \,\middle|\, \Pi\right), \quad \psi(P) = \Theta\begin{bmatrix}\mathbf{p} \\ \mathbf{q} + \frac{1}{2}\mathbf{e}_1\end{bmatrix}\left(\int_{\lambda_1}^P dV \,\middle|\, \Pi\right); \tag{3.22}$$

$\mathbf{p}, \mathbf{q} \in \mathbb{C}$ *are arbitrary constant vectors such that* $p_1 = 0$. *Then the function* Φ *is holomorphic and invertible on* $\mathcal{L}$ *outside the branch points* λ_j *and transforms as follows with respect to analytic continuation along the basic cycles of* $\mathcal{L}$:

$$T_{a_1}[\Phi(P)] = \sigma_1 \Phi(P) \tag{3.23}$$

$$T_{b_1}[\Phi(P)] = \sigma_3 \Phi(P) e^{-2\pi i q_1 \sigma_3} e^{-\pi i \Pi_{11} - 2\pi i \int_{\lambda_1}^P dV_1} \tag{3.24}$$

$$T_{a_j}[\Phi(P)] = \Phi(P) e^{2\pi i p_j \sigma_3} \tag{3.25}$$

$$T_{b_j}[\Phi(P)] = \Phi(P) e^{-2\pi i q_j \sigma_3} e^{-\pi i \Pi_{jj} - 2\pi i \int_{\lambda_1}^P dV_j} \tag{3.26}$$

for $j = 2, \dots, g$.

PROOF. Taking into account the definition of the Prym differentials dV_j (3.19) we see that

$$T_{a_1}\left[\int_{\lambda_1}^P dV\right] = \int_{\lambda_1}^P dV + \frac{\mathbf{e}_1}{2}$$

$$T_{a_j}\left[\int_{\lambda_1}^P dV\right] = \int_{\lambda_1}^P dV + \mathbf{e}_1, \quad j = 2, \dots, g,$$

$$T_{b_j}\left[\int_{\lambda_1}^P dV\right] = \int_{\lambda_1}^P dV + \Pi\mathbf{e}_j, \quad j = 1, \dots, g.$$

Substituting these expressions into the formulas for φ and ψ and taking into account the behaviour of the 1-forms dU_j under the action of the involution $*$ (3.18) we derive the following transformation properties of the functions φ and ψ:

$$T_{a_1}[\varphi(P)] = \psi(P), \qquad T_{a_1}[\psi(P)] = \varphi(P) \tag{3.27}$$

$$T_{a_1}[\varphi(P^*)] = \psi(P^*), \qquad T_{a_1}[\psi(P^*)] = \varphi(P^*) \tag{3.28}$$

$$T_{b_1}[\varphi(P)] = e^{-2\pi i q_1} e^{-\pi i \Pi_{11} - 2\pi i \int_{\lambda_1}^P dV_1} \varphi(P) \tag{3.29}$$

$$T_{b_1}[\psi(P)] = -e^{-2\pi i q_1} e^{-\pi i \Pi_{11} - 2\pi i \int_{\lambda_1}^P dV_1} \psi(P) \tag{3.30}$$

$$T_{b_1}[\varphi(P^*)] = e^{2\pi i q_1} e^{-\pi i \Pi_{11} - 2\pi i \int_{\lambda_1}^P dV_1} \varphi(P^*) \tag{3.31}$$

$$T_{b_1}[\psi(P^*)] = -e^{2\pi i q_1} e^{-\pi i \Pi_{11} - 2\pi i \int_{\lambda_1}^P dV_1} \psi(P^*) \tag{3.32}$$

and

$$T_{a_j}[\varphi(P)] = e^{2\pi i p_j} \varphi(P) \tag{3.33}$$

$$T_{a_j}[\varphi(P^*)] = e^{-2\pi i p_j} \varphi(P^*) \tag{3.34}$$

$$T_{b_j}[\varphi(P)] = e^{-2\pi i q_j} e^{-\pi i \Pi_{jj} - 2\pi i \int_{\lambda_1}^P dV_j} \varphi(P) \tag{3.35}$$

$$T_{b_j}[\varphi(P^*)] = e^{2\pi i q_j} e^{-\pi i \Pi_{jj} - 2\pi i \int_{\lambda_1}^P dV_j} \varphi(P^*). \tag{3.36}$$

The transformation laws of ψ along the cycles a_j, b_j for $j > 1$ coincide with the transformation laws of φ.

Combining the above relations into matrix form, we come to the transformation laws (3.5)–(3.5).

It remains to verify non-degeneracy of $\Phi(P)$ outside the singularities λ_j. We know that $\det \Phi(P)$ has at least simple zeros at the points λ_j (at these points the columns of $\Phi(P)$ are proportional to each other). To check that $\det \Phi(P)$ does not vanish on $\hat{\mathcal{L}}$ outside of λ_j let us first observe that

$$\begin{aligned} T_{a_1}[\det \Phi(P)] &= -\det \Phi(P) \\ T_{a_j}[\det \Phi(P)] &= \det \Phi(P), \quad j = 2, \dots, g, \end{aligned} \tag{3.37}$$

$$T_{b_j}[\det \Phi(P)] = e^{-2\pi i \Pi_{jj} - 4\pi i \int_{\lambda_1}^P dV_j} \det \Phi(P) \quad j = 1, \dots, g. \tag{3.38}$$

Now let us calculate the integral

$$\begin{aligned} \oint_{\partial \hat{\mathcal{L}}} d \ln \det \Phi(P) = &\oint_{a_1} d\{\ln \det \Phi(P) - \ln \det T_{b_1}[\Phi(P)]\} \\ &+ \oint_{a_{n+1}} d\{\ln \det \Phi(P) - \ln \det T_{b_{n+1}}[\Phi(P)]\} \\ &+ \sum_{j=2}^{g} \oint_{a_j} d\{\ln \det \Phi(P) - \ln \det T_{b_1}[\Phi(P)]\}. \end{aligned} \tag{3.39}$$

Taking into account (3.3) and (3.3) as well as the normalization of the basic integrals dU_j, we see that the first two terms of the right hand side of this expression equal $2\pi i$, whereas each term in the sum equals $4\pi i$. Altogether, we get $4\pi i g$, and, therefore, $\det \Phi(P)$ has in $\hat{\mathcal{L}}$ exactly $2g$ zeros which coincide with λ_j. □

Let us also choose some domain $\Omega \subset E$ which does not overlap with the projections of any basic cycle onto E. Then the domain Ω^* does not overlap with the boundary of $\hat{\mathcal{L}}$, and the functions $\varphi^*(P)$ and $\psi^*(P)$ are uniquely defined in $\hat{\mathcal{L}}$ by (3.22). Let us now choose some sheet of the universal covering X of the torus E with punctures $\{\lambda_1, \dots, \lambda_{2g}\}$, and define a new function $\Psi(\lambda)$ on the subset Ω of this sheet by the formula

$$\Psi(\lambda \in \Omega) = \frac{1}{\sqrt{\det \Phi(\lambda)}} \Phi(\lambda). \tag{3.40}$$

Then we extend the function $\Psi(\lambda)$ to the rest of X by analytic continuation.

The following theorem shows that the function Ψ satisfies conditions of isomonodromy, and therefore generates a class of solutions of the elliptic Schlesinger system (3.6):

THEOREM 3.6. *The function $\Psi(\lambda \in X)$ defined by formulas* (3.21), (3.22) *and* (3.40) *is holomorphic and invertible on X outside the points λ_j, $j = 1, \dots, 2g$. Moreover, it transforms as follows with respect to analytic continuation along the basic cycles of E:*

$$T_a[\Psi(\lambda)] = i\sigma_1 \Psi(\lambda) \quad T_b[\Psi(\lambda)] = i\sigma_3 \Psi(\lambda) e^{-2\pi i q_1 \sigma_3} \tag{3.41}$$

and around closed cycles surrounding the points λ_j:

$$T_{\lambda_j}[\Psi(\lambda)] = \Psi(\lambda) M_j, \tag{3.42}$$

where T_b, T_b and T_{λ_j} denote the corresponding operators of analytic continuation;

$$M_j = \begin{pmatrix} 0 & -m_j \\ m_j^{-1} & 0 \end{pmatrix}, \tag{3.43}$$

and

$$m_1 = i, \qquad m_2 = -i \exp\left\{-2\pi i \sum_{j=2}^{g} p_j\right\} \tag{3.44}$$

$$m_{2l} = -i \exp\left\{2\pi i q_l - 2\pi i \sum_{k=l+1}^{g} p_k\right\} \tag{3.45}$$

$$m_{2l-1} = i \exp\left\{2\pi i q_l - 2\pi i \sum_{k=l}^{g} p_k\right\} \tag{3.46}$$

for $l = 2, \dots, g$.

PROOF. Holomorphy and invertibility of the function Ψ follow from the same statements concerning the function Φ (3.21). The relations (3.41) directly follow from (3.5), (3.5). To calculate m_j let us observe that, according to (3.5)–(3.5), monodromies of Ψ are related to the constants $\mathbf{p}$ and $\mathbf{q}$ as follows:

$$M_{2j} M_{2j-1} = e^{2\pi i p_j \sigma_3}, \qquad j = 2, \dots, g, \tag{3.47}$$

$$M_{2j-1} M_{2j-2} = e^{2\pi i (q_j - q_{j-1}) \sigma_3}, \quad j = 3, \dots, g, \tag{3.48}$$

$$M_3 M_2 = e^{2\pi i q_2}. \tag{3.49}$$

Moreover, we order the monodromies in such a way that

$$M_a M_b M_a^{-1} M_b^{-1} M_{2g} \cdots M_1 = I \tag{3.50}$$

(since $M_a = I$ the first four factors in this relation drop out). The monodromy M_1 corresponding to our choice of basic cycles equals $i\sigma_1$. Altogether, these relations lead to (3.43), (3.6)–(3.6) after elementary calculations. □

COROLLARY 3.7. *The residues*

$$A_j(\{\lambda_j\},\mu) \equiv \operatorname{res}|_{\lambda=\lambda_j} \Psi_\lambda \Psi^{-1} \tag{3.51}$$

satisfy the elliptic Schlesinger system (3.6).

4. Outlook

Let us mention several applications of the mathematical results described above. Recently [**22**] a close relationship was established between the Schlesinger system and the Ernst equation of general relativity, which allows to apply to the Ernst equation all results of Section 2. In particular, one can get in this way a class of algebro-geometric solutions of the Ernst equation [**21**], which turns out to coincide with the class of algebro-geometric solutions of the Ernst equation known since 1988 [**19**]. It is rather satisfactory that a certain subclass of genus 2 algebro-geometric solutions of the Ernst equation recently found a realistic physical application in the problem of the description of different kinds of dust discs [**25, 18**]. Another application of the construction of Section 2 is the theory of $SU(2)$-invariant gravitational instantons [**1**], where it allows to considerably simplify the results of Hitchin [**9**].

So far we don't know about physical applications of the elliptic version of the Schlesinger system, and all results of Section 3 have at the moment purely mathematical significance; however, we strongly believe that such applications will be found in the near future.

Appendix A. Some Elliptic Functions

The elliptic theta-function with characteristic $[p,q]$ $(p,q\in\mathbb{C})$ on a torus E is defined by the series

$$\vartheta[p,q](\lambda\mid\mu) = \sum_{m\in\mathbb{Z}} e^{\pi i\mu(m+p)^2+2\pi i(m+p)(\lambda+q)}. \tag{A.1}$$

Let us introduce on the torus E the standard Jacobi theta-functions:

$$\begin{cases} \vartheta_1(\lambda)\equiv -\vartheta\left[\frac12,\frac12\right](\lambda\mid\mu), \\ \vartheta_2(\lambda)\equiv \vartheta\left[\frac12,0\right](\lambda\mid\mu), \\ \vartheta_3(\lambda)\equiv\vartheta(\lambda)\equiv \vartheta[0,0](\lambda\mid\mu), \\ \vartheta_4(\lambda)\equiv \vartheta\left[0,\frac12\right](\lambda\mid\mu), \end{cases} \tag{A.2}$$

and the corresponding theta-constants

$$\vartheta_j\equiv\vartheta_j(0),\quad j=2,3,4.$$

We define the following three combinations of Jacobi theta-functions:

$$w_1(\lambda)=\pi\vartheta_2\vartheta_3\frac{\vartheta_4(\lambda)}{\vartheta_1(\lambda)},\quad w_2(\lambda)=\pi\vartheta_2\vartheta_4\frac{\vartheta_3(\lambda)}{\vartheta_1(\lambda)},\quad w_3(\lambda)=\pi\vartheta_3\vartheta_4\frac{\vartheta_2(\lambda)}{\vartheta_1(\lambda)}. \tag{A.3}$$

All these functions have simple poles at $\lambda=0$ with residue 1. Moreover, they possess the following periodicity properties:

$$\begin{cases} w_1(\lambda+1)=-w_1(\lambda) & w_1(\lambda+\mu)=w_1(\lambda), \\ w_2(\lambda+1)=-w_2(\lambda) & w_2(\lambda+\mu)=-w_2(\lambda), \\ w_3(\lambda+1)=w_3(\lambda) & w_3(\lambda+\mu)=-w_3(\lambda). \end{cases} \tag{A.4}$$

Let us also define the following functions $\mathcal{Z}_\alpha$:

$$\mathcal{Z}_1 = \frac{w_1}{2\pi i}\frac{\vartheta_4'(\lambda)}{\vartheta_4(\lambda)}, \quad \mathcal{Z}_2 = \frac{w_2}{2\pi i}\frac{\vartheta_3'(\lambda)}{\vartheta_3(\lambda)}, \quad \mathcal{Z}_3 = \frac{w_3}{2\pi i}\frac{\vartheta_2'(\lambda)}{\vartheta_2(\lambda)} \tag{A.5}$$

which have the periodicity properties:

$$\begin{cases} \mathcal{Z}_1(\lambda+1) = -\mathcal{Z}_1(\lambda) & \mathcal{Z}_1(\lambda+\mu) = \mathcal{Z}_1(\lambda) - w_1, \\ \mathcal{Z}_2(\lambda+1) = -\mathcal{Z}_2(\lambda) & \mathcal{Z}_2(\lambda+\mu) = -\mathcal{Z}_2(\lambda) + w_2, \\ \mathcal{Z}_3(\lambda+1) = \mathcal{Z}_3(\lambda) & \mathcal{Z}_3(\lambda+\mu) = -\mathcal{Z}_3(\lambda) + w_3. \end{cases} \tag{A.6}$$

It is easy to verify the identity

$$\frac{dw_\alpha}{d\mu}(\lambda) = \frac{d\mathcal{Z}_\alpha}{d\lambda}(\lambda), \tag{A.7}$$

which follows from the analyticity and twist properties of both sides.

Notice also that the functions w_α may be represented as ratios of Jacobi's elliptic functions as follows:

$$w_1(\lambda) = \frac{1}{\mathrm{sn}(\lambda)}, \quad w_2(\lambda) = \frac{\mathrm{dn}(\lambda)}{\mathrm{sn}(\lambda)}, \quad w_3(\lambda) = \frac{\mathrm{cn}(\lambda)}{\mathrm{sn}(\lambda)}. \tag{A.8}$$

In calculating the transformation of the elliptic τ-function under the action of elliptic Schlesinger transformations one has to use also the summation theorem and some integral relations for the functions w_α [**20**].

References

1. M. Babich and D. Korotkin, *Self-dual* SU(2)*-invariant Einstein metrics and modular dependence of theta functions*, Lett. Math. Phys. **46** (1998), 323–337.
2. E. Belokolos, A. Bobenko, V. Enolskii, A. Its, and V. Matveev, *Algebro-geometric approach to nonlinear integrable equations*, Springer Ser. Nonlinear Dynamics, Berlin-Heidelberg-New York, 1994.
3. R. Bikbaev, A. Bobenko, and A. Its, *Theory of exact solutions of Landau-Lifshitz equation*, Preprint DonFTI 84-6(81) I, II (1984).
4. E. Date, M. Jimbo, M. Kashiwara, and T. Miwa, *Landau-Lifshitz equation: solitons, quasiperiodic solutions and infinite-dimensional Lie algebras*, J. Phys. A **16** (1983), 221–236.
5. P. Deift, A. Its, A. Kapaev, and X. Zhou, *On the algebro-geometric integration of the Schlesinger equations and on elliptic solutions of the Painlevé 6 equation*, Comm. Math. Phys. **203** (1999), 613–634.
6. P. I. Etingof, *Representations of affine Lie algebras, elliptic r-matrix systems, and special functions*, Comm. Math. Phys. **159** (1994), 471–502.
7. J. D. Fay, *Theta functions on Riemann surfaces*, Lecture Notes in Math., vol. 352, Springer, 1973.
8. J. Harnad and A. Its, *Integrable Fredholm operators and dual isomonodromic deformations*, preprint CRM-2477, Montréal; solv-int/9706002.
9. N. Hitchin, *Twistor spaces, Einstein metrics and isomonodromics deformations*, J. Differential Geom. **42** (1995), 30–112.
10. D. Ivanov, *Knizhnik-Zamolodchikov-Bernard equations as a quantization of nonstationary Hitchin system*, hep-th/9610207.
11. K. Iwasaki, *Moduli and deformation for Fuchsian projective connections on a Riemann surface*, J. Fac. Sci. Univ. Tokyo Sect. IA Math. **38** (1991), 431–531; *Fuchsian moduli on a Riemann surface—its Poisson structure and Poincaré-Lefschetz duality*, Pacific J. Math. **155** (1992), 319–340.
12. M. Jimbo and T. Miwa, *Monodromy preserving deformations of linear ordinary differential equations with rational coefficients*. II, Physica D **2** (1981), 407–448.
13. M. Jimbo and T. Miwa, *Monodromy preserving deformations of linear ordinary differential equations with rational coefficients*. III, Physica D **2** (1981), 26–46.

14. M. Jimbo, T. Miwa, Y. Môri, and M. Sato, *Density matrix of an impenetrable Bose gas and the fifth Painlevé transcendent*, Physica D **1** (1980), 80–158.
15. M. Jimbo, T. Miwa, and K. Ueno, *Monodromy preserving deformations of linear ordinary differential equations with rational coefficients*, Physica D **2** (1981), 306–352.
16. A. Kitaev, *Special functions of the isomonodromy type*, Preprint SFB-288-272, Berlin, 1997.
17. A. Kitaev and D. Korotkin, *On solutions of the Schlesinger system in terms of theta-functions*, Internat. Math. Res. Notices **17** (1998), 877–905.
18. C. Klein and O. Richter, *Explicit solution of Riemann-Hilbert problems for the Ernst equation*, Phys. Rev. D (3) **57** (1998), 857–862.
19. D. Korotkin, *Finite-gap solutions of the stationary axially symmetric Einstein equation in vacuo*, Theor. Math. Phys. **77** (1988), 1018–1031.
20. D. Korotkin, N. Manojlović, and H. Samtleben, *Schlesinger transformations for elliptic isomonodromic deformations*, solv-int/9910010.
21. D. Korotkin and V. Matveev, *On theta-functional solutions of Schlesinger system and Ernst equation*, gr-qc/9810041.
22. D. Korotkin, H. Nicolai, *Separation of variables and Hamiltonian formulation for the Ernst equation*, Phys. Rev. Lett. **74** (8) (1995), 1272–1275.
23. D. Korotkin and H. Samtleben, *On the quantization of isomonodromic deformations on the torus*, Internat. J. Modern Phys. A **12** (1997), 2013–2029.
24. A. Levin and M. Olshanetskii, *Hierarchies of isomonodromic deformations and Hitchin systems*, Preprint ITEP-TH 45/9; hep-th/9709207.
25. G. Neugebauer and R. Meinel, *General relativistic gravitational field of a rigidly rotating disk of dust: solution in terms of ultraelliptic functions*, Phys. Rev. Lett. **75** (1995), no. 17, 3046–3047.
26. K. Okamoto, *On Fuchs' problem on a torus.* I, Funkcial. Ekvac. **14** (1971), 137–152; *Sur le problème de Fuchs sur un tore.* II, J. Fac. Sci. Univ. Tokyo Sect. IA Math. **24** (1977), 357–372.
27. K. Okamoto, *Déformation d'une équation différentielle linéaire avec une singularité irrégulière sur un tore*, J. Fac. Sci. Univ. Tokyo Sect. IA Math. **26** (1979), 501–518.
28. L. Schlesinger, *Über eine Klasse von Differentialsystemen beliebiger Ordnung mit festen kritischen Punkten*, J. Reine Angew. Math. **141** (1912), 96–145.
29. K. Takasaki, *Gaudin model, KZB equation, and isomonodromic problem on torus*, Lett. Math. Phys. **44** (1998), 143–156.

MAX-PLANCK-INSTITUT FÜR GRAVITATIONSPHYSIK, AM MÜHLENBERG 1, 14476 GOLM, GERMANY

Current address: Department of Mathematics and Statistics, Concordia University, 7141 Sherbrooke Str. W., Montréal, Qué., H4B 1R6, Canada

E-mail address: korotkin@discrete.concordia.ca

Centre de Recherches Mathématiques
CRM Proceedings and Lecture Notes
Volume **26**, 2000

Quantum Inverse Scattering Problem and Correlation Functions of Integrable Models

J.-M. Maillet

Dedicated to the memory of Anatoly Izergin.

ABSTRACT. A general approach to the computation of exact correlation functions of lattice quantum integrable models proposed recently is reviewed. It is based on the resolution of the quantum inverse scattering problem. The application of this method to the XXZ Heisenberg spin-$\frac{1}{2}$ chain in a magnetic field leads to multiple integral representations of its n-point correlation functions. For zero magnetic field it gives a proof, in both the massless and massive (anti-ferromagnetic) regimes, of the formulas obtained from the q-deformed KZ equations (massless regime) and the representation theory of the quantum affine algebra $\mathcal{U}_q(\widehat{\mathrm{sl}}_2)$ together with the corner transfer matrix approach (massive regime).

1. Introduction

During the last twenty years, considerable progress has been achieved in our understanding of classical and quantum integrable models of field theory and statistical mechanics [**5, 12, 15, 22, 31, 38**]. New concepts and methods have emerged from this domain, leading to major breakthroughs not only in theoretical physics but also in mathematics. In this context, one of the main challenges of the theory of quantum integrable models remained largely open until now, despite the great number of remarkable works which have been devoted to its resolution: computing exact and manageable expressions for their correlation functions. This is a fundamental problem, both to enlarge the range of applications of these models, for example in the realm of condensed matter physics, and also to understand in a deeper way their underlying mathematical structures.

The aim of this lecture is to review an approach proposed recently in [**19, 28, 29**] (see also [**33, 43**] for related works) to compute correlation functions of lattice quantum integrable models using the algebraic Bethe ansatz method [**5, 12, 13, 31**] and the explicit resolution of the quantum inverse scattering problem [**28, 34**].

2000 *Mathematics Subject Classification.* 82B23, 82B20, 81R50, 81U40, 37K15, 34A55.
Supported by CNRS (France) and the EC-TMR contract FMRX-CT96-0012.
This is the final form of the paper.

For a long time, only very few models were known for which correlation functions can be computed in an exact and explicit way. The typical examples are the Ising model [**24, 46, 47**], related to free fermions, and conformal field theories dealing with critical or massless systems in the continuum [**6, 30**]. Besides these models (for which series of formidable works have been necessary to obtain exact correlation functions), in the class of integrable models solvable by means of the Bethe ansatz, and in particular for integrable lattice models, two different but complementary approaches have been considered to solve this problem.

One of them relies on the study of infinite dynamical symmetries of integrable models directly in the infinite volume limit. It started with the solution of bootstrap equations for S-matrices and form factors of (massive) relativistic quantum field theories [**26,27,38,51**] and merged with the representation theory of the associated Yangians or quantum affine symmetries [**11,14,20,39–41,45**]. Using the additional ideas coming from two-dimensional statistical models resolution (like the Baxter corner transfer matrix method) [**3–5, 44**], it leads (using however some hypothesis) to multiple integral representations of correlation functions of the XXZ Heisenberg spin chains (at zero temperature and zero magnetic field) in the massive regime [**21, 22, 25**]. Since in that approach these functions are related to the trace of products of q-deformed vertex operators satisfying q-deformed KZ-equations, it also led to conjectures for other regimes (gapless region) for which the corner transfer matrix approach is not available. More general models have been considered since then using this method (see the lectures by T. Miwa in this volume).

Another approach, described essentially in the book [**31**], is based on the algebraic Bethe ansatz analysis [**5, 12, 13, 31**] of these lattice models in the finite volume. However, due to the complicated combinatorial structure of Bethe eigenstates and of the scalar products of general states, it did not succeed in obtaining explicit and manageable expressions for the correlation functions for generic models. Instead, this method seems to give insights into the long distance asymptotics of correlation functions [**18**] (see also the lectures by V. Korepin in this volume).

In [**19,28,29**], a new approach to the exact computation of correlation functions of lattice quantum integrable models was given. It is based on the actual resolution of the so-called quantum inverse scattering problem, namely, on the reconstruction of the local quantum operators (the correlation functions of which we want to calculate) in terms of the elements of the monodromy matrix of the model satisfying a Yang-Baxter algebra and containing in particular the creation/annihilation operators of the Bethe eigenstates. Although this method has been developed first in the example of the Heisenberg spin-$\frac{1}{2}$ XXZ chain [**28**], it is clear already that the basic tools and techniques used there can be applied in principle to generic quantum integrable lattice models (see [**34**] for a general solution of the quantum inverse scattering problem for a large class of integrable lattice models).

The aim of this lecture is to give a description of this new approach. I will first review the problem to be solved and the general features of our method in Section 2. In the next sections, the spin-$\frac{1}{2}$ XXZ Heisenberg chain will be used as a representative example. The basics of this model and the corresponding Bethe ansatz approach are recalled in Section 3. Sections 4 and 5 are devoted to the construction of the main tools (solution of the quantum inverse scattering problem, scalar product formulas, etc.) to be used in Section 6 for the computation of correlation functions. In the last section, some comments and final remarks are given.

2. General Approach to Correlation Functions

Correlation functions of a certain product of local operators denoted generically by $\mathcal{O}$ are defined by

$$\frac{\operatorname{tr}_{\mathcal{H}}\left(\mathcal{O}\, e^{-\mathbf{H}/kT}\right)}{\operatorname{tr}_{\mathcal{H}}\left(e^{-\mathbf{H}/kT}\right)}, \tag{2.1}$$

where T is the temperature, k the Boltzmann constant, and H is the Hamiltonian of the model acting in the quantum space $\mathcal{H}$. In the zero temperature limit, only the groundstate $|\psi_g\rangle$ of H contributes to the above trace, and the correlation function reduces to

$$\frac{\langle\psi_g|\mathcal{O}|\psi_g\rangle}{\langle\psi_g|\psi_g\rangle}. \tag{2.2}$$

Let us first review the two main difficulties in trying to compute such a quantity. The first problem to solve is of course the determination of the ground state $|\psi_g\rangle$. For interacting theories, this is already a highly non-trivial (and in general non-perturbative) problem. For integrable models solvable by means of the algebraic Bethe ansatz [**5, 12, 13, 31**], it admits a solution in terms of creation (resp. annihilation) operators which are elements of the associated quantum monodromy matrix $T_i(\lambda) \in \operatorname{End}(V_i) \otimes \mathcal{H}$ satisfying the Yang-Baxter algebra [**2, 48**],

$$R_{12}(\lambda,\mu)T_1(\lambda)T_2(\mu) = T_2(\mu)T_1(\lambda)R_{12}(\lambda,\mu), \tag{2.3}$$

with the usual tensor notations $T_1(\lambda) = T(\lambda) \otimes \mathrm{Id}$ and $T_2(\mu) = \mathrm{Id} \otimes T(\mu)$, R being a linear operator in the tensor product of two two-dimensional linear spaces $V_1 \otimes V_2$ and obeying the Yang-Baxter equation,

$$R_{12}(\lambda_1,\lambda_2)R_{13}(\lambda_1,\lambda_3)R_{23}(\lambda_2,\lambda_3) = R_{23}(\lambda_2,\lambda_3)R_{13}(\lambda_1,\lambda_3)R_{12}(\lambda_1,\lambda_2). \tag{2.4}$$

In the simplest cases the monodromy matrix is a 2×2 matrix in the auxiliary space V, with operator-valued entries acting in $\mathcal{H}$,

$$T(\lambda) = \begin{pmatrix} A(\lambda) & B(\lambda) \\ C(\lambda) & D(\lambda) \end{pmatrix}. \tag{2.5}$$

The ground state is then constructed by the multiple action of the B operators on a known reference state $|0\rangle$ as $B(\lambda_1)\cdots B(\lambda_N)|0\rangle$ (resp. $\langle 0|C(\lambda_1)\cdots C(\lambda_N)\rangle$) which is a common eigenstate of the transfer matrices if the set of spectral parameters $\{\lambda_j\}_{1\le j\le N}$ is a solution of the Bethe equations [**7**],

$$\frac{a(\lambda_j)}{d(\lambda_j)} \prod_{\substack{k=1\\k\neq j}}^{N} \frac{b(\lambda_j,\lambda_k)}{b(\lambda_k,\lambda_j)} = 1, \quad 1 \le j \le N, \tag{2.6}$$

where $a(\lambda)$ and $d(\lambda)$ are the eigenvalues of the operators $A(\lambda)$ and $D(\lambda)$, respectively, on the reference state $|0\rangle$.

The second problem is to express in a compact way the action of any product of local operators $\mathcal{O}$ on such Bethe eigenstates. This question remained open for many years. The reason is the following. Although the operators $B(\lambda)$ satisfy together with the other elements of the monodromy matrix a nice quadratic algebra (2.3), they are highly non-local quantities in terms of the local operators defining the model. Hence, there was until recently no way to obtain the needed algebra between local operators and the monodromy matrix operator entries. It turns out that for almost all known lattice quantum integrable models, this problem has a simple and explicit solution: it consists of reconstructing the local operators in terms of the monodromy matrix entries, or in other words, to solve explicitly the so-called quantum inverse scattering problem. We will give in the following the formulas corresponding to the particular case of the XXZ spin-$\frac{1}{2}$ Heisenberg finite chain [**28**]. However, it will be made clear that the simplicity of the result and of its proof already open the way for generalisations to almost all known lattice integrable models [**34**]. This is quite unexpected since for the classical integrable models, the solution to the corresponding classical inverse scattering problem (which enables the determination of soliton solutions of the associated non-linear equations of motion) is much more involved and far less explicit compared to our quantum formulas.

However, the parallel with the classical situation is quite instructive. The interpretation is the following. The algebraic Bethe ansatz which leads to the construction of eigenvectors and eigenvalues of the transfer matrix is the quantum analogue of the *direct* classical scattering problem giving the action-angle variables in terms of the original local ones. It is the inverse scattering problem part, namely the reconstruction of the local fields in terms of the monodromy matrix elements (and hence in terms of the action-angle variables), which really solves the initial dynamical problem. At the quantum level, the reconstruction of the local operators will give the key to the computation of the correlation functions.

To achieve this goal, our method will be decomposed into the following main steps:

(1) Construction of the groundstate $|\psi_g\rangle$ in the algebraic Bethe ansatz framework as, $\langle\psi_g| = \langle 0|\prod_{i\in I} C(\lambda_i)$ and $|\psi_g\rangle = \prod_{i\in I} B(\lambda_i)|0\rangle$, $|0\rangle$ being some reference state, and the parameters λ_i, $i \in I$, satisfying the Bethe equations.

(2) Algebraic solution of the quantum inverse scattering problem for local operators, namely their reconstruction in terms of the monodromy matrix operator entries.

(3) Expression of the action on the ground state of any product of local operators using their expression in terms of the monodromy matrix and the Yang-Baxter algebra it satisfies, as $\langle\psi_g|\mathcal{O} = \sum_{j\in J}\alpha_j\langle 0|\prod_{k\in K_j} C(\lambda_k)$, with some computable coefficients α_j and sets J and K_j depending on $\mathcal{O}$ and $\langle\psi_g|$.

(4) Computation the correlation function as (multiple) sums of scalar products of states, one of them being the ground state,

$$\langle\psi_g|\mathcal{O}|\psi_g\rangle = \sum_{j\in J}\alpha_j\langle 0|\prod_{k\in K_j} C(\lambda_j)\prod_{i\in I} B(\lambda_i)|0\rangle.$$

These scalar products can be computed using for example a new basis of the space of states induced by the so-called factorizing F-matrices (see [**33**]), in which the operators B and C simplify drastically, hence allowing a direct computation (for the XXZ model it is given as a ratio of two determinants [**28**]).

(5) Take the thermodynamic limit (the limit of infinite lattice) of the above

results, where multiple sums lead to multiple integral representations of the correlation functions.

The above steps 1, 2, 3 can be achieved for a very general class of lattice integrable models. Steps 4 and 5 need more development at the moment for the general models. However, the F basis has been determined recently for higher spin Heisenberg models in [**43**], using the functional Bethe ansatz technique of Sklyanin [**36**], which gives some hope for a general solution of step 4.

In the following sections, I will outline the application of this general method to the representative example of the XXZ Heisenberg spin-$\frac{1}{2}$ in a magnetic field. Multiple integral representations of its correlation functions have been obtained in [**29**]. This approach gives for this model a proof of the zero magnetic field expressions given by Jimbo, Miwa and their collaborators [**22**].

3. The XXZ Spin-$\frac{1}{2}$ Heisenberg Chain

The Hamiltonian of the XXZ spin-$\frac{1}{2}$ Heisenberg chain of finite length M [**7,17**] is given by

$$H_{\mathrm{XXZ}} = \sum_{m=1}^{M} \{\sigma_m^x \sigma_{m+1}^x + \sigma_m^y \sigma_{m+1}^y + \Delta(\sigma_m^z \sigma_{m+1}^z - 1)\}, \tag{3.1}$$

and we impose periodic boundary conditions. Here σ_m^a, $a = x, y, z$, are the Pauli spin operators acting in the local quantum spin-$\frac{1}{2}$ space $\mathcal{H}_m$ at site m. The anisotropy parameter Δ defines the physical nature of the model: when $\Delta \leq -1$, the ground state of the Hamiltonian is ferromagnetic, whereas its magnetization is equal to zero when $\Delta > -1$. We shall focus our attention on this last domain, which decomposes into a massive regime (for $\Delta > 1$), and a gapless regime (for $-1 < \Delta < 1$) in the thermodynamic limit ($M \to \infty$).

The R-matrix of the XXZ model is

$$R(\lambda, \mu) = \begin{pmatrix} 1 & 0 & 0 & 0 \\ 0 & b(\lambda,\mu) & c(\lambda,\mu) & 0 \\ 0 & c(\lambda,\mu) & b(\lambda,\mu) & 0 \\ 0 & 0 & 0 & 1 \end{pmatrix} \tag{3.2}$$

where the functions $b(\lambda, \mu)$ and $c(\lambda, \mu)$ are defined as

$$b(\lambda,\mu) = \frac{\sinh(\lambda-\mu)}{\sinh(\lambda-\mu+\eta)}, \quad c(\lambda,\mu) = \frac{\sinh\eta}{\sinh(\lambda-\mu+\eta)}.$$

The parameter η is here related to the anisotropy parameter Δ of the Hamiltonian by

$$\Delta = \frac{1}{2}(q + q^{-1}), \quad \text{with } q = e^{\eta}.$$

The R-matrix is a linear operator in the tensor product of two two-dimensional linear spaces $V_1 \otimes V_2$, where each V_i is isomorphic to $\mathbb{C}^2$, and depends generically on two spectral parameters λ_1 and λ_2 associated to these two vector spaces. It is denoted by $R_{12}(\lambda_1, \lambda_2)$. Such an R-matrix satisfies the Yang-Baxter equation

$$R_{12}(\lambda_1,\lambda_2)\, R_{13}(\lambda_1,\lambda_3)\, R_{23}(\lambda_2,\lambda_3) = R_{23}(\lambda_2,\lambda_3)\, R_{13}(\lambda_1,\lambda_3)\, R_{12}(\lambda_1,\lambda_2). \tag{3.3}$$

Moreover, it is unitary, except for a finite number of values of the spectral parameter, namely if $b(\lambda_1, \lambda_2) \neq \pm c(\lambda_1, \lambda_2)$ we have,

$$R_{12}(\lambda_1 - \lambda_2)\, R_{21}(\lambda_2 - \lambda_1) = \mathbf{1}, \tag{3.4}$$

and reduces to the permutation operator for a particular value of the spectral parameter, $R_{12}(0) = P_{12}$.

Identifying one of the two vector spaces of the R-matrix with the quantum space $\mathcal{H}_m$, one defines the quantum L-operator of the inhomogeneous chain at site m as

$$L_m(\lambda, \xi_m) = R_{0m}(\lambda - \xi_m), \tag{3.5}$$

where ξ_m is an arbitrary inhomogeneity parameter attached to the site m. Here R_{0m} acts in $V_0 \otimes \mathcal{H}_m$, where V_0 is an auxiliary space isomorphic to $\mathbb{C}^2$. The monodromy matrix is constructed as an ordered product of such L-operators:

$$T_0(\lambda) \equiv T_{0,1\dots N}(\lambda; \xi_1, \dots, \xi_N) = R_{0N}(\lambda - \xi_N)\dots R_{01}(\lambda - \xi_1), \tag{3.6}$$

and can be represented in the auxiliary space V_0 as a 2×2 matrix,

$$T(\lambda) = \begin{pmatrix} A(\lambda) & B(\lambda) \\ C(\lambda) & D(\lambda) \end{pmatrix}, \tag{3.7}$$

whose matrix elements A, B, C, D are linear operators on the quantum space of states of the chain $\mathcal{H} = \bigotimes_{n=1}^{N} \mathcal{H}_n$.

The transfer matrix $\mathcal{T}(\lambda)$ is defined as the trace $A(\lambda) + D(\lambda)$ of the monodromy matrix. Transfer matrices commute with each other for different values of the spectral parameter λ. They commute also with the Hamiltonian (3.1) in the homogeneous case where all ξ_m are equal, as the Hamiltonian can be reconstructed in terms of the transfer matrix by means of the following "trace identity":

$$H_{\mathrm{XXZ}} = 2 \sinh \eta \partial_\lambda \log \mathcal{T}(\lambda)\big|_{\lambda = \xi_j} + \text{const.} \tag{3.8}$$

Common eigenstates of the transfer matrices (and thus of the Hamiltonian (3.1) in the homogeneous case) can be constructed by successive actions of operators $B(\lambda)$ on the reference state $|0\rangle$, which is the ferromagnetic state with all the spins up. More precisely, the state $B(\lambda_1) \cdots B(\lambda_N)|0\rangle$ is a common eigenstate of the transfer matrices if the set of spectral parameters $\{\lambda_j\}_{1 \le j \le N}$ is a solution of the Bethe equations,

$$\frac{a(\lambda_j)}{d(\lambda_j)} \prod_{\substack{=1 \\ k \ne j}}^{N} \frac{b(\lambda_j, \lambda_k)}{b(\lambda_k, \lambda_j)} = 1, \quad 1 \le j \le N, \tag{3.9}$$

where $a(\lambda) = 1$ and $d(\lambda) = \prod_{i=1}^{M} b(\lambda, \xi_i)$ are the eigenvalues of operators $A(\lambda)$ and $D(\lambda)$, respectively, on the reference state $|0\rangle$. The corresponding eigenvalue for the transfer matrix $\mathcal{T}(\mu)$ is then

$$\tau(\mu, \{\lambda_j\}) = a(\mu) \prod_{j=1}^{n} b^{-1}(\lambda_j, \mu) + d(\mu) \prod_{j=1}^{n} b^{-1}(\mu, \lambda_j). \tag{3.10}$$

The Bethe equations can also be written in a logarithmic form:

$$M p_{0_{\mathrm{tot}}}(\lambda_j) + \sum_{k=1}^{N} \theta(\lambda_j - \lambda_k) = 2\pi n_j, \quad 1 \le j \le N, \tag{3.11}$$

where n_j are integers for N odd and half integers for N even. The bare momentum $p_{0_{\rm tot}}(\lambda)$ and the scattering phase $\theta(\lambda)$ are defined as

$$p_{0_{\rm tot}}(\lambda) = \frac{i}{M}\ln\frac{d(\lambda)}{a(\lambda)} = \frac{1}{M}\sum_{k=1}^{M} p_0\Big(\lambda-\xi_k+\frac{\eta}{2}\Big),$$
$$p_0(\lambda) = i\ln\frac{\sinh(\lambda-\eta/2)}{\sinh(\lambda+\eta/2)},$$
$$\theta(\lambda) = i\ln\frac{\sinh(\eta+\lambda)}{\sinh(\eta-\lambda)}.$$

In the thermodynamic limit ($M\to\infty$), these Bethe equations for the ground state become an integral equation for the quasi-particle density ρ in the rapidity representation (Lieb equation) [**32, 49**]:

(3.12) $$\rho_{\rm tot}(\alpha) + \int_{-\Lambda}^{\Lambda} K(\alpha-\beta)\rho_{\rm tot}(\beta)\,d\beta = \frac{p'_{0_{\rm tot}}(\alpha)}{2\pi},$$

where the new real variables α are defined in terms of general spectral parameters λ differently in the two domains:

$$\alpha = \lambda \quad \text{for } -1<\Delta<1,$$
$$\alpha = i\lambda \quad \text{for } \Delta>1.$$

The density ρ is defined as the limit of the quantity $1/M(\alpha_{j+1}-\alpha_j)$, and the functions $K(\alpha)$ and $p'_{0_{\rm tot}}(\alpha)$ are the derivatives with respect to α of the functions $-\theta(\lambda(\alpha))/2\pi$ and $p_{0_{\rm tot}}(\lambda(\alpha))$:

(3.13) $$K(\alpha) = \frac{\sin 2\zeta}{2\pi\sinh(\alpha+i\zeta)\sinh(\alpha-i\zeta)} \qquad \text{for } -1<\Delta<1, \text{ with } \zeta = i\eta,$$
$$p'_0(\alpha) = \frac{\sin\zeta}{\sinh(\alpha+i\frac{\zeta}{2})\sinh(\alpha-i\frac{\zeta}{2})}$$

(3.14) $$K(\alpha) = \frac{\sinh 2\zeta}{2\pi\sin(\alpha+i\zeta)\sin(\alpha-i\zeta)} \qquad \text{for } \Delta>1, \text{ with } \zeta=-\eta,$$
$$p'_0(\alpha) = \frac{\sinh\zeta}{\sin(\alpha+i\frac{\zeta}{2})\sin(\alpha-i\frac{\zeta}{2})}$$

with

(3.15) $$p'_{0_{\rm tot}}(\alpha) = \frac{1}{M}\sum_{i=1}^{M} p'_0(\alpha-\beta_k-i\frac{\zeta}{2}),$$

where $\beta_k=\xi_k$ in the domain $-1<\Delta<1$, and $\beta_k=i\xi_k$ in the domain $\Delta>1$. The integration limit Λ is equal to $\pi/2$ for $\Delta>1$, and to $+\infty$ for $-1<\Delta<1$.

The solution for the Lieb equation (3.12) in the homogeneous model where all parameters ξ_k are equal to $\eta/2$, that is, the density for the ground state of the Hamiltonian (3.1) in the thermodynamic limit, is given by the following function [**49**]:

(3.16) $$\rho(\alpha) = \frac{1}{2\zeta\cosh(\pi\alpha/\zeta)} \qquad \text{for } -1<\Delta<1,$$

$$\rho(\alpha) = \frac{1}{2\pi} \sum_{n=-\infty}^{+\infty} \frac{e^{2in\alpha}}{\cosh(n\zeta)} \quad \text{for } \Delta > 1. \tag{3.17}$$

For technical convenience, we will also use in the following the solution of the inhomogeneous Lieb equation, that is, the function

$$\rho_{\text{tot}}(\alpha) = \frac{1}{M} \sum_{i=1}^{M} \rho\Big(\alpha - \beta_k - i\frac{\zeta}{2}\Big). \tag{3.18}$$

It will also be convenient to assume, without any loss of generality, that the inhomogeneity parameters are contained in the region $-\zeta < \operatorname{Im}\beta_j < 0$.

Let us mention at last that the ground state of the XXZ model in the region $\Delta > 1$ is degenerate in the thermodynamic limit ($M \to \infty$), namely there are two states with the same energy (and characterized by the same density (3.17)), which we will call the ground state $|\Psi_1\rangle$ and the quasi-ground state $|\Psi_2\rangle$ (on the finite lattice, these states possess different energy). In this domain, the correlation function at zero temperature is thus half of the trace on these two states, that is, of the sum of the two corresponding matrix elements. In the domain $-1 < \Delta < 1$, the ground state is not degenerate.

4. The Quantum Inverse Scattering Problem

The solution of this problem was a dream for most of us working in the domain of quantum integrable models. It was shown for the first time in [**28**] that the explicit solution of the quantum inverse scattering problem (namely the reconstruction of any local spin operator at any site of the chain in terms of the elements of the quantum monodromy matrix satisfying a Yang-Baxter algebra, and containing creation/annihilation operators of Bethe eigenstates of the Hamiltonian) can be obtained for the XXZ spin-$\frac{1}{2}$ model in a very simple and purely algebraic way.

The elementary nature of the answer in this very representative example among the models solvable by means of the algebraic Bethe ansatz method, is quite unexpected. The algebraic Bethe ansatz method, also called quantum inverse scattering method, appeared twenty years ago as a quantum analogue of the classical inverse scattering problem approach to non-linear wave equations having soliton solutions, in order to solve quantum integrable models in two dimensions. The essential tools of this method are the quantum monodromy matrix satisfying quadratic commutation relations (Yang-Baxter algebra) whose structure constants are given by an R-matrix solving the Yang-Baxter (cubic) equation [**5, 48**]. It has been designed to diagonalize the corresponding Hamiltonians simultaneously with their associated commuting family of integrals of motion [**13, 42**]. This method is in fact the quantum analogue of the *direct part* of the classical inverse scattering problem method in its Hamiltonian formulation [**50**], in which the Lax matrix is used to construct the monodromy matrix containing the action-angle variables which linearize the dynamics. However, the *inverse scattering problem part* of the classical theory, namely, the reconstruction of the local classical field variables contained in the Lax matrix in terms of the elements of the monodromy matrix (and hence in terms of the action-angle variables) using the Gel'fand-Levitan-Marchenko equations [**1, 16, 35**], being already a quite difficult problem to solve, it was not at all obvious to find a direct way to extend it to the quantum situation, although the motivations (form factors and correlation functions) were clear from the very beginning [**8–10**].

The very remarkable feature of the solution to this problem given in [**28**] was not only that the quantum inverse scattering problem can be solved explicitly, but also that both its resulting expressions (reconstruction of the local spin operators at any site of the chain in terms of a simple product of the quantum monodromy matrix elements) and their proofs are very elementary. In turn, it essentially relies on the fact that the quantum R-matrix $R(\lambda, \mu)$ solving the Yang-Baxter equation reduces to the permutation operator when the two spectral parameters λ and μ are equal.

This fact being almost a consequence of the Yang-Baxter equation itself, and hence satisfied for very generic cases, it immediatly suggested that the quantum inverse scattering problem can indeed be solved for almost all known quantum integrable (lattice) models. This is achieved for a large class of integrable lattice models in [**34**]. It is to be compared to the attempts at solving the corresponding problem directly for the continuum quantum integrable field theories such as the Sine-Gordon model, using a quantum version of the Gel'fand-Levitan-Marchenko equations, where it appeared to be an extremely difficult problem to handle, due in particular to the presence of boundstates [**38**].

Let us now describe the solution and its proof for the XXZ spin-$\frac{1}{2}$ model [**28,34**]. This proof can be easily extended to very general lattice models, including the so-called fundamental lattice models, but also the models with impurities and the fusion models (see [**34**]).

THEOREM 4.1. *The local spin operators at a given site i of the inhomogeneous XXZ Heisenberg chain are given by*

$$\sigma_i^- = \prod_{\alpha=1}^{i-1} (A+D)(\xi_\alpha) \cdot B(\xi_i) \cdot \prod_{\alpha=i+1}^{N} (A+D)(\xi_\alpha), \tag{4.1}$$

$$\sigma_i^+ = \prod_{\alpha=1}^{i-1} (A+D)(\xi_\alpha) \cdot C(\xi_i) \cdot \prod_{\alpha=i+1}^{N} (A+D)(\xi_\alpha), \tag{4.2}$$

$$\sigma_i^z = \prod_{\alpha=1}^{i-1} (A+D)(\xi_\alpha) \cdot (A-D)(\xi_i) \cdot \prod_{\alpha=i+1}^{N} (A+D)(\xi_\alpha). \tag{4.3}$$

Note that the case of homogeneous models where all ξ are equal is also included in this result by just taking the desired specific equal values for the ξ_j.

The proof of this theorem is a straightforward consequence of the following lemma when x_0 is respectively equal to σ_0^-, σ_0^+ and σ_0^z:

LEMMA 4.2. *Let x_i be an operator acting on the quantum spin space $\mathcal{H}_i$, and denote by x_0 the corresponding 2×2 matrix acting on the auxiliary space V_0. They are related by the identity*

$$\operatorname{tr}_0\big(x_0 \; R_{0,1\ldots N}(\xi_i)\big) = \prod_{\alpha=1}^{i-1} (A+D)^{-1}(\xi_\alpha) \cdot x_i \cdot \prod_{\alpha=1}^{i} (A+D)(\xi_\alpha), \tag{4.4}$$

where the trace on the left hand side is taken on the matrix acting in V_0.

PROOF. Let us first prove this for the local operators at site 1 of the chain. We have:

$$\begin{aligned}\mathrm{tr}_0\big(x_0\ R_{0,1\ldots N}(\xi_1)\big) &= \mathrm{tr}_0(x_0\ R_{0N}(\xi_1)\ldots R_{02}(\xi_1)\ P_{01}),\\ &= x_1\ R_{1N}(\xi_1)\ldots R_{12}(\xi_1),\\ &= x_1\ R_{1,2\ldots N}(\xi_1).\end{aligned}$$

The reconstruction at any other site i of the chain is then obtained by acting with the propagator (shift operator) from site 1 to site i as described in the following lemma. □

LEMMA 4.3. *Let U_1^i be the propagator from site 1 to site i of the chain. It can be written in the following two forms*:

$$U_1^i = R_{i-1,i\ldots N1\ldots i-2}\ldots R_{2,3\ldots N1}\ R_{1,2\ldots N} = \prod_{\alpha=1}^{i-1}\big(A_{1\ldots N}(\xi_\alpha)+D_{1\ldots N}(\xi_\alpha)\big). \tag{4.5}$$

PROOF. The action of the propagator consists in shifting the beginning of the chain from site 1 to site i. For an operator entry $X_{1\ldots N}$ of the monodromy matrix ($X = A$, B, C or D), this means that

$$U_1^i\ X_{1\ldots N} = X_{i\ldots N1\ldots i-1}\ U_1^i, \tag{4.6}$$

which, in terms of the monodromy matrix, can be written

$$U_1^i\ T_{0,1\ldots N} = T_{0,i\ldots N1\ldots i-1}\ U_1^i. \tag{4.7}$$

leading to the first expression (4.5)

The second equality (4.5) follows from the identity

$$A_{1\ldots N}(\xi_\alpha) + D_{1\ldots N}(\xi_\alpha) = R_{\alpha,\alpha+1\ldots N1\ldots\alpha-1}(\xi_\alpha;\xi_{\alpha+1},\ldots,\xi_{\alpha-1}). \tag{4.8}$$

The proof of (4.8) is based on the observation that $R_{0\alpha}(\xi_\alpha)$ is the permutation matrix $P_{0\alpha}$ of the spaces 0 and α. Writing $A_{1\ldots N}(\xi_\alpha)+D_{1\ldots N}(\xi_\alpha)$ as a trace in the auxiliary space V_0, and making $P_{0\alpha}$ act on every factor, we obtain, thanks to the cyclicity of the trace,

$$\begin{aligned}A_{1\ldots N}(\xi_\alpha)+D_{1\ldots N}(\xi_\alpha) &= \mathrm{tr}_0\big(R_{0N}(\xi_\alpha)\ldots R_{0\alpha+1}(\xi_\alpha)\ P_{0\alpha}\ R_{0\alpha-1}(\xi_\alpha)\ldots R_{01}(\xi_\alpha)\big),\\ &= R_{\alpha\alpha-1}(\xi_\alpha)\ldots R_{\alpha 1}(\xi_\alpha)\ R_{\alpha N}(\xi_\alpha)\ldots R_{\alpha\alpha+1}(\xi_\alpha),\\ &= R_{\alpha,\alpha+1\ldots N1\ldots\alpha-1}(\xi_\alpha;\xi_{\alpha+1},\ldots,\xi_{\alpha-1}),\end{aligned}$$

which ends the proof of Lemma 4.3. Note that in (4.5), all the factors commute with each other. □

REMARK 4.4. Due to the unitarity of the R matrix, the propagator U_1^1 through the whole chain is the identity, hence,

$$\prod_{\alpha=1}^{N}\big(A_{1\ldots N}(\xi_\alpha)+D_{1\ldots N}(\xi_\alpha)\big) = R_{N,1\ldots N-1}\ldots R_{1,2\ldots N} = 1. \tag{4.9}$$

Thus, the inverse $(U_1^i)^{-1}$ of the propagator on a part of the chain is merely the propagator $U_i^1 = \prod_{\alpha=i}^{N}(A+D)(\xi_\alpha)$ on the remaining part.

This result is extended in [**34**] to a very large class of quantum integrable lattice models.

5. From the Inverse Scattering Problem to Correlation Functions

Using the above solution to the quantum inverse scattering problem for local spins, we can compute the correlation functions. For this we need three more ingredients, which are the computation of generic actions of local operators on the ground state using Yang-Baxter commutation relations, the scalar product of states, one of them being a Bethe state, and finally a simple procedure to take the thermodynamic limit. These tools are given in the next subsections.

5.1. Action of Operators A, B, C, D on a General State. We want now to express the action of any local operator on a state constructed by action of C operators on the reference state. This is obtained in turn from the solution of the quantum inverse scattering problem by the successive action of some product of A, B, C, D operators. The actions of A, B, C, D on such a state are well known (see for example [**31**]), but we recast them here in a form that is more convenient for our purpose.

The action of the operators $A(\lambda)$ and $D(\lambda)$ on the states constructed by successive actions of operators $C(\lambda)$ can be written in the following form:

$$\langle 0|\prod_{k=1}^{N} C(\lambda_k)A(\lambda_{N+1}) = \sum_{a'=1}^{N+1} a(\lambda_{a'})\frac{\prod_{k=1}^{N}\sinh(\lambda_k-\lambda_{a'}+\eta)}{\prod_{\substack{k=1\\k\neq a'}}^{N+1}\sinh(\lambda_k-\lambda_{a'})}\langle 0|\prod_{\substack{k=1\\k\neq a'}}^{N+1} C(\lambda_k); \tag{5.1}$$

$$\langle 0|\prod_{k=1}^{N} C(\lambda_k)D(\lambda_{N+1}) = \sum_{a=1}^{N+1} d(\lambda_{a})\frac{\prod_{k=1}^{N}\sinh(\lambda_a-\lambda_{k}+\eta)}{\prod_{\substack{k=1\\k\neq a}}^{N+1}\sinh(\lambda_a-\lambda_{k})}\langle 0|\prod_{\substack{k=1\\k\neq a}}^{N+1} C(\lambda_k). \tag{5.2}$$

The action of the operator $B(\lambda)$ is more complicated:

$$\begin{aligned}\langle 0|\prod_{k=1}^{N} C(\lambda_k)B(\lambda_{N+1}) &= \sum_{a=1}^{N+1} d(\lambda_{a})\frac{\prod_{k=1}^{N}\sinh(\lambda_a-\lambda_{k}+\eta)}{\prod_{\substack{k=1\\k\neq a}}^{N+1}\sinh(\lambda_a-\lambda_{k})}\\ &\times\sum_{\substack{a'=1\\a'\neq a}}^{N+1}\frac{a(\lambda_{a'})}{\sinh(\lambda_{N+1}-\lambda_{a'}+\eta)}\frac{\prod_{\substack{j=1\\j\neq a}}^{N+1}\sinh(\lambda_j-\lambda_{a'}+\eta)}{\prod_{\substack{j=1\\j\neq a,a'}}^{N+1}\sinh(\lambda_j-\lambda_{a'})}\langle 0|\prod_{\substack{k=1\\k\neq a,a'}}^{N+1} C(\lambda_k),\end{aligned} \tag{5.3}$$

but in the case which is interesting for the computation of the correlation functions when $\lambda_{N+1}=\xi_k$ and hence $d(\lambda_{N+1})=0$, we obtain a simpler result:

$$\begin{aligned}\langle 0|\prod_{k=1}^{N} C(\lambda_k)B(\lambda_{N+1}) &= \sum_{a=1}^{N} d(\lambda_{a})\frac{\prod_{k=1}^{N}\sinh(\lambda_a-\lambda_{k}+\eta)}{\prod_{\substack{k=1\\k\neq a}}^{N+1}\sinh(\lambda_a-\lambda_{k})}\\ &\times\sum_{\substack{a'=1\\a'\neq a}}^{N+1} a(\lambda_{a'})\frac{\prod_{\substack{j=1\\j\neq a}}^{N}\sinh(\lambda_j-\lambda_{a'}+\eta)}{\prod_{\substack{j=1\\j\neq a,a'}}^{N+1}\sinh(\lambda_j-\lambda_{a'})}\langle 0|\prod_{\substack{k=1\\k\neq a,a'}} C(\lambda_k).\end{aligned} \tag{5.4}$$

It should be mentioned that the action of B is similar to the successive action of D and A. Using these formulae, one can reduce expressions for general correlation functions to (multiple) sums of scalar products of a Bethe state with an arbitrary state constructed by successive actions of B operators on the reference state.

5.2. Scalar Products. The next step of the computation is to find an explicit and convenient expression for scalar products of two states, one of them being a Bethe eigenstate. The usual Bethe ansatz techniques, based only on the use of commutation relations, generally generate huge sums which are difficult to sum up. In [**28**], a direct computation in a new basis (F-basis) [**33**] has been performed and leads to an explicit expression for such scalar products as a determinant of usual functions of the model:

THEOREM 5.1. *Let $\{\lambda_1, \ldots, \lambda_N\}$ be a solution of the Bethe equations* (2.6) *and let $\{\mu_1, \ldots, \mu_N\}$ be an arbitrary set of parameters. Then the scalar product*

$$S_N(\{\mu_j\}, \{\lambda_k\}) = \langle 0| \prod_{j=1}^{N} C(\mu_j) \prod_{k=1}^{N} B(\lambda_k) |0\rangle \tag{5.5}$$

can be represented as a ratio of two determinants,

$$S_N(\{\mu_j\}, \{\lambda_k\}) = S_N(\{\lambda_k\}, \{\mu_j\}) = \frac{\det T(\{\mu_j\}, \{\lambda_k\})}{\det V(\{\mu_j\}, \{\lambda_k\})}, \tag{5.6}$$

of the following $N \times N$ matrices T and V:

$$T_{ab} = \partial_{\lambda_a} \tau(\mu_b, \{\lambda_k\}), \qquad V_{ab} = \frac{1}{\sinh(\mu_b - \lambda_a)}, \quad 1 \le a, b \le N, \tag{5.7}$$

where $\tau(\mu_b, \{\lambda_k\})$ is the eigenvalue of the transfer matrix $\mathcal{T}(\mu_b)$ corresponding to the Bethe state $\prod_{k=1}^{N} B(\lambda_k) |0\rangle$ given by (3.10).

This result is equivalent to the scalar product formula obtained in [**37**].

When particularizing this formula to the case when the two states are equal, one obtains the Gaudin formula for the norm of a Bethe state:

$$\langle 0| \prod_{j=1}^{N} C(\lambda_j) \prod_{k=1}^{N} B(\lambda_k) |0\rangle = \sinh^N \eta \prod_{\alpha \neq \beta} \frac{\sinh(\lambda_\alpha - \lambda_\beta + \eta)}{\sinh(\lambda_\alpha - \lambda_\beta)} \det \Phi'(\{\lambda_a\}), \tag{5.8}$$

where Φ' is a $N \times N$ matrix the elements of which are given by:

$$\Phi'_{ab} = -\partial_{\lambda_b} \ln\left(\frac{a(\lambda_a)}{d(\lambda_a)} \prod_{\substack{k=1 \\ k \neq a}}^{N} \frac{b(\lambda_a, \lambda_k)}{b(\lambda_k, \lambda_a)} \right). \tag{5.9}$$

By means of this expression for the scalar product, general correlation functions for the finite chain can now be expressed as sums of determinants.

5.3. Thermodynamic Limit. The last step of our method, to obtain the general correlation functions in the infinite volume limit, is to take the thermodynamic limit of the expressions obtained for the finite chain. In the thermodynamic limit $M \to \infty$, the Bethe equations for the ground state become the integral Lieb equation (3.12) for the density. In a more general way, for any $\mathcal{C}^\infty$ function f (π-periodic in the domain $\Delta > 1$), sums over all the values of f at the point α_j, $1 \le j \le N$, parametrizing the ground state, can be replaced in the thermodynamic limit by an integral involving the density ρ solution of the Lieb equation [**19**]:

$$\frac{1}{M} \sum_{j=1}^{N} f(\alpha_j) = \int_{-\Lambda}^{\Lambda} f(\alpha) \rho_{\text{tot}}(\alpha)\, d\alpha + O(M^{-\infty}). \tag{5.10}$$

Thus, sums over determinants will become multiple integrals.

This property enabled us in [**19**] to obtain the expression of the matrix elements of the Gaudin matrix (5.9) in the thermodynamic limit:

$$\Phi'_{ab}(\alpha) = -2i\pi M\left\{\delta_{ab}\rho_{\text{tot}}(\alpha_a) + \frac{1}{M}K(\alpha_a - \alpha_b)\right\} + O(M^{-\infty}) \quad \text{for } -1 < \Delta < 1, \tag{5.11}$$

$$\Phi'_{ab}(\alpha) = 2\pi M\left\{\delta_{ab}\rho_{\text{tot}}(\alpha_a) + \frac{1}{M}K(\alpha_a - \alpha_b)\right\} + O(M^{-\infty}) \quad \text{for } \Delta > 1. \tag{5.12}$$

These expressions will be useful in the following to compute the determinants which appear in the formulae for the correlation functions in the thermodynamic limit. Finally, we will obtain correlation functions as multiple integrals of usual functions of the model.

6. Correlation Functions

The general formulae for the correlation functions of the XXZ Heisenberg spin-$\frac{1}{2}$ chain in a magnetic field were given in [**29**]. Here, I will simply give an explicit formula for a simple example (the probability to observe m spins aligned in the antiferromagnetic ground state), and then describe only the structure of such a formula in terms of multiple integrals for the generic building blocks of correlation functions defined by

$$F_m(\{\epsilon_j, \epsilon'_j\}) = \frac{\langle\psi_g|\prod_{j=1}^m E_j^{\epsilon'_j,\epsilon_j}|\psi_g\rangle}{\langle\psi_g|\psi_g\rangle}, \tag{6.1}$$

where $E_m^{\epsilon'_m,\epsilon_m}$ are the elementary operators acting on $\mathcal{H}_m$ at site m as the 2×2 matrices $E_{lk}^{\epsilon',\epsilon} = \delta_{l,\epsilon'}\delta_{k,\epsilon}$, and $\prod_{j=1}^m E_j^{\epsilon'_j,\epsilon_j}$ is any product of such elementary operators from site one to m. Any n-point correlation function can be obtained from these building blocks.

6.1. Emptiness Formation Probability. We consider first the simplest m-point correlation function: the emptiness formation probability, i.e. the probability to detect a ferromagnetic domain of length m in the antiferromagnetic ground state of the XXZ model. This probability can be expressed in the following form:

$$\tau(m) = \frac{\langle\psi_g|\prod_{j=1}^m 1/2(1-\sigma_j^z)|\psi_g\rangle}{\langle\psi_g|\psi_g\rangle}, \tag{6.2}$$

where $|\psi_g\rangle$ is the ground state in the massless case and any one of two ground states constructed by the algebraic Bethe ansatz in the massive regime.

Using the solution of the quantum inverse scattering problem one can express the operators $1/2(1-\sigma_j^z)$ in terms of the monodromy matrix elements:

$$\frac{1}{2}(1-\sigma_j^z) = \prod_{k=1}^{j-1}\Big(A(\xi_k) + D(\xi_k)\Big)D(\xi_j)\prod_{k=j+1}^{M}\Big(A(\xi_k) + D(\xi_k)\Big).$$

So the emptiness formation probability can be written uniquely in terms of the monodromy matrix elements:

$$(6.3) \qquad \tau(m) = \phi_m(\{\lambda\}) \frac{\langle 0| \prod_{a=1}^N C(\lambda_a) \prod_{j=1}^m D(\xi_j) \prod_{a=1}^N B(\lambda_a)|0\rangle}{\langle 0| \prod_{a=1}^N C(\lambda_a) \prod_{a=1}^N B(\lambda_a)|0\rangle},$$

where $\phi_m(\{\lambda\})$ is the ground state eigenvalue of the corresponding product of the transfer matrices:

$$\phi_m(\{\lambda\}) = \prod_{j=1}^m \prod_{a=1}^N \frac{\sinh(\lambda_a - \xi_j)}{\sinh(\lambda_a - \xi_j + \eta)}.$$

Using the relation (5.2) we obtain the following action of a product of the operators $D(\lambda)$ on a state constructed by the action of the operators $C(\lambda)$,

$$(6.4) \quad \langle 0| \prod_{k=1}^N C(\lambda_k) \prod_{j=1}^m D(\lambda_{N+j})$$
$$= \sum_{a_1=1}^{N+1} \sum_{\substack{a_2=1 \\ a_2 \neq a_1}}^{N+2} \cdots \sum_{\substack{a_m=1 \\ a_m \neq a_1,\dots,a_{m-1}}}^{N+m} G_{a_1\dots a_m}(\lambda_1 \dots \lambda_{N+m}) \langle 0| \prod_{\substack{k=1 \\ k\neq a_1,\dots,a_m}}^{N+m} C(\lambda_k),$$

where the function G can be written as:

$$(6.5) \qquad G_{a_1\dots a_m}(\lambda_1, \dots, \lambda_{N+m}) = \prod_{j=1}^m d(\lambda_{a_j}) \frac{\prod_{\substack{b=1 \\ b\neq a_1,\dots,a_{j-1}}}^{N+j-1} \sinh(\lambda_{a_j} - \lambda_b + \eta)}{\prod_{\substack{b=1 \\ b\neq a_1,\dots,a_j}}^{N+j} \sinh(\lambda_{a_j} - \lambda_b)}.$$

To compute the emptiness formation probability one should take the parameters λ_a for $a > N$ equal to ξ_{a-N}. This means, in particular, that the sums should be taken up to $a_j = N$ as $d(\xi_k) = 0$.

Now we calculate the scalar products in each term of the sum. In fact we have to calculate the following "normalized" product:

$$(6.6) \quad \mathbb{S}(\{\lambda_1, \dots, \lambda_{N-m}, \xi_1, \dots, \xi_m\}, \{\lambda\})$$
$$= \frac{\langle 0| \prod_{b=1}^{N-m} C(\lambda_b) \prod_{k=1}^m C(\xi_k) \prod_{k=1}^N B(\lambda_k)|0\rangle}{\langle 0| \prod_{k=1}^N C(\lambda_k) \prod_{k=1}^N B(\lambda_k)|0\rangle}.$$

This quantity can be easily calculated using the representation for the scalar products (5.6) (as one of the states in the numerator is a Bethe state) and the Gaudin formula (5.8) for the norm of Bethe vectors. Finally we obtain,

$$(6.7) \quad \mathbb{S}(\{\lambda_1, \dots, \lambda_{N-m}, \xi_1, \dots, \xi_m\}, \{\lambda\}) =$$
$$\prod_{\substack{j,k=1 \\ j>k}}^m \frac{\sinh(\lambda_{N-m+k} - \lambda_{N-m+j})}{\sinh(\xi_k - \xi_j)} \prod_{j=1}^m \prod_{k=1}^{N-m} \frac{\sinh(\lambda_k - \lambda_{N-m+j})}{\sinh(\lambda_k - \xi_j)}$$
$$\times \prod_{a=1}^N \prod_{k=1}^m \frac{\sinh(\lambda_a - \xi_k + \eta)}{\sinh(\lambda_a - \lambda_{N-m+k} + \eta)} \frac{\det \Psi'(\{\lambda\}, \{\xi\})}{\det \Phi'(\{\lambda\})},$$

where the $N \times N$ matrix Φ' is the Gaudin matrix (5.9). The first $N-m$ columns of the $N \times N$ matrix Ψ' are the same as in the Gaudin matrix but the other columns are different:

$$\Psi'_{ab} = \Phi'_{ab}, \qquad b \le N-m,$$

$$\Psi'_{ab} = \frac{\sinh\eta}{\sinh(\lambda_a - \xi_{b+m-N})\sinh(\lambda_a - \xi_{b+m-N}+\eta)}, \qquad b > N-m.$$

As the Gaudin matrix is invertible, the fraction of the two determinants in (6.7) can be represented as one determinant:

$$\frac{\det\Psi'(\{\lambda\},\{\xi\})}{\det\Phi'(\{\lambda\})} = \det\big(\Phi'^{-1}(\{\lambda\})\Psi'(\{\lambda\},\{\xi\})\big).$$

The first $N-m$ columns of the matrix $\Phi'^{-1}\Psi'$ are those of the unit matrix columns:

$$(\Phi'^{-1}\Psi')_{ab} = \delta_{ab}, \quad b \le N-m.$$

The action of the inverse Gaudin matrix on the other columns of the matrix Ψ' can be calculated in the thermodynamic limit. Using the representation for the Gaudin matrix in the thermodynamic limit (5.11), (5.12) and the Lieb equation (3.12) one concludes that

$$\frac{1}{M}\sum_{b=1}^{N}\Phi'_{ab}\frac{\tilde\rho(\lambda_b - \xi_k + \eta/2)}{\tilde\rho_{\mathrm{tot}}(\lambda_b)} = \frac{\sinh\eta}{\sinh(\lambda_a-\xi_k)\sinh(\lambda_a-\xi_k+\eta)} + O(M^{-\infty}),$$

where $\tilde\rho(\lambda)$ is defined differently for the two regimes:

$$\tilde\rho(\lambda) = \rho(\lambda), \quad -1<\Delta\le 1,$$
$$\tilde\rho(\lambda) = i\rho(i\lambda), \quad \Delta>1.$$

Thus for the corresponding matrix elements we obtain

$$(\Phi'^{-1}\Psi')_{ab} = \frac{\tilde\rho(\lambda_a - \xi_{b+m-N}+\frac{\eta}{2})}{M\tilde\rho_{\mathrm{tot}}(\lambda_a)} + O(M^{-\infty}), \quad b>N-m.$$

Finally, the fraction of the two determinants in (6.7) can be written in a very simple form in the thermodynamic limit:

$$(6.8)\quad \frac{\det\Psi'(\{\lambda\},\{\xi\})}{\det\Phi'(\{\lambda\})} = \prod_{a=N-m+1}^{N}\frac{1}{M\tilde\rho_{\mathrm{tot}}(\lambda_a)}\det S(\{\lambda_{N-m+1},\dots,\lambda_N\},\{\xi\}) + O(M^{-\infty}),$$

in terms of the $m\times m$ matrix S,

$$(6.9)\quad S_{ab} = \tilde\rho\left(\lambda_{N-m+a} - \xi_b + \frac{\eta}{2}\right).$$

Now, using (6.3), (6.4), (6.5), (6.7) and (6.8), we obtain the following representation for the emptiness formation probability:

$$(6.10)\quad \tau(m) = \frac{1}{M^m\prod_{k<l}\sinh(\xi_k-\xi_l)} \times \sum_{a_m=1}^{N}\sum_{a_{m-1}=1}^{N}\cdots\sum_{a_1=1}^{N} H(\{\lambda_{a_1},\dots,\lambda_{a_m}\},\{\xi\})\prod_{j=1}^{m}\tilde\rho_{\mathrm{tot}}^{-1}(\lambda_{a_j}),$$

where the function H is

$$(6.11)\quad H(\{\lambda_{a_1},\dots,\lambda_{a_m}\},\{\xi\}) = \frac{1}{\prod_{k>l}\sinh(\lambda_{a_k}-\lambda_{a_l}+\eta)}\det S(\{\lambda_{a_1},\dots,\lambda_{a_m}\},\{\xi\}) \times \prod_{j=1}^{m}\Big(\prod_{k=1}^{j-1}\sinh(\lambda_{a_j}-\xi_k+\eta)\prod_{k=j+1}^{m}\sinh(\lambda_{a_j}-\xi_k)\Big)+O(M^{-\infty}).$$

Here we take the sums over all the values of a_j, as if two indices coincide ($a_j = a_k$, $j \neq k$) the determinant of the matrix S vanishes.

In the thermodynamic limit for the ground state the sums can be replaced by the integrals (5.10) and the emptiness formation probability can be expressed as a multiple integral,

$$(6.12)\qquad \tau(m) = \frac{1}{\prod_{k<l}\sinh(\xi_k-\xi_l)}\int_{-\tilde{\Lambda}}^{\tilde{\Lambda}} d\lambda_1\cdots\int_{-\tilde{\Lambda}}^{\tilde{\Lambda}} d\lambda_m H(\{\lambda_1,\dots,\lambda_m\},\{\xi\}),$$

where $\tilde{\Lambda} = \Lambda$ for $-1 < \Delta \leq 1$ and $\tilde{\Lambda} = -i\Lambda$ for $\Delta > 1$.

Thus we obtain explicit results for both regimes of the XXZ model. In both cases the determinant of the matrix S can be calculated explicitly. In the massless case it is the Cauchy determinant:

$$(6.13)\qquad \det S = \left(\frac{i}{2\zeta}\right)^m \frac{\prod_{k<l}\sinh\pi/\zeta(\xi_k-\xi_l)\prod_{a>b}\sinh\pi/\zeta(\lambda_a-\lambda_b)}{\prod_{a=1}^{m}\prod_{k=1}^{m}\sinh\pi/\zeta(\lambda_a-\xi_k)}.$$

The emptiness formation probability has in this case the following form:

$$(6.14)\quad \tau(m) = \prod_{k<l}\frac{\sinh\pi/\zeta(\xi_k-\xi_l)}{\sinh(\xi_k-\xi_l)} \times \int_{-\infty}^{\infty} i\frac{d\lambda_1}{2\zeta}\cdots\int_{-\infty}^{\infty} i\frac{d\lambda_m}{2\zeta}\prod_{a>b}\frac{\sinh\pi/\zeta(\lambda_a-\lambda_b)}{\sinh(\lambda_a-\lambda_b-i\zeta)} \times \prod_{a=1}^{m}\prod_{k=1}^{m}\frac{1}{\sinh\pi/\zeta(\lambda_a-\xi_k)}\prod_{j=1}^{m}\Big(\prod_{k=1}^{j-1}\sinh(\lambda_j-\xi_k-i\zeta)\prod_{k=j+1}^{m}\sinh(\lambda_j-\xi_k)\Big).$$

In the homogeneous limit ($\xi_j = -i\zeta/2$, $\forall j$) we obtain the following result for the emptiness formation probability:

$$(6.15)\quad \tau(m) = (-1)^m\Big(-\frac{\pi}{\zeta}\Big)^{m(m+1)/2} \times \int_{-\infty}^{\infty}\frac{d\lambda_1}{2\pi}\cdots\int_{-\infty}^{\infty}\frac{d\lambda_m}{2\pi}\prod_{a>b}\frac{\sinh\pi/\zeta(\lambda_a-\lambda_b)}{\sinh(\lambda_a-\lambda_b-i\zeta)} \times \prod_{j=1}^{m}\frac{\sinh^{j-1}(\lambda_j-i\zeta/2)\sinh^{m-j}(\lambda_j+i\zeta/2)}{\cosh^m(\pi/\zeta)\lambda_j}.$$

In the massive case the determinant of the matrix S is more complicated but can be expressed in terms of the Theta functions:

$$(6.16)\quad \det S = g_m\Big(-\frac{1}{2\pi}\Big)^m\frac{\prod_{j<k}\theta_1(i\lambda_j-i\lambda_k)\theta_1(i\xi_k-i\xi_j)}{\prod_{j,k=1}^{m}\theta_1(i\lambda_j-i\xi_k)}\theta_2\Big(\sum_{j=1}^{m}(i\lambda_j-i\xi_j)\Big),$$

where

$$g_m = \prod_{n=1}^{\infty}\left(\frac{1-q^{2n}}{1+q^{2n}}\right)^2\left[2q^{1/4}\prod_{n=1}^{\infty}(1-q^{2n})^3\right]^{m-1}.$$

As usual in this regime, we change the variables to more convenient ones: $\beta = i\xi$, $\zeta = -\eta$. For the emptiness formation probability we obtain:

$$\begin{aligned}(6.17)\quad \tau(m) = g_m \prod_{k<l}\frac{\theta_1(\beta_k-\beta_l)}{\sin(\beta_k-\beta_l)}\int_{-\pi/2}^{\pi/2} i\,\frac{d\lambda_1}{2\pi}\cdots\int_{-\pi/2}^{\pi/2} i\,\frac{d\lambda_m}{2\pi}\theta_2\\ \times\left(\sum_{j=1}^{m}(\lambda_j-\beta_j)\right)\prod_{a>b}\frac{\theta_1(\lambda_a-\lambda_b)}{\sin(\lambda_a-\lambda_b-i\zeta)}\\ \times\prod_{a=1}^{m}\left(\prod_{k=1}^{m}\frac{1}{\theta_1(\lambda_a-\beta_k)}\prod_{k=1}^{a-1}\sin(\lambda_a-\beta_k-i\zeta)\prod_{k=a+1}^{m}\sin(\lambda_a-\beta_k)\right).\end{aligned}$$

In the homogenous limit $\beta_j = -i\zeta/2$ we have the following result:

$$\begin{aligned}(6.18)\quad \tau(m) = \prod_{n=1}^{\infty}\left(\frac{1-q^{2n}}{1+q^{2n}}\right)^2\left[2q^{1/4}\prod_{n=1}^{\infty}(1-q^{2n})^3\right]^{m(m+1)/2-1}\\ \times\int_{-\pi/2}^{\pi/2} i\,\frac{d\lambda_1}{2\pi}\cdots\int_{-\pi/2}^{\pi/2} i\,\frac{d\lambda_m}{2\pi}\theta_2\left(\sum_{j=1}^{m}\left(\lambda_j+i\frac{\zeta}{2}\right)\right)\\ \times\prod_{a>b}\frac{\theta_1(\lambda_a-\lambda_b)}{\sin(\lambda_a-\lambda_b-i\zeta)}\prod_{j=1}^{m}\frac{\sin^{j-1}(\lambda_j-i\zeta/)\sin^{m-j}(\lambda_j+i\zeta/2)}{\theta_1^m(\lambda_j+i\zeta/2)}.\end{aligned}$$

We have shown here that the simplest m-point correlation function for the XXZ model can be expressed using the algebraic Bethe ansatz as multiple integrals of the elementary or elliptic functions. These results reproduce for this particular case the formulae obtained by M. Jimbo, T. Miwa, and their collaborators [**21–23**].

6.2. General Correlation Functions. In this section we consider a more general case of correlation functions : the ground state mean value of any product of the local elementary 2×2 matrices $E_{lk}^{\epsilon',\epsilon} = \delta_{l,\epsilon'}\delta_{k,\epsilon}$:

$$(6.19)\qquad F_m(\{\epsilon_j,\epsilon'_j\}) = \frac{\langle\psi_g|\prod_{j=1}^m E_j^{\epsilon'_j,\epsilon_j}|\psi_g\rangle}{\langle\psi_g\mid\psi_g\rangle}.$$

It should be mentioned that an arbitrary n-point correlation function can be obtained as a sum of such mean values.

To calculate this product we use the solution of the quantum inverse scattering problem, representing the local elementary matrices in terms of the corresponding monodromy matrix elements:

$$E_j^{\epsilon'_j,\epsilon_j} = \prod_{k=1}^{j-1}(A(\xi_k)+D(\xi_k))T_{\epsilon_j,\epsilon'_j}(\xi_j)\prod_{k=j+1}^{M}(A(\xi_k)+D(\xi_k)).$$

Thus we reduce the problem to the computation of the ground state mean value of an arbitrary ordered product of the monodromy matrix elements,

$$(6.20)\qquad F_m(\{\epsilon_j,\epsilon'_j\}) = \phi_m(\{\lambda\})\frac{\langle\psi_g|T_{\epsilon_1,\epsilon'_1}(\xi_1)\cdots T_{\epsilon_m,\epsilon'_m}(\xi_m)|\psi_g\rangle}{\langle\psi_g\mid\psi_g\rangle},$$

where $\phi_m(\{\lambda\})$ is the ground state eigenvalue of the corresponding product of the transfer matrices:

$$\phi_m(\{\lambda\}) = \prod_{j=1}^{m}\prod_{a=1}^{N} \frac{\sinh(\lambda_a - \xi_j)}{\sinh(\lambda_a - \xi_j + \eta)}.$$

Now to calculate these mean values we use the commutation relations of the monodromy matrix elements.

An arbitrary product of the monodromy matrix elements can be treated in a rather general way. One should consider the following two sets of indices:

$$\begin{aligned}
&\boldsymbol{\alpha}^{+} = \{j : 1 \le j \le m, \epsilon_j = 1\}, \quad \mathrm{card}(\boldsymbol{\alpha}^{+}) = s',\\
&\max_{j\in\boldsymbol{\alpha}^{+}}(j) \equiv j'_{\max}, \quad \min_{j\in\boldsymbol{\alpha}^{+}}(j) \equiv j'_{\min},\\
&\boldsymbol{\alpha}^{-} = \{j : 1 \le j \le m, \epsilon'_j = 2\}, \quad \mathrm{card}(\boldsymbol{\alpha}^{-}) = s,\\
&\max_{j\in\boldsymbol{\alpha}^{-}}(j) \equiv j_{\max}, \quad \min_{j\in\boldsymbol{\alpha}^{-}}(j) \equiv j_{\min}.
\end{aligned} \tag{6.21}$$

It should be mentioned that in a general case the intersection of these two sets is not empty and corresponds to the operators $B(\xi_j)$.

Consider now the action of an arbitrary product on a state constructed by the action of the operators $C(\lambda)$,

$$\langle 0| \prod_{k=1}^{N} C(\lambda_k) T_{\epsilon_1,\epsilon'_1}(\lambda_{N+1}) \dots T_{\epsilon_m,\epsilon'_m}(\lambda_{N+m}),$$

applying one by one the formulae (5.1)–(5.3). For all the indices j from the sets $\boldsymbol{\alpha}^{+}$ and $\boldsymbol{\alpha}^{-}$ one obtains a summation on the corresponding indices a'_j (for $j \in \boldsymbol{\alpha}^{+}$, corresponding to the action of the operators $A(\lambda)$ or $B(\lambda)$) or a_j (for $j \in \boldsymbol{\alpha}^{-}$, corresponding to the action of the operators $D(\lambda)$ or $B(\lambda)$). As the product of the monodromy matrix elements is ordered, these summations are also ordered and the corresponding indices should be taken from the following sets:

$$\begin{aligned}
\mathbf{A}_j &= \{b : 1 \le b \le N+m,\ b \ne a_k, a'_k,\ k < j\},\\
\mathbf{A}'_j &= \{b : 1 \le b \le N+m,\ b \ne a'_k,\ k < j,\ b \ne a_k,\ k \le j\}.
\end{aligned}$$

Thus the action of a product of the monodromy matrix elements can be written as the following sum:

$$\begin{aligned}
\langle 0|\prod_{k=1}^{N} C(\lambda_k)\, T_{\epsilon_1,\epsilon'_1}(\lambda_{N+1}) \dots T_{\epsilon_m,\epsilon'_m}(\lambda_{N+m}) &=\\
\sum_{\{a_j,a'_j\}} G_{\{a_j,a'_j\}}(\lambda_1,\dots,\lambda_{N+m})\langle 0| &\prod_{b\in\mathbf{A}_{m+1}} C(\lambda_b).
\end{aligned} \tag{6.22}$$

The summation is taken over the indices a_j for $j \in \boldsymbol{\alpha}^{-}$ and a'_j for $j \in \boldsymbol{\alpha}^{+}$ such that:

$$\begin{aligned}
1 \le a_j \le N+j, \quad a_j \in \mathbf{A}_j,\\
1 \le a'_j \le N+j, \quad a'_j \in \mathbf{A}'_j.
\end{aligned}$$

The functions $G_{\{a_j,a'_j\}}(\lambda_1,\dots\lambda_{N+m})$ can be easily obtained from the formulae (5.1)–(5.3) taking into acount that $\lambda_a = \xi_{N-a}$ for $a > N$:

$$(6.23) \qquad G_{\{a_j,a'_j\}}(\lambda_1,\dots,\lambda_{N+m}) = \prod_{j\in\boldsymbol{\alpha}^-} d(\lambda_{a_j}) \frac{\prod_{\substack{b=1\\ b\in\mathbf{A}_j}}^{N+j-1} \sinh(\lambda_{a_j}-\lambda_b+\eta)}{\prod_{\substack{b=1\\ b\in\mathbf{A}'_j}}^{N+j} \sinh(\lambda_{a_j}-\lambda_b)}$$
$$\times \prod_{j\in\boldsymbol{\alpha}^+} a(\lambda_{a'_j}) \frac{\prod_{\substack{b=1\\ b\in\mathbf{A}'_j}}^{N+j-1} \sinh(\lambda_b-\lambda_{a'_j}+\eta)}{\prod_{\substack{b=1\\ b\in\mathbf{A}_{j+1}}}^{N+j} \sinh(\lambda_b-\lambda_{a'_j})}.$$

Now to calculate the normalized mean value (6.20) we apply the representation for the scalar product (5.6) and the Gaudin formula (5.8). It should be mentioned that the number of operators $C(\lambda)$ has to be equal to the number of the operators $B(\lambda)$, as otherwise the mean value is zero, and hence the total number of elements in the sets $\boldsymbol{\alpha}^+$ and $\boldsymbol{\alpha}^-$ is $s+s'=m$. Taking into account that in (6.20), for $b>N$, $\lambda_b = \xi_{b-N}$, one can consider the scalar products appearing in the representation for the ground state mean values,

$$\frac{\langle 0|\prod_{b\in\mathbf{A}_{m+1}} C(\lambda_b)\prod_{k=1}^N B(\lambda_k)|0\rangle}{\langle 0|\prod_{k=1}^N C(\lambda_k)\prod_{k=1}^N B(\lambda_k)|0\rangle},$$

for all the permitted values of a_j, a'_j, using the same method as for the emptiness formation probability. Finally we obtain:

$$(6.24) \qquad F_m(\{\epsilon_j,\epsilon'_j\}) = \frac{1}{\prod_{k<l}\sinh(\xi_k-\xi_l)} \sum_{\{a_j,a'_j\}} H_{\{a_j,a'_j\}}(\lambda_1,\dots,\lambda_{N+m}),$$

the sum being taken over the same set of indices a_j, a'_j as in (6.22). The functions $H_{\{a_j,a'_j\}}(\{\lambda\})$ can be obtained using (6.23) and the representations for the scalar products. It is convenient to introduce the following set of indices:

$$\{b_1,\dots,b_m\} = \{a'_{j'_{\max}},\dots,a'_{j'_{\min}},a_{j_{\min}},\dots,a_{j_{\max}}\}.$$

One should also take into account that for the XXZ model $a(\lambda)=1$ and $d(\xi_k)=0$. Then one obtains that $a_j \le N$, for all $j\in\boldsymbol{\alpha}^-$ (otherwise the corresponding term is zero):

$$(6.25) \qquad H_{\{a_j,a'_j\}}(\{\lambda\}) = \frac{(-1)^{s'}}{\prod_{k>l}\sinh(\lambda_{b_k}-\lambda_{b_l}+\eta)}$$
$$\times \prod_{j\in\boldsymbol{\alpha}^+}\left(\prod_{k=1}^{j-1}\sinh(\lambda_{a'_j}-\xi_k-\eta)\prod_{k=j+1}^{m}\sinh(\lambda_{a'_j}-\xi_k)\right)$$
$$\times \prod_{j\in\boldsymbol{\alpha}^-}\left(\prod_{k=1}^{j-1}\sinh(\lambda_{a_j}-\xi_k+\eta)\prod_{k=j+1}^{m}\sinh(\lambda_{a_j}-\xi_k)\right)$$
$$\times \left(\det M(\{b_k\}) + O(M^{-\infty})\right),$$

where the $m\times m$ matrix $M(\{b_k\})$ is slightly different in comparison with the case of the emptiness formation probability,

$$b_l > N, \quad M_{lk} = -\delta_{b_l-N,k},$$

$$b_l \le N, \quad M_{lk} = \frac{\tilde{\rho}(\lambda_{b_l} - \xi_k + \eta/2)}{\tilde{\rho}_{\mathrm{tot}}(\lambda_{b_l})}.$$

The sum in (6.24) can be rewritten in a simpler way if one takes into account that the function $H_{\{a_j,a'_j\}}(\{\lambda\})$ defined by (6.25) is equal to zero if $b_j = b_k$, $j \ne k$ (as the determinant vanishes in this case), or if $a'_j > N + j$:

$$(6.26) \quad F_m(\{\epsilon_j, \epsilon'_j\}) = \frac{1}{\prod_{k<l} \sinh(\xi_k - \xi_l)} \sum_{b_1=1}^{N+m} \cdots \sum_{b_{s'}=1}^{N+m} \sum_{b_{s'+1}=1}^{N} \cdots \sum_{b_m=1}^{N} H_{\{a_j,a'_j\}}(\{\lambda\}).$$

The sum over $1 \le b_j \le N$ is just a sum over the rapidities in the ground state and can be replaced by integrals as in the case of the emptiness formation probability:

$$\sum_{a=1}^{N} f(\lambda_a) = \int_{-\tilde{\Lambda}}^{\tilde{\Lambda}} d\lambda\, \tilde{\rho}_{\mathrm{tot}}(\lambda) f(\lambda) + O(M^{-\infty}).$$

The contributions of the terms with $a'_j > N$ can be rewritten as integrals over the contours $\Gamma_{a'_j-N}$ surrounding the pole of the corresponding density function $\tilde{\rho}(\lambda - \xi_{a'_j-N} + \eta/2)$ at the point $\lambda = \xi_{a'_j-N}$. The residues of the density function $\rho(\lambda - \xi + \eta/2)$ (for both regimes) at these points are:

$$2\pi i \,\mathrm{Res}\left(\rho\left(\lambda - \xi + \frac{\eta}{2} \right) \right)\Bigg|_{\lambda=\xi} = -1.$$

The other points ξ_l should be outside the contour $\Gamma_{a'_j-N}$. The matrix $M(\{b_k\})$ then should be replaced by the same matrix $S(\{\lambda\},\{\xi\})$ as in the representation for the emptiness formation probability. Finally, for the correlation function one obtains:

$$(6.27) \quad F_m(\{\epsilon_j, \epsilon'_j\}) = \frac{1}{\prod_{k<l} \sinh(\xi_k - \xi_l)} \left(\int_{-\tilde{\Lambda}}^{\tilde{\Lambda}} + \sum_{j=1}^{m} \oint_{\Gamma_j} \right) d\lambda_1 \cdots \left(\int_{-\tilde{\Lambda}}^{\tilde{\Lambda}} + \sum_{j=1}^{m} \oint_{\Gamma_j} \right) d\lambda_{s'}$$
$$\times \int_{-\tilde{\Lambda}}^{\tilde{\Lambda}} d\lambda_{s'+1} \cdots \int_{-\tilde{\Lambda}}^{\tilde{\Lambda}} d\lambda_m \widetilde{H}_{\{\epsilon_j,\epsilon'_j\}}(\lambda_1, \ldots, \lambda_m) + O(M^{-\infty}),$$

where the function $\widetilde{H}_{\{\epsilon_j,\epsilon'_j\}}(\lambda_1, \ldots, \lambda_m)$ is defined as

$$(6.28) \quad \widetilde{H}_{\{\epsilon_j,\epsilon'_j\}}(\{\lambda\}) = \frac{(-1)^{s'}}{\prod_{k>l} \sinh(\lambda_k - \lambda_l + \eta)}$$
$$\times \prod_{j\in\boldsymbol{\alpha}^-} \left(\prod_{k=1}^{j-1} \sinh(\mu_j - \xi_k + \eta) \prod_{k=j+1}^{m} \sinh(\mu_j - \xi_k) \right)$$
$$\times \prod_{j\in\boldsymbol{\alpha}^+} \left(\prod_{k=1}^{j-1} \sinh(\mu'_j - \xi_k - \eta) \prod_{k=j+1}^{m} \sinh(\mu'_j - \xi_k) \right) \det S(\{\lambda\},\{\xi\}),$$

where

$$S_{lk} = \tilde{\rho}\left(\lambda_l - \xi_k + \frac{\eta}{2} \right),$$

and the parameters of integration are ordered in the following way:

$$\{\lambda_1, \dots, \lambda_m\} = \{\mu'_{j'_{\max}}, \dots, \mu'_{j'_{\min}}, \mu_{j_{\min}}, \dots, \mu_{j_{\max}}\}.$$

Finally we obtain the expression of these correlation functions in the homogeneous limit where all ξ are equal. It is given as an m-fold integral over contours C_j^h which depend on the value of j, on the regime considered, and also on the value of the magnetic field. The general answer can be written as [**29**],

$$F_m(h, \{\epsilon_j, \epsilon'_j\}) = \prod_{j=1}^{m} \int_{C_j^h} d\lambda_j \, \Omega_m(\{\lambda_j\}, \{\epsilon_j, \epsilon'_j\}) \det S_h(\{\lambda\}),$$

where $\Omega_m(\{\lambda_j\}, \{\epsilon_j, \epsilon'_j\})$ is a purely algebraic quantity, in particular not depending on the regime or on the magnetic field, while $S_h(\{\lambda\})$ is a matrix depending on the density function $\rho_h(\lambda)$ describing the ground state, which is a solution of the Lieb equation, and hence depends both on the regime and on the value of the magnetic field h.

7. Conclusions

We have described a general approach to compute correlation functions of lattice quantum integrable models, with the XXZ Heisenberg spin-$\frac{1}{2}$ chain in a magnetic field as a typical example. The main ingredients are the solution of the quantum inverse scattering problem [**28, 34**] and the scalar product formula [**28, 37**].

Our method should now be extended to temperature and time dependent correlation functions. Another very important problem concerns the long distance asymptotics and the scaling limit (the limit in which the lattice spacing goes to zero, while the number of sites N goes to infinity, their product being kept fixed). This would open the possibility to consider also the quantum integrable field theory models. Another route towards this goal would be to solve the quantum inverse problem directly for the lattice versions of quantum integrable field theory models, along the lines described in [**34**] for higher spin Heisenberg chains. In this context, we will have to deal with quantum Lax operators having infinite dimensional representation auxiliary space. I expect the resulting reconstruction expressions for the local quantum fields, at least for the simplest cases, to be given by formulas very similar to the infinite spin limit of the result for spin-s Heisenberg chains [**34**].

Acknowledgements

I would like to thank here my collaborators on this difficult subject, J. Sanchez de Santos, N. Kitanine, V. Terras, and of course A. Izergin.

References

1. M. J. Ablowitz, D. J. Kaup, A. C. Newell, and H. Segur, *Method for solving the sine-Gordon equation*, Phys. Rev. Lett. **30** (1973), 1262–1264.
2. R. J. Baxter, *Partition function of the eight-vertex lattice model*, Ann. Physics **70** (1972), 193–228.
3. R. J. Baxter, *Corner transfer matrices of the eight-vertex model.* I. *Low temperature expansions and conjectured properties*, J. Statist. Phys. **15** (1976), 485–503.
4. R. J. Baxter, *Corner transfer matrices of the eight-vertex model.* II. *The Ising model case*, J. Statist. Phys. **17** (1977), 1–14.
5. R. J. Baxter, *Exactly solved models in statistical mechanics*, Academic Press, London-New York, 1982.

6. A. A. Belavin, A. M. Polyakov, and A. B. Zamolodchikov, *Infinite conformal symmetry in two-dimensional quantum field theory*, Nuclear Phys. B **241** (1984), 333–380.
7. H. Bethe, *Zur Theorie der Metalle. I. Eigenwerte und Eigenfunktionen der linearen Atomkette*, Z. Phys. **71** (1931), 205–226.
8. D. B. Creamer, H. B. Thacker, and D. Wilkinson, *Gel'fand-Levitan method for operator fields*, Phys. Rev. D **21** (1980), 1523–1528.
9. D. B. Creamer, H. B. Thacker, and D. Wilkinson, *Quantum Gel'fand Levitan method as a generalized Jordan-Wigner transformation*, Phys. Lett. B **92** (1980), 144–148.
10. D. B. Creamer, H. B. Thacker, and D. Wilkinson, *Some exact results for the two-point function of an integrable quantum field theory*, Phys. Rev. D **23** (1981), 3081–3084.
11. E. Date, M. Jimbo, and M. Okado, *Crystal base and q-vertex operators*, Comm. Math. Phys. **155** (1993), 47–69.
12. L. D. Faddeev, *Integrable models in* (1+1)*-dimensional quantum field theory*, Recent Advances in Field Theory and Statistical Mechanics (Les Houches 1982) (J. B. Zuber and R. Stora, eds.), Elsevier, 1984, pp. 561–608.
13. L. D. Faddeev, E. K. Sklyanin, and L. A. Takhtajan, *Quantum inverse problem method.* I., Teoret. Mat. Fiz. **40** (1979), 194–220; English transl. Theoret. and Math. Phys. **40** (1980), 688.
14. I. B. Frenkel and N. Y. Reshetikhin, *Quantum affine algebras and holonomic difference equations*, Comm. Math. Phys. **146** (1992), 1–60.
15. M. Gaudin, *La fonction d'onde de Bethe*, Masson, Paris, 1983.
16. I. M. Gel'fand and B. M. Levitan, *On the determination of a differential equation from its spectral function*, Amer. Math. Soc. Transl. Ser. 2 **1** (1955), 253–304.
17. W. Heisenberg, *Zur Theorie des Ferromagnetismus*, Z. Phys. **49** (1928), 619–636.
18. A. R. Its and N. A. Slavnov, *On the Riemann-Hilbert approach to the asymptotic analysis of the correlation functions of the quantum nonlinear Schrödinger equation. Non-free fermionic case*, Preprint MI-98-76, math-ph/9811009, 1998.
19. A. G. Izergin, N. Kitanine, J. M. Maillet, and V. Terras, *Spontaneous magnetisation of the XXZ Heisenberg spin-$\frac{1}{2}$ chain*, Nuclear Phys. B **554** (1999), 679–696.
20. M. Jimbo, T. Kojima, T. Miwa, and Y.-H. Quano, *Smirnov's integrals and the quantum Knizhnik-Zamolodchikov equation of level* 0, J. Phys. A **27** (1994), 3267–3283.
21. M. Jimbo, K. Miki, T. Miwa, and A. Nakayashiki, *Correlation functions of the XXZ model for* $\Delta < -1$, Phys. Lett. A **168** (1992), 256–263.
22. M. Jimbo and T. Miwa, *Algebraic analysis of solvable lattice models*, Amer. Math. Soc., Providence, RI, 1995.
23. M. Jimbo and T. Miwa, *Quantum KZ equation with* $|q| = 1$ *and correlation functions of the XXZ model in the gapless regime*, J. Phys. A **29** (1996), 2923–2958.
24. M. Jimbo, T. Miwa, Y. Môri, and M. Sato, *Density matrix of an impenetrable Bose gas and the fifth Painlevé transcendent*, Phys. D **1** (1980), 80–158.
25. M. Jimbo, T. Miwa, and A. Nakayashiki, *Difference equations for the correlation functions of the eight-vertex model*, J. Phys. A **26** (1993), 2199–2209.
26. M. Karowski and H. J. Thun, *Complete S-matrix of the* $O(2N)$ *Gross-Neveu model*, Nuclear Phys. B **190** (1981), 61–92.
27. M. Karowski and P. Weisz, Nuclear Phys. B **139** (1978), 445.
28. N. Kitanine, J.-M. Maillet, and V. Terras, *Form factors of the XXZ Heisenberg spin-$\frac{1}{2}$ finite chain*, Nuclear Phys. B **554** (1999), 647–678.
29. N. Kitanine, J.-M. Maillet, and V. Terras, *Correlation functions of the XXZ Heisenberg spin-$\frac{1}{2}$ chain in a magnetic field*, Nuclear Phys. B **567** (2000), 554–582.
30. V. G. Knizhnik and A. B. Zamolodchikov, *Current algebra and Wess-Zumino model in two dimensions*, Nuclear Phys. B **247** (1984), 83–103.
31. V. E. Korepin, N. M. Bogoliubov, and A. G. Izergin, *Quantum inverse scattering method and correlation functions*, Cambridge University Press, 1993.
32. E. H. Lieb and W. Liniger, *Exact analysis of an interacting Bose gas.* I. *The general solution and the ground state*, Phys. Rev. **130** (1963), 1605–1616.
33. J.-M. Maillet and J. Sanchez de Santos, *Drinfel'd twists and algebraic Bethe Ansatz*, q-alg/9612012, 1996; L. D. Fadeev's Seminar on Mathematical Physics (M. Semenov-Tian-Shansky, ed.), Amer. Math. Soc. Transl. Ser. 2, vol. 201, Amer. Math. Soc., Providence, RI, (to appear).

34. J.-M. Maillet and V. Terras, *On the quantum inverse scattering problem*, Nuclear Phys. B (2000), (to appear); Preprint LPENSL-TH-19/99, hep-th/9911030.
35. V. A. Marchenko, *On reconstruction of the potential energy from phases of the scattered waves*, Dokl. Akad. Nauk SSSR (N.S.) **104** (1955), 695–698 (Russian).
36. E. K. Sklyanin, *Functional Bethe Ansatz*, Integrable and Superintegrable Systems (B. Kupershmidt, ed.), World Scientific, Singapore, 1990, pp. 8–33.
37. N. A. Slavnov, *Calculation of scalar products of wave functions and form-factors in the framework of the algebraic Bethe ansatz*, Theoret. and Math. Phys. **79** (1989), 502–508.
38. F. A. Smirnov, *Form factors in completely integrable models of quantum field theory*, World Scientific, Singapore, 1992.
39. F. A. Smirnov, *Remarks on deformed and undeformed Knizhnik-Zamolodchikov equations*, RIMS preprint 860, 1992.
40. F. A. Smirnov, *Form factors, deformed Knizhnik-Zamolodchikov equations and the finite-gap integration*, Comm. Math. Phys. **155** (1993), 459–487.
41. F. A. Smirnov, *What are we quantizing in integrable field theory?*, RIMS preprint 935, 1993.
42. L. A. Takhtajan and L. D. Faddeev, *The quantum method of the inverse problem and the Heisenberg XXZ model*, Russian Math. Surveys **34** (1979), no. 5, 11–68.
43. V. Terras, *Drinfel'd twists and functional Bethe Ansatz*, Lett. Math. Phys. **48** (1999), 263–276.
44. H. B. Thacker, *Corner transfer matrices and Lorentz invariance on a lattice*, Phys. D **18** (1986), 348–359.
45. A. Varchenko, *Quantized Knizhnik-Zamolodchikov equations, quantum Yang-Baxter equation, and difference equations for q-hypergeometric functions*, Comm. Math. Phys. **162** (1994), 499–528.
46. T. T. Wu, B. M. McCoy, C. A. Tracy, and E. Barouch, *Spin-spin correlation functions for the two-dimensional Ising model: Exact theory in the scaling region*, Phys. Rev. B **13** (1976), 316–374.
47. C. N. Yang, *The spontaneous magnetization of a two-dimensional Ising model*, Phys. Rev. **85** (1952), 808–815.
48. C. N. Yang, *Some exact results for the many-body problem in one dimension with repulsive delta-function interaction*, Phys. Rev. Lett. **19** (1967), 1312–1315.
49. C. N. Yang and C. P. Yang, *One-dimensional chain of anisotropic spin-spin interactions.* II. *Properties of the ground state energy per lattice site for an infinite system*, Phys. Rev. **150** (1966), 327–339.
50. V. E. Zakharov and L. D. Faddeev, *Korteweg-de Vries equation: a completely integrable Hamiltonian system*, Funct. Anal. Appl. **5** (1971), 280–287.
51. A. B. Zamolodchikov and A. B. Zamolodchikov, *Factorized S-matrices in two dimensions as the exact solutions of certain relativistic quantum field theories*, Ann. Physics **120** (1979), 253–291.

Laboratoire de Physique[1] Groupe de Physique Théorique, ENS Lyon, 46 allée d'Italie, 69364 Lyon Cedex 07, France
E-mail address: jean-michel.maillet@ens-lyon.fr

[1]UMR 5672 du CNRS, associée à l'École Normale Supérieure de Lyon.

Centre de Recherches Mathématiques
CRM Proceedings and Lecture Notes
Volume **26**, 2000

Multiseparability and Superintegrability for Classical and Quantum Systems

W. Miller, Jr.

ABSTRACT. It has long been known that there are potentials on n-dimensional constant curvature spaces for which a given Hamiltonian system in classical mechanics, and Schrödinger equation in quantum mechanics, admits solutions via separation of variables in more than one coordinate system. Smorodinsky, Winternitz et al. initiated the methodical search for such potentials in two and three dimensions, and there has been a considerable amount of work for various examples. Such a system is called *maximal* in dimension n if there exist $2n-1$ functionally independent integrals of motion. In some papers, these systems are called *superintegrable*.

In the first part of this paper we outline the basic ideas relating to the notion of superintegrable potentials. The energy observable is degenerate for potentials of this type and the corresponding integrals of motion that arise from the simultaneous separability close quadratically under repeated commutation. We give examples of these systems and indicate how superintegrability can be of use, particularly in relation to bound states. Virtually all of the special functions of mathematical physics (in one and several variables) arise in this study and formulas expanding one type of special function as a series in another type emerge as a byproduct.

Finally, we describe how one can, in principle, classify all such systems and deduce the structure of the quadratic algebra. Many of the results reported here were obtained in collaboration with E. G. Kalnins and G. S. Pogosyan.

1. Introduction

It has long been known that Schrödinger's equation with certain special potentials can admit (multiplicative) separation of variables in more than one coordinate system. This is intimately related to the notion of superintegrability, [**11, 10, 37**]. This subject has been studied by a number of authors, based on the use of the corresponding differential equations that that are implied by the requirement of simultaneous separability, [**8, 12, 13, 28, 2, 5, 29, 14, 15, 27, 21, 33**]. Specifically, superintegrability means that for a Schrödinger equation in dimension n there exist $2n-1$ functionally independent quantum mechanical second-order observables (i.e., second-order self-adjoint operators that commute with the Hamiltonian). There is an analogous concept of superintegrability for classical mechanical systems. This

1991 *Mathematics Subject Classification.* 22E70, 35C05, 35Q40, 37J15.
This is the final form of the paper.

relates to the corresponding additive separation of variables of the Hamilton-Jacobi equation. Furthermore, one observes that if we do have simultaneous separability then the resulting constants of the motion close quadratically under repeated application of the Poisson bracket, [**14**]. We also know that for spaces of constant curvature separable coordinate systems of the free motion are described by quadratic elements of the corresponding first order symmetries, [**31, 18, 16**].

Although concrete examples of superintegrable systems are easily at hand, a complete classification of all such systems has presented major difficulties. How can one be sure that all systems for free motion have been found? (For example, Rañada's classification [**33**] omits our system 5 below.) Once these are determined, how can one be sure that the most general additive potential term has been calculated?

Here we will present the background information to understand the problem, and its importance, and present a new approach to its solution, with details for two dimensional complex Euclidean space.

Consider an n-dimensional Riemannian manifold R_n. (In most of the following we will assume that R_n is a space of constant curvature, for that is the case where the most interesting and rich applications arise.) In local coordinates $q_1, \dots, q_n$ the contravariant metric tensor is $\big(g^{jk}(\mathbf{q})\big)$. Let $V(\mathbf{q})$ be a potential function on R_n. The corresponding Hamilton-Jacobi equation is

$$\mathcal{H}(\mathbf{q}, \mathbf{p}) = E \tag{1.1}$$

where

$$\mathcal{H}(\mathbf{q}, \mathbf{p}) = \sum_{j,k=1}^{n} g^{jk}(\mathbf{q}) p_j p_k + V(\mathbf{q}) = \sum_{j,k=1}^{n} g^{jk}(\mathbf{q}) \frac{\partial S}{\partial q_j} \frac{\partial S}{\partial q_k} + V(\mathbf{q}), \tag{1.2}$$

and $S(\mathbf{q})$ is the action function, [**26**].

The quantum analog of this classical system is given by the Schrödinger equation

$$H\Psi(\mathbf{q}) = E\Psi(\mathbf{q}) \tag{1.3}$$

where in local coordinates

$$H = \Delta_n + V(\mathbf{q}), \quad \Delta_n = \frac{1}{\sqrt{g}} \sum_{j,k=1}^{n} \frac{\partial}{\partial q_j} \cdot (\sqrt{g} g^{jk}) \frac{\partial}{\partial q_k} \tag{1.4}$$

and $g = \det(g^{jk})^{-1}$.

Recall that a *complete integral* $S(\mathbf{q}, \lambda_1, \dots, \lambda_n)$ of the Hamilton-Jacobi equation solves the associated classical mechanical system [**26, 1**]. (A complete integral is a solution of (1.1) such that locally

$$\det\left(\frac{\partial^2 S}{\partial q_j \partial \lambda_k}\right) \neq 0.$$

Any solution of the Hamilton-Jacobi equation via (additive) separation of variables

$$S(\mathbf{q}, \lambda_1, \dots, \lambda_n) = \sum_{j=1}^{n} S^{(j)}(q_j, \boldsymbol{\lambda})$$

where $\lambda_1 = E, \lambda_2, \dots, \lambda_n$ are the separation constants, yields a complete integral.

Similarly, in the quantum case, if the Schrödinger equation $H\Psi = E\Psi$ (multiplicatively) separates in the coordinates $\mathbf{q}$ then we can write

$$\Psi(\mathbf{q}) = \prod_{j=1}^{n} \Psi^{(j)}(q_j, \boldsymbol{\lambda}) \tag{1.5}$$

and this ansatz allows the decomposition of (1.3) into n ordinary differential equations, one for each of the factors $\Psi^{(j)}$. Many of the special functions of mathematical physics occur as solutions of these ordinary differential equations.

For orthogonal coordnates $\mathbf{q}$ on an n-dimensional constant curvature space (i.e., such that $g^{jk} = 0$ for $j \neq k$) one can show that the Hamilton-Jacobi equation is additively separable if and only if the Shrödinger equation is multiplicatively separable. (See [**32**] for a discussion of the relationship for general Riemannian manifolds.) We shall see that superintegrability is closely linked to symmetry properties of (1.1) and (1.3), and to separation of variables (special function) solutions of (1.3).

At this point it is useful to summarize briefly the history of the symmetry/special function approach to solving the Schrödinger equation (1.3). (Superintegrability is just one of the latest chapters.) Beginning with the introduction of the Schrödinger equation in the 1930's, and continuing until the 1960's, the main emphasis was on the study of Lie symmetry groups of unitary operators that commuted with the Hamiltonian H, hence mapped solutions of (1.3) to solutions. At the Lie algebra level, one looked for algebras of first-order differential operators L that commuted with H: $[L, H] \equiv LH - HL = 0$. (Again L maps solutions Ψ of (1.3) to solutions.) This led to studies of rotationally invariant potentials and the theory of spherical harmonics [**35**].

A related concept was that of *dynamical symmetry groups* or Lie algebras. The idea was to imbed H as an element of a Lie algebra of first and second-order differential operators. The representation theory of the Lie algebra could then be used to derive information about the eigenvalues and eigenvectors of H. The harmonic oscillator and the Morse potential were treated in this way. The so-called factorization method for solving the Schrödinger equation is related to this approach [**30**]. Among the special functions that arise and whose properties can be studied from this connection are Bessel functions and (more generally) hypergeometric functions. (More recently, q-analogs of the dynamical symmetry algebra approach have led to q-hypergeometric functions [**20**].)

Since the 1960's we have been in the "Cheshire Cat" era, [**6**]. In the most recent theories relating integrabilty, superintegrability and variable separation, the Lie groups and algebras have disappeared, but their *grin* persists, $\smile$. The focus here is on second order constants of the motion (symmetry operators that are built out of products of first order Lie symmetries for the zero potential problem) and their connection with variable separation for both the Hamilton-Jacobi and Schrödinger equations. One of the seminal papers in this regard was by Smorodinsky et al. [**12**]. Some other contributions are, for example, by Shapovalov, Kalnins and Miller, and Winternitz [**36, 34, 17**].These approaches also exploit the maximum symmetry of the physical system, but no longer in terms of Lie algebras of operators.

To examine these ideas, let us start with a classical system

$$\mathcal{H} = \sum g^{jk} p_j p_k + V(\mathbf{q}) \tag{1.6}$$

where the p_j are the momenta conjugate to the coordintes q_j. Recall that the *Poisson bracket* of two functions $f_h(\mathbf{q},\mathbf{p})$, $h=1,2$, is the function

$$\{f_1,f_2\}(\mathbf{q},\mathbf{p}) = \sum_{j=1}^{n}\left(\frac{\partial f_1}{\partial q_j}\frac{\partial f_2}{\partial p_j} - \frac{\partial f_1}{\partial p_j}\frac{\partial f_2}{\partial q_j}\right), \tag{1.7}$$

[**1**]. A *second-order constant of the motion* for (1.6) is a function

$$\mathcal{L} = \sum a^{jk}(\mathbf{q})p_jp_k + W(\mathbf{q}), \quad a^{jk} = a^{kj}, \tag{1.8}$$

such that $\{\mathcal{L},\mathcal{H}\} = 0$.

Note that the null space of the map

$$T: \quad df(\mathbf{q},\mathbf{p}) \to \{f,\mathcal{H}\}(\mathbf{q},\mathbf{p})$$

is $2n-1$ dimensional. Thus (locally) there are $2n-1$ functionally independent constants of the motion (but not necessarily second-order). For the purposes of this paper we adopt the following definitions. We say that the classical system $\mathcal{H} = E$ is *superintegrable* or *maximal* if there are $2n-1$ functionally independent second-order constants of the motion:

$$\begin{aligned} \mathcal{L}_\ell &= \sum a^{jk}(\mathbf{q})p_jp_k + W_\ell(\mathbf{q}) \\ \mathcal{L}_0 &= \mathcal{H}, \quad \ell = 0,1,\ldots,2n-2 \\ \{\mathcal{L}_\ell,\mathcal{H}\} &= 0. \end{aligned} \tag{1.9}$$

We say that the quantum system $H\Psi = E\Psi$ is *superintegrable* or *maximal* if there are $2n-1$ linearly independent second-order symmetry operators:

$$\begin{aligned} L_\ell &= \sum \frac{1}{\sqrt{g}}\partial_{q_j}(\sqrt{g}a^{jk}(\mathbf{q}))\partial_{q_k} + W_\ell(\mathbf{q}) \\ L_0 &= H, \quad \ell = 0,1,\ldots,2n-2 \\ [L_\ell,H] &\equiv L_\ell H - HL_\ell = 0. \end{aligned} \tag{1.10}$$

To clarify the connection between these ideas and variable separation we assume that the coordinates $\mathbf{q}$ are orthogonal, i.e., the covariant metric tensor is diagonal:

$$ds^2 = \sum g_{jk}\, dq_j dq_k = \sum H_j^2(\mathbf{q})\, dq_j^2,$$

so that the Hamilton-Jacobi equation is given by

$$\mathcal{H} = \sum_j H_j^{-2}\left(\frac{\partial S}{\partial q_j}\right)^2 + V(\mathbf{q}) = E. \tag{1.11}$$

Set $\partial S/\partial q_j = S_j = p_j$ and assume additive separation in the $\mathbf{q}$ coordinates, so that $\partial_j S_i = \partial_j\partial_i S = 0$ for $i \neq j$. The separation equations are postulated to be

$$S_i^2 - \sum_{j=1}^{n} u_{ij}(q_i)\lambda^j + f_i(q_i) = 0, \quad i = 1,\ldots,n, \quad \lambda^1 = E. \tag{1.12}$$

Here $\partial_k u_{ij}(q_i) = 0$ for $k \neq i$ and $\det(u_{ij}) \neq 0$. We say that $U = (u_{ij})$ is a *Stäckel matrix.*

Then (1.11) can be recovered from (1.12) provided $H_j^{-2} = (U^{-1})^{1j}$. The quadratic forms

$$\mathcal{L}_\ell = \sum_{j=1}^{n}(U^{-1})^{\ell j}(p_j^2 + f_j(q_j)) = \sum_{j=1}^{n}(U^{-1})^{\ell j}p_j^2 + W_\ell(\mathbf{q})$$

satisfy

$$\mathcal{L}_\ell = -\lambda_\ell, \quad \mathcal{H} = \mathcal{L}_1 = \sum H_j^2 p_j^2 + V(\mathbf{q})$$

for a separable solution. Furthermore, we have

$$\{\mathcal{L}_\ell, \mathcal{L}_j\} = 0, \quad \ell \neq j.$$

Thus the $\mathcal{L}_\ell$, $2 \leq \ell \leq n$, are *constants of the motion* for the *Hamiltonian* $\mathcal{L}_1$.

An analogous construction, replacing (1.12) by n second-order linear ODE's for factors $\Psi^{(i)}(q_i)$ leads to second order linear partial differential operators $L_1 = H, L_2, \ldots, L_n$ such that

$$H\Psi = E\Psi, \quad L_\ell \Psi = \lambda_\ell \Psi, \quad \ell = 2, \ldots, n \tag{1.13}$$

and

$$\Psi(\mathbf{q}) = \prod_{k=1}^{n} \Psi^{(i)}(q_i).$$

Then one can verify that

$$[L_k, H] = 0, \quad [L_k, L_j] = 0.$$

How does one find all orthogonal separable coordinate systems $\mathbf{q}$ for a given space R_n for zero potential, $V \equiv 0$? This is a difficult problem in differential geometry. The answer is known for some constant curvature spaces. In real Euclidean 2-space there are four separable systems: Cartesian, polar, parabolic and elliptic. For complex Euclidean 2-space, including real Euclidean space and real Minkowski space, there are six [**8, 31, 16**]: Cartesian, polar, parabolic, elliptic, hyperbolic and semi-hyperbolic. We describe these coordinate systems and their corresponding free particle constants of the motion L. (We adopt the basis $p_x, p_y, M = xp_y - yp_x$ for the Lie algebra $e(2, C)$ and define $p_\pm = p_x \pm ip_y$, $\smile$.) There is one orbit of constants of the motion, with representative Mp_+, that is not associated with variable separation [**32**]. The separable systems are:

Cartesian coordinates:

$$x, y, \quad L = p_x^2 \tag{1.14}$$

Polar Coordinates:

$$x = r\cos\theta, \quad y = r\sin\theta, \quad L = M^2 \tag{1.15}$$

Parabolic Coordinates:

$$x_P = \frac{1}{2}(\xi^2 - \eta^2), \quad y_P = \xi\eta, \quad L = Mp_y \tag{1.16}$$

Elliptic Coordinates (in algebraic form)**:**

$$x_E^2 = c^2(u-1)(v-1), \quad y_E^2 = -c^2 uv, \quad L = M^2 + c^2 p_x^2 \tag{1.17}$$

Hyperbolic Coordinates:

$$x_H = \frac{r^2 + r^2 s^2 + s^2}{2rs}, \quad y_H = i\frac{r^2 - r^2 s^2 + s^2}{2rs}, \quad L = M^2 + p_+^2 \tag{1.18}$$

Semi-Hyperbolic Coordinates:

$$\begin{gathered} x_{SH} = -\frac{1}{4}(w-u)^2 + \frac{1}{2}(w+u), \quad iy_{SH} = -\frac{1}{4}(w-u)^2 - \frac{1}{2}(w+u), \\ L = 2Mp_+ + p_-^2 \end{gathered} \tag{1.19}$$

TABLE 1. Separable coordinates in 3-D real Euclidean space.

	Coordinate System	Coordinates
I.	Cartesian $x, y, z \in \mathbf{R}$	$x,\ y,\ z$
II.	Cylindrical polar $\rho > 0,\ \varphi \in [0, 2\pi)$	$x = \rho\cos\varphi,\ y = \rho\sin\varphi,\ z$
III.	Cylindrical elliptic $z \in \mathbf{R},\ e_1 < \mu_1 < e_2 < \mu_2$	$x^2 = \frac{(\mu_1-e_1)(\mu_2-e_1)}{(e_2-e_1)}$, $y^2 = \frac{(\mu_1-e_2)(\mu_2-e_2)}{(e_1-e_2)}$, z
IV.	Cylindrical parabolic $\xi, x \in \mathbf{R},\ \eta \geq 0$	$x,\ y = \xi\eta,\ z = \frac{1}{2}(\xi^2 - \eta^2)$
V.	Spherical $r > 0,\ \theta \in [0, \pi],\ \varphi \in [0, 2\pi)$	$x = r\cos\theta\cos\varphi$, $y = r\sin\theta\sin\varphi$, $z = r\cos\theta$
VI.	Prolate spheroidal $e_1 < u_1 < e_2 < u_2$, $\varphi \in [0, 2\pi)$	$x^2 = \frac{(u_1-e_2)(u_2-e_2)}{(e_1-e_2)}\cos^2\varphi$, $y^2 = \frac{(u_1-e_2)(u_2-e_2)}{(e_1-e_2)}\sin^2\varphi$, $z^2 = \frac{(u_1-e_1)(u_2-e_1)}{(e_2-e_1)}$
VII.	Oblate spheroidal $e_1 < u_1 < e_2 < u_2$, $\varphi \in [0, 2\pi)$	$x^2 = \frac{(u_1-e_1)(u_2-e_1)}{(e_2-e_1)}\cos^2\varphi$, $y^2 = \frac{(u_1-e_1)(u_2-e_1)}{(e_2-e_1)}\sin^2\varphi$, $z^2 = \frac{(u_1-e_2)(u_2-e_2)}{(e_1-e_2)}$
VIII.	Sphero-conical $r \geq 0$, $e_1 < \rho_1 < e_2 < \rho_2 < e_3$	$x^2 = r^2\frac{(\rho_1-e_1)(\rho_2-e_1)}{(e_1-e_2)(e_1-e_3)}$, $y^2 = r^2\frac{(\rho_1-e_2)(\rho_2-e_2)}{(e_2-e_1)(e_2-e_3)}$, $z^2 = r^2\frac{(\rho_1-e_3)(\rho_2-e_3)}{(e_3-e_2)(e_3-e_1)}$
IX.	Parabolic $\xi, \eta \geq 0,\ \varphi \in [0, 2\pi)$	$x = \xi\eta\cos\varphi$, $y = \xi\eta\sin\varphi$, $z = \frac{1}{2}(\xi^2 - \eta^2)$
X.	Ellipsoidal $a_1 < u_1 < a_2 < u_2 < a_3 < u_3$	$x^2 = \frac{(u_1-a_1)(u_2-a_1)(u_3-a_1)}{(a_3-a_1)(a_2-a_1)}$, $y^2 = \frac{(u_1-a_2)(u_2-a_2)(u_3-a_2)}{(a_1-a_2)(a_3-a_2)}$, $z^2 = \frac{(u_1-a_3)(u_2-a_3)(u_3-a_3)}{(a_1-a_3)(a_2-a_3)}$
XI.	Paraboloidal $0 < \eta_1 < a_2 < \eta_2 < a_3 < \eta_3$	$x^2 = \frac{(\eta_1-a_3)(\eta_2-a_3)(\eta_3-a_3)}{(a_3-a_2)}$, $y^2 = \frac{(\eta_1-a_2)(\eta_2-a_2)(\eta_3-a_2)}{(a_2-a_3)}$, $z^2 = \frac{1}{2}(\eta_1 + \eta_2 + \eta_3 - a_2 - a_3)$

In real Euclidean 3-space there are 11 separable systems see Table 1, [**8, 31**]. On the real 2-sphere there are two: spherical and ellipsoidal. For real n-dimensional Euclidean space and the n-sphere Kalnins and the author have a graphical procedure to classify and construct all possibilities, [**18, 16**]. On the 2-hyperboloid there are 9 separable systems, [**16**]. For the n-hyperboloid of two sheets there is again a graphical procedure to construct all possibilites, [**16**].

In each case above, the symmetries $\mathcal{L}_j$ are second order elements in the enveloping algebra of the symmetry Lie algebra of the corresponding manifold, e.g., the Lie algebra $e(n, C)$ for Euclidean n-space and $so(n+1, C)$ for the n-sphere, $\smile$.

We see that for zero potential, each of the constant curvature spaces listed above is separable in multiple coordinate systems. Indeed we can veryfy that the zero potential is superintegrable on each of these spaces. However, a potential $V \neq 0$ "breaks the symmetry" and reduces the number of separable systems, usually to zero. (See [**32**] for conditions that must be satisfied by a potential in order to permit separation in a given coordinate system.)

How does one determine which constants of the motion lead to variable separation?

THEOREM 1. *Necessary and sufficient conditions for the existence of an orthogonal separable coordinate system $\{q_i\}$ for the Hamilton-Jacobi equation $\mathcal{H}^1 = E$ on an n-dimensional Riemannian manifold are that there exist n quadratic forms $\mathcal{L}_k = \sum_{i,j=1}^n L_{ij}^{(k)} p_i p_j + W_k$ on the manifold such that:*

(1) $\{\mathcal{L}_k, \mathcal{L}_\ell\} = 0$, $1 \leq k, i \leq n$.
(2) *The set $\{\mathcal{L}_k\}$ is linearly independent (as n quadratic forms).*
(3) *There is a basis $\{\omega_{(j)} : 1 \leq j \leq n\}$ of simultaneous eigenforms for the n matrices $\{L_{ij}^{(k)}\}$.*

If conditions 1–3 are satisfied then there exist functions $g_i(\mathbf{q})$ such that:

$$\omega_{(j)} = g_j dq_j, \quad j = 1, \ldots, n.$$

THEOREM 2. *Necessary and sufficient conditions for the existence of an orthogonal separable coordinate system $\{q_i\}$ for the Schrödinger equation $(\Delta_n + V)\Psi = E\Psi$ on an n-dimensional constant curvature space are that there exists a linearly independent set $\{L_1 = H = \Delta_n + V, L_2, \ldots, L_n\}$ of second-order differential operators on the manifold such that:*

(1) $[L_k, L_\ell] = 0$, $1 \leq k, \ell \leq n$.
(2) *Each L_k is in self-adjoint form.*
(3) *There is a basis $\{\omega_{(j)} : 1 \leq j \leq n\}$ of simultaneous eigenforms for the $\{L_k\}$.*

If conditions 1–3 are satisfied then there exist functions $g_i(\mathbf{q})$ such that:

$$\omega_{(j)} = g_j dq_j, \quad j = 1, \ldots, n.$$

See [**17, 32**] for proofs and discussions of these theorems. The main point of the theorems is that, under the required hypotheses the eigenforms ω^ℓ of the quadratic forms L^{ij} are normalizable, i.e., that up to multiplication by a nonzero function, ω^ℓ is the differential of a coordinate. This fact permits us to compute the coordinates directly from a knowledge of the symmetry operators. For general Riemannian manifolds Theorem 1 remains true, but Theorem 2 is false unless separation is replaced by the more general concept of R-separation [**32**].

We expect a superintegrable system to separate in multiple coordinate systems, though the above remarks do not constitute a proof of this. Thus, one way to find superintegrable systems is to search for potentials $V(\mathbf{q})$ that permit separation in multiple coordinate systems.

2. Examples for the Euclidean Plane

To illustrate the basic ideas we can consider the example of the Schrödinger equation with potential

$$V(x,y)=\frac{1}{2}\left(\omega^2(x^2+y^2)+\frac{k_1^2-\frac{1}{4}}{x^2}+\frac{k_2^2-\frac{1}{4}}{y^2}\right),$$

i.e.,

$$\left(\frac{\partial^2}{\partial x^2}+\frac{\partial^2}{\partial y^2}\right)\Psi-\left(\omega^2(x^2+y^2)+\frac{k_1^2-\frac{1}{4}}{x^2}+\frac{k_2^2-\frac{1}{4}}{y^2}\right)\Psi=-2E\Psi.$$

This equation separates in three coordinate systems: *Cartesian* coordinates (x,y); *polar* coordinates $x=r\cos\theta$, $y=r\sin\theta$, and *elliptical* coordinates

$$x^2=c^2\frac{(u_1-e_1)(u_2-e_1)}{(e_1-e_2)},\quad y^2=c^2\frac{(u_1-e_2)(u_2-e_2)}{(e_2-e_1)}.$$

The bound state energies are given by

$$E_n=\omega(2n+2+k_1+k_2)$$

for integer n. The wave functions for each of these coordinate systems are:

(1) *Cartesian* coordinates

$$\Psi_{n_1,n_2}(x,y)=2\omega^{(k_1+k_2+2)/2}\sqrt{\frac{n_1!n_2!}{\Gamma(n_1+k_1+1)\Gamma(n_2+k_2+1)}}\times x^{(k_1+1/2)}y^{(k_2+1/2)}e^{-\omega/2(x^2+y^2)}L_{n_1}^{k_1}(\omega x^2)L_{n_2}^{k_2}(\omega y^2)$$

where $n=n_1+n_2$, and the $L_n^k(x)$ are Laguerre polynomials, [**9**];

(2) *polar* coordinates

$$\Psi(r,\theta)=\Phi_q^{(k_1,k_2)}(\theta)\omega^{(2q+k_1+k_2+1)/2}\sqrt{\frac{2m!}{\Gamma(m+2q+k_1+k_2+1)}}\times e^{(-\omega r^2/2)}r^{(2q+k_1+k_2+1)}L_m^{2q+k_1+k_2+1}(\omega r^2)$$

where $n=m+q$,

$$\Phi_q^{(k_1,k_2)}(\theta)=\sqrt{2(2q+k_1+k_2+1)\frac{q!\Gamma(k_1+k_2+q+1)}{\Gamma(k_2+q+1)\Gamma(k_1+q+1)}}\times(\cos\theta)^{k_1+(1/2)}(\sin\theta)^{k_2+(1/2)}P_q^{(k_1,k_2)}(\cos 2\theta), \tag{2.1}$$

and the $P_q^{(k_1,k_2)}(\cos 2\theta)$ are Jacobi polynomials, [**9**];

(3) *elliptical* coordinates

$$\Psi=e^{-\omega(x^2+y^2)}x^{k_1+1/2}y^{k_2+1/2}\prod_{m=1}^{n}\left(\frac{x^2}{\theta_m-e_1}+\frac{y^2}{\theta_m-e_2}-c^2\right)$$

where use has been made of the identity

$$\frac{x^2}{\theta-e_1}+\frac{y^2}{\theta-e_2}-c^2=-c^2\frac{(u_1-\theta)(u_2-\theta)}{(\theta-e_1)(\theta-e_2)}.$$

The zeros θ_j satisfy the relations

$$\frac{k_1+1}{\theta_m-e_1}+\frac{k_2+1}{\theta_m-e_2}+\sum_{j\neq m}\frac{2}{(\theta_m-\theta_j)}-\omega=0.$$

Associated with the separability of the Schrödinger equation in these coordinate systems there are second order symmetry operators. A basis for such operators is

$$L_1=\partial_x^2+\frac{(\frac{1}{4}-k_1^2)}{x^2}-\omega^2x^2,\quad L_2=\partial_y^2+\frac{(\frac{1}{4}-k_2^2)}{y^2}-\omega^2y^2$$

$$M=(x\partial_y-y\partial_x)^2+\left(\frac{1}{4}-k_1^2\right)\frac{y^2}{x^2}+\left(\frac{1}{4}-k_2^2\right)\frac{x^2}{y^2}-\frac{1}{2}.$$

(Note that $H=L_1+L_2$.) The separable eigenfunctions already given are eigenfunctions of the symmetry operators L_1, M and $M+e_2L_1+e_1L_2$ with eigenvalues

$$\lambda_c=-\omega(2n_1+k_1+1),\quad \lambda_p=(2q+k_1+k_2+1)^2+(1+k_1^2+k_2^2),$$

$$\lambda_e=2(1-k_1)(1-k_2)-2e_2\omega(k_1+1)-2e_1\omega(k_2+1)-\omega^2e_1e_2$$

$$-4\sum_{m=1}^{q}\left[e_2\frac{k_1+1}{\theta_m-e_1}+e_1\frac{k_2+1}{\theta_m-e_2}\right].$$

The algebra constructed by repeated commutators is (R is defined by the first relation)

$$[L_1,M]=[M,L_2]=R,\quad [L_i,R]=-4\{L_i,L_j\}+16\omega^2M,\quad i\neq j,$$

$$[M,R]=4\{L_1,M\}-4\{L_2,M\}+8(1-k_2^2)L_1-8(1-k_1^2)L_2,$$

$$R^2=\frac{8}{3}\{M,L_1,L_2\}+\frac{64}{3}\{L_1,L_2\}+16\omega^2M^2-16(1-k_2^2)L_1^2$$

$$-16(1-k_1^2)L_2^2-\frac{128}{3}\omega^2M-64\omega^2(1-k_1^2)(1-k_2^2).$$

These relations are quadratic.

In real Euclidean two-space there are precisely four potentials that have the multiseparation property, [**21**]. The second potential is

$$V(x,y)=\omega^2(4x^2+y^2)-\frac{(\frac{1}{4}-k_2^2)}{y^2}.$$

The corresponding Schrödinger equation is separable in two coordinate systems: *Cartesian* coordinates and *parabolic* coordinates

$$x=\frac{1}{2}(\xi^2-\eta^2),\quad y=\xi\eta.$$

The third potential is

$$V(x,y)=-\frac{\alpha}{\sqrt{x^2+y^2}}+\frac{1}{4\sqrt{x^2+y^2}}\left(\frac{(k_1^2-\frac{1}{4})}{\sqrt{x^2+y^2}+x}+\frac{(k_2^2-\frac{1}{4})}{\sqrt{x^2+y^2}-x}\right).$$

The corresponding Schrödinger equation is separable in two coordinate systems: *polar*, *parabolic* and *modified elliptic* coordinates, where

$$\xi^2=c^2\frac{(u_1-e_1)(u_2-e_1)}{(e_1-e_2)},\quad \eta^2=c^2\frac{(u_1-e_2)(u_2-e_2)}{(e_2-e_1)}.$$

This last coordinate system can be written as

$$x=\sqrt{\frac{(U_1-E_1)(U_2-E_1)}{4(E_1-E_2)}},\quad y=\sqrt{\frac{(U_1-E_2)(U_2-E_2)}{4(E_2-E_1)}}-2\sqrt{E_1-E_2},$$

where $E_1=-e_1e_2$, $E_2=-\frac{1}{4}(e_1+e_2)^2$ and $U_j=u_j^2-u_j(e_1+e_2)$.

The fourth potential is

$$V(x,y)=-\frac{\alpha}{\sqrt{x^2+y^2}}+\frac{B_1}{4}\frac{\sqrt{\sqrt{x^2+y^2}+x}}{\sqrt{x^2+y^2}}+\frac{B_2}{4}\frac{\sqrt{\sqrt{x^2+y^2}-x}}{\sqrt{x^2+y^2}}.$$

Separation occurs here in *parabolic* and *parabolic coordinates of the second type*

$$x=\mu\nu,\quad y=\frac{1}{2}(\mu^2-\nu^2).$$

As an illustration of the utility of the notion of a quadratic algebra consider the last potential given. A basis for the quadratic algebra consists of L_1, L_2 and H with defining relations

$$[R,L_1]=-4L_2H+B_1B_2,\quad [R,L_2]=4L_1H+\frac{1}{2}(B_1^2-B_2^2)$$
$$R^2=4L_1^2H+4L_2^2H-16\alpha^2H+(B_2^2-B_1^2)L_1-2B_1B_2L_2-2\alpha^2(B_1^2+B_2^2)$$

with $R=[L_1,L_2]$. If we look for eigenfunctions of the operators L_1, L_2 respectively, we have

$$L_1\varphi_m=\lambda_m\varphi_m,\quad L_2\psi_n=\rho_n\psi_n.$$

If we write

$$L_1\psi_n=\sum_\tau C_{n\tau}\psi_\tau$$

then the quadratic algebra relations imply

$$[(\rho_n-\rho_\tau)^2+8E]C_{n\tau}=-\left[\frac{1}{2}(B_1^2-B_2^2)-16\alpha E\right]\delta_{n\tau}$$
$$\sum_\tau C_{n\tau}C_{\tau\sigma}(2\rho_\tau-\rho_n-\rho_\sigma)=(8E\rho_n+B_1B_2+16\alpha E)\delta_{n\sigma}.$$

These relations in turn imply that

$$C_{nn}=-\frac{\frac{1}{2}(B_1^2-B_2^2)+16\alpha E}{8E}$$

and $C_{nn+1}=C^*_{n+1n}$ are the only nonzero coefficents. Indeed they can essentially be determined by the relation

$$4\sqrt{-2E}(|C_{n,n+1}|^2-|C_{n-1,n}|^2)=8E\rho_n+B_1B_2+16\alpha E$$

where the eigenvalues λ_m and ρ_n are given by

$$\lambda_m=2\alpha-\frac{B_1^2}{8E}-(2m+1)\sqrt{-2E},\quad \rho_n=2\alpha-\frac{(B_1+B_2)^2}{16E}-(2n+1)\sqrt{-2E}$$

and the quantisation condition for E is

$$4\alpha-\frac{B_1^2+B_2^2}{8E}=-(q+2)\sqrt{-2E}$$

for integer q.

3. More Examples: The Real 2-Sphere

These ideas work also for separable coordinates on the real two dimensional sphere, [**21**]. An important example is the potential

$$V = \frac{1}{2}\left[\frac{k_1^2 - \frac{1}{4}}{s_1^2} + \frac{k_2^2 - \frac{1}{4}}{s_2^2} + \frac{k_3^2 - \frac{1}{4}}{s_3^2}\right]$$

where $s_1^2 + s_2^2 + s_3^2 = 1$. The corresponding Schrödinger equation has the form

$$\left[\left(s_1\frac{\partial}{\partial s_2} - s_2\frac{\partial}{\partial s_1}\right)^2 + \left(s_1\frac{\partial}{\partial s_3} - s_3\frac{\partial}{\partial s_1}\right)^2 + \left(s_3\frac{\partial}{\partial s_2} - s_2\frac{\partial}{\partial s_3}\right)^2\right]\Psi$$
$$+\left[\frac{(k_1^2 - \frac{1}{4})}{s_1^2} + \frac{(k_2^2 - \frac{1}{4})}{s_2^2} + \frac{(k_3^2 - \frac{1}{4})}{s_3^2}\right]\Psi = -2E\Psi.$$

This equation admits solution via separation of variables in two coordinate systems: *spherical* coordinates

$$s_1 = \sin\theta\cos\varphi, \quad s_2 = \sin\theta\sin\varphi, \quad s_3 = \cos\theta,$$

and *elliptical* coordinates

$$s_i^2 = \frac{(u_1 - e_i)(u_2 - e_i)}{(e_1 - e_j)(e_i - e_j)}, \quad i, j, k = 1, 2, 3, \quad i \neq j \neq k.$$

Indeed a basis for the second-order symmetries of Schrödinger's equation with this potential is

$$L_{ij} = (s_i\partial_{s_j} - s_j\partial_{s_i})^2 + \left(\frac{1}{4} - k_i^2\right)\frac{s_j^2}{s_i^2} + \left(\frac{1}{4} - k_j^2\right)\frac{s_i^2}{s_j^2} - \frac{1}{2}, \quad i \neq j.$$

These symmetries satisfy the quadratic algebra relations

$$[L_{ij}, R] = 4\{L_{ij}, L_{jk}\} - 4\{L_{ij}, L_{ik}\} + 8(1 - k_i^2)L_{jk} - 8(1 - k_j)^2L_{ik},$$
$$R^2 = -\frac{4}{3}\{L_{ij}, L_{ik}, L_{jk}\} + \frac{64}{3}\{L_{ij}, L_{ik}\} + \frac{64}{3}\{L_{ij}, L_{jk}\} + \frac{64}{3}\{L_{ik}, L_{jk}\}$$
$$- 16(1 - k_k^2)L_{ij}^2 - 16(1 - k_j^2)L_{ik}^2 - 16(1 - k_i^2)L_{jk}^2$$
$$+ \frac{128}{3}(1 - k_i^2)L_{jk} + \frac{128}{3}(1 - k_j^2)L_{ik} + \frac{128}{3}(1 - k_k^2)L_{ij}, \quad i \neq j \neq k.$$

The eigenfunctions with bound state energy eigenvalues

$$E_p = \frac{1}{2}(2p + 2 + k_1 + k_2 + k_3)^2 - \frac{1}{8}$$

in these coordinate systems are *polar*:

$$\Psi = (\sin\theta)^{-1}\Phi_n^{(k_2,k_1)}(\varphi)\Phi_m^{(2n+k_1+k_2,k_3)}(\theta)$$

where $p = m + n$, (here $\Phi_n^{(k_2,k_1)}$ is given by (2.1)) and *elliptical*

$$\Psi = \left(\prod_{\ell=1}^{3} s_\ell^{(k_\ell+1/2)}\right)\prod_{j=1}^{q}\left(\frac{s_1^2}{\theta_j - e_1} + \frac{s_2^2}{\theta_j - e_2} + \frac{s_3^2}{\theta_j - e_3}\right)$$

where

$$\frac{k_1 + 1}{\theta_m - e_1} + \frac{k_2 + 1}{\theta_m - e_2} + \frac{k_3 + 1}{\theta_m - e_3} + \sum_{j\neq m}\frac{2}{\theta_m - \theta_j} = 0$$

and $q = p$. Here we have made use of the identity

$$\frac{s_1^2}{\theta_j - e_1} + \frac{s_2^2}{\theta_j - e_2} + \frac{s_3^2}{\theta_j - e_3} = \frac{\prod_{\ell=1}^2 (u_\ell - \theta_j)}{\prod_{m=1}^3 (\theta_j - e_m)}.$$

The separable eigenfunctions already given are eigenfunctions of the symmetry operators L_{12} and $e_3 L_{12} + e_2 L_{13} + e_1 L_{23}$:

$$\lambda_S = (2n + k_1 + k_2 + 1)^2,$$

$$\lambda_E = -2[k_1(e_2+e_3) + k_2(e_1+e_3) + k_3(e_2+e_1) + e_3k_1k_2 + e_1k_2k_3 + e_2k_1k_3] - \frac{3}{2}(e_1+e_2+e_3) - 4e_2e_3(k_1+1)\sum_{m=1}^{q} \frac{1}{\theta_m - e_1} - 4e_2e_1(k_3+1)\sum_{m=1}^{q} \frac{1}{\theta_m - e_3} - 4e_1e_3(k_2+1)\sum_{m=1}^{q} \frac{1}{\theta_m - e_2}.$$

A second multiseparable potential on the sphere is

$$V = -\frac{\alpha s_3}{\sqrt{s_1^2+s_2^2}} + \frac{1}{4\sqrt{s_1^2+s_2^2}}\left[\frac{(k_1^2 - \frac14)}{\sqrt{s_1^2+s_2^2}+s_1} + \frac{(k_2^2-\frac14)}{\sqrt{s_1^2+s_2^2}-s_1}\right].$$

Schrödinger's equation for this potential has the form

$$\left[\left(s_1\frac{\partial}{\partial s_2} - s_2\frac{\partial}{\partial s_1}\right)^2 + \left(s_1\frac{\partial}{\partial s_3} - s_3\frac{\partial}{\partial s_1}\right)^2 + \left(s_3\frac{\partial}{\partial s_2} - s_2\frac{\partial}{\partial s_3}\right)^2\right]\Psi + \left(-\frac{\alpha s_3}{\sqrt{s_1^2+s_2^2}} + \frac{1}{4\sqrt{s_1^2+s_2^2}}\left[\frac{(k_1^2-\frac14)}{\sqrt{s_1^2+s_2^2}+s_1} + \frac{(k_2^2-\frac14)}{\sqrt{s_1^2+s_2^2}-s_1}\right]\right)\Psi = -2E\Psi.$$

This equation admits solution via separation of variables in two coordinates systems: *spherical* and *elliptical coordinates of modified type*

$$s_1' = \cos f\, s_1 + \sin f\, s_3, \quad s_2' = s_2, \quad s_3' = -\sin f\ s_1 + \cos f\ s_3$$

where

$$s_i^2 = \frac{(y_1 - e_i)(y_2 - e_i)}{(e_1 - e_j)(e_i - e_j)}, \quad i,j,k = 1,2,3 \quad i \neq j \neq k,$$

and

$$\sin f = \sqrt{\frac{(e_2 - e_1)}{(e_3 - e_1)}}, \quad e_1 = e_2 + \frac14 (E_+ + E_-)^2, \quad e_3 = e_2 + \frac14(E_+ - E_-)^2,$$

$$y_j = e_2 + \frac14(E_+^2 + E_-^2) + \frac14 E_+E_-\left(Z_j + \frac{1}{Z_j}\right), \quad j = 1,2.$$

Indeed if we use the variables

$$U_3^2 = Z_1Z_2, \quad U_1^2 = \frac{(Z_1 + \Omega_-)(Z_2+\Omega_-)}{(\Omega_-^2 - 1)}, \quad U_2^2 = \frac{(Z_1+\Omega_+)(Z_2+\Omega_+)}{(\Omega_+^2 - 1)}$$

where

$$\Omega_+ = \frac{E_+}{E_-}, \quad \Omega_- = \frac{E_-}{E_+},$$

then, putting $k_3 = \sqrt{2(E-i\alpha)+\frac{1}{4}}$ and $\hat{E} = i\alpha + E$ and multiplying the Schrödinger equation by $(Z_1Z_2)^{-1} - 1$, we see that the resulting equation has the form

(3.1)

$$H\Psi = \left(\left(U_1\frac{\partial}{\partial U_2} - U_2\frac{\partial}{\partial U_1}\right)^2 + \left(U_1\frac{\partial}{\partial U_3} - U_3\frac{\partial}{\partial U_1}\right)^2 + \left(U_3\frac{\partial}{\partial U_2} - U_2\frac{\partial}{\partial U_3}\right)^2\right)\Psi$$
$$+\left[\frac{(\frac{1}{4}-k_1^2)}{U_1^2} + \frac{(\frac{1}{4}-k_2^2)}{U_2^2} + \frac{(\frac{1}{4}-k_3^2)}{U_3^2}\right]\Psi + 2\hat{E}\Psi = 0$$

which is essentially the same form as for the first potential. The bound state quantisation condition has the form

$$E = \frac{1}{2}\left[2q + 2 + k_1 + k_2 + \sqrt{\frac{1}{4} + 2(E - i\alpha)}\right]^2 - \frac{1}{8}.$$

4. Features of Superintegrability ($n = 2$)

Based on the examples of the last two sections, we can point out some basic features of superintegrability in two dimensions.

(1) The potential V permits separability of the Hamilton-Jacobi equation $\mathcal{H} = E$ and the Schrödinger equation $H\Psi = E\Psi$ in at least two coordinate systems, characterized by symmetry conditions $\mathcal{L}_1 = \lambda_1$, $\mathcal{L}_2 = \lambda_2$ in the first case and $L_1\Psi = \lambda_1\Psi$, $L_2\Psi = \lambda_2\Psi$ in the second.

(2) One can obtain alternate spectral resolutions $\{\Psi_j^{(1)}\}, \{\Psi_k^{(2)}\}$ for the multiply degenerate eigenspaces of H,
$$L_1\Psi_j^{(1)} = \lambda_1^{(1)}\Psi_j^{(1)}, \quad L_2\Psi_k^{(2)} = \lambda_2^{(2)}\Psi_k^{(2)}.$$
These alternate resolutions resolve the degeneracy problem.

(3) The interbasis expansions
$$\Psi_k^{(2)} = \sum_j a_{jk}\Psi_j^{(1)}$$
yield important special function identities. In many cases, these become expansions of one set of multivariable orthogonal polynomials in terms of another set.

(4) The operators H, L_1, L_2 generate a quadratic algebra. Indeed, with $R = [L_1, L_2]$ we have that R^2 is a polynomial of order 3 in H, L_1, L_2, whereas $[L_1, R]$ and $[L_2, R]$ are polynomials of order 2 in H, L_1, L_2. A corresponding statement is true for algebra generated by the symmetries $\mathcal{H}$, $\mathcal{L}_1$, $\mathcal{L}_2$ under the Poisson bracket. (Note: This is a remarkable property, and is false for general symmetries. Consider Euclidean 2-space with Hamiltonian $\mathcal{H} = p_x^2 + p_y^2$. The algebra of all symmetries of $\mathcal{H}$ is generated by $p_x, p_y, M = xp_y - yp_x$. Let
$$\mathcal{L}_1 = M^2 + p_xp_y, \quad \mathcal{L}_2 = p_x^2.$$
Then we have $\mathcal{R} = \{\mathcal{L}_1, \mathcal{L}_2\} = 4Mp_xp_y$ and
$$\mathcal{R}^2 = \mathcal{F}(\mathcal{L}_0, \mathcal{L}_1, \mathcal{L}_2) = 16\mathcal{L}_1\mathcal{L}_2(\mathcal{L}_0 - \mathcal{L}_2) - 16\mathcal{L}_2^{3/2}(\mathcal{L}_0 - \mathcal{L}_2)^{3/2}.$$
Since $\mathcal{H}$, $\mathcal{L}_1$, $\mathcal{L}_2$ are functionally independent $\mathcal{R}^2$ must be a function of these symmetries. However, although $\mathcal{F}$ is defined and bounded at the

point $(\mathcal{L}_0, \mathcal{L}_1, \mathcal{L}_2) = (0,0,0)$, it is not a polynomial, and not even analytic at this point. Thus it has no power series expansion about the origin.)

(5) The quadratic algebra structure can be used to compute the interbase expansion coefficients.

5. Examples in Higher Dimensions

An extreme case, superintegrability in n dimensions, occurs for the Schrödinger equation

$$(5.1)\qquad \big(\Delta_n + V_n(\mathbf{q})\big)\Psi = -M(M+G-1)\Psi$$

where

$$(5.2)\qquad V_n = -\frac{1}{4}\sum_{i=1}^{n}\frac{(\gamma_i - \frac{1}{2})(\gamma_i - 3/2)}{q_i^2} - \frac{1}{4}\frac{(\gamma_{n+1} - \frac{1}{2})(\gamma_{n+1} - 3/2)}{q_0^2} + \alpha,$$

$$\alpha = \frac{1}{4}\left[(1-G)^2 - 1 - \frac{(n-3)(n+1)}{4}\right],$$

$G = \sum_{j=1}^{n+1}\gamma_j$, and Δ_n is the Laplace-Beltrami operator on the n-sphere. Here

$$q_0^2 + q_1^2 + \ldots + q_n^2 = 1.$$

One can transfer this Schrödinger equation with a scalar potential V_n to one with vector potential Λ_n through the use of a multiplier transformation ρ. Setting $\Phi(\mathbf{x}) = \rho(\mathbf{x})\Psi(\mathbf{x})$ for a nonzero scalar function ρ we find

$$(5.3)\qquad H\Phi \equiv (\Delta_n + \Lambda_n)\Phi = -M(M+G-1)\Phi$$
$$\Longleftrightarrow\ \big(\Delta_n + V_n(\mathbf{x})\big)\Psi = -M(M+G-1)\Psi,$$

provided

$$\rho^{-1} = x_1^{\gamma_1/2-1/4}\cdots x_n^{\gamma_n/2-1/4}(1-x)^{\gamma_{n+1}/2-1/4},$$

and

$$\Lambda_n = \sum_{j=1}^{n}\left[\gamma_j - \frac{1}{2} + \left(\frac{n+1}{2} - G\right)x_j\right]\partial_{x_j}.$$

Here

$$H = \sum_{i,j=1}^{n}(x_i\delta_{ij} - x_ix_j)\partial_{x_ix_j} + \sum_{i=1}^{n}(\gamma_i - Gx_i)\partial_{x_i}$$

in the coordinates

$$\begin{aligned} q_0^2 &= 1 - \sum_{i=1}^{n} x_i = 1 - x \\ q_1^2 &= x_1 \\ q_2^2 &= x_2 \\ &\vdots \\ q_n^2 &= x_n. \end{aligned}$$

In the paper [**18**] and the book [**16**] all separable coordinates for the equation $\Delta_n\Psi = \lambda\Psi$ are constructed, where Δ_n is the Laplace-Beltrami operator on S^n. It is shown that all separable coordinates are orthogonal and that for each separable coordinate system the corresponding separated solutions are characterized as simultaneous eigenfunctions of a set of n second order commuting symmetry operators for Δ_n. Moreover, the equation $(\Delta_n + V_n)\Psi = \lambda\Psi$ where the scalar potential takes the form

$$V_n = \sum_{i=1}^{n} \frac{\alpha_i}{q_i^2} + \frac{\alpha_0}{q_0^2}, \qquad \alpha_0, \alpha_1, \dots, \alpha_n$$

is separable in *all* the coordinate systems in which the Laplace-Beltrami eigenvalue equation is separable. These results can easily be extended to solutions of

$$(\Delta_n + \Lambda_n)\Phi = \lambda\Phi \tag{5.4}$$

through the mappings

$$\begin{aligned} \Delta_n + \Lambda_n &= \rho(\Delta_n + V_n)\rho^{-1} \\ S_{ij} &= \rho S'_{ij}\rho^{-1} \\ \Phi &= \rho\Psi. \end{aligned} \tag{5.5}$$

Indeed all separable solutions Ψ map to R-separable solutions Φ of (5.4), [**32**]. Finally, since $H = \Delta_n + \Lambda_n$ maps polynomials of maximum order m_k in x_k to polynomials of the same type, it follows that a basis of separated solutions can be expressed as *polynomials* in the x_i. The second-order symmetry operators for this operator can be chosen to be self-adjoint, so the basis of simultaneous eigenfunctions can be chosen to be *orthogonal* with respect to the inner product

$$\begin{aligned} (\Phi_1, \Phi_2) \equiv \langle \Psi_1, \Psi_2 \rangle &= \int \cdots \int_{x_i > 0, x < 1} \Phi_1(\mathbf{x})\overline{\Phi_2}(\mathbf{x})\rho^{-2}(\mathbf{x})\, d\omega \\ &= \int \cdots \int_{x_i > 0, x < 1} \Phi_1\overline{\Phi_2}\, d\tilde{\omega}, \\ d\tilde{\omega} = x_1^{\gamma_1 - 1} &\dots x_n^{\gamma_n - 1}(1 - x)^{\gamma_{n+1} - 1} dx_1 \dots dx_n. \end{aligned}$$

Thus every separable coordinate system for the Laplace-Beltrami eigenvalue equation on the n-sphere yields an orthogonal basis of polynomial solutions of equation (5.4), hence an orthogonal basis for all n-variable polynomials with inner product (5). For details about the bases that can occur and the interbasis expansion coefficients, see [**25, 19**]. Among the special functions that arise are polyspherical harmonics, products of Jacobi polynomials, Heun polynomials, Lame' polynomials, ellipsoidal polynomials, and Lauricella polynomials.

We will look at one more example, the generalized isotropic oscillator in Euclidean 3-space. This is the Schrödinger equation

$$H\Psi = -\frac{1}{2}\left(\frac{\partial^2}{\partial x^2} + \frac{\partial^2}{\partial y^2} + \frac{\partial^2}{\partial z^2}\right)\Psi + V(x, y, z)\Psi = E\Psi, \tag{5.6}$$

with potential

$$V(x, y, z) = \frac{\omega^2}{2}(x^2 + y^2 + z^2) + \frac{1}{2}\left[\frac{(k_1^2 - \frac{1}{4})}{x^2} + \frac{(k_2^2 - \frac{1}{4})}{y^2} + \frac{(k_3^2 - \frac{1}{4})}{z^2}\right] \tag{5.7}$$

where the constant $k_i \geq \frac{1}{2}$. For $k_i = \frac{1}{2}$ we have the ordinary isotropic oscillator potential. The corresponding Schrödinger equation admits solutions via a separation of variables in eight coordinate systems: Cartesian, spherical, sphero-conical, cylindrical polar, cylindrical elliptic, prolate and oblate spheroidal, ellipsoidal. See [**22**] for the complete details. Here we consider only the quadratic algebra struture.

Note that a basis for the symmetries of Schrödinger's equation with the potential (5.7) consists of the six operators:

$$M_i = -D_{ii} - \frac{k_i^2 - \frac{1}{4}}{x_i^2}, \qquad -\mathcal{H} = M_1 + M_2 + M_3 \tag{5.8}$$

$$J_{ij} = L_{ij}^2 - \left(k_i^2 - \frac{1}{4}\right)\frac{x_j^2}{x_i^2} - \left(k_j^2 - \frac{1}{4}\right)\frac{x_i^2}{x_j^2} - \frac{1}{2} \quad i,j = 1,2,3 \tag{5.9}$$

where $L_{ij} = x_i\partial_{x_j} - x_j\partial_{x_i}$, $D_{ii} = -\partial_{x_i}^2 + \omega^2 x_i^2$ is a diagonal components of the Demkov tensor [**7**] and we have the notation $x_1 = x$, $x_2 = y$, $x_3 = z$.

The commutators of the operators (5.8)–(5.9) can be closed to form a quadratic algebra as follows

$$[M_i, M_j] = 0, \quad [M_i, J_{jk}] = 0,$$
$$[M_i, J_{ij}] = Q_{ij} = Q_{[ij]}, \quad [J_{ij}, J_{ik}] = R_{[ijk]} = R$$

where Q_{ij} is totally antisymmetric and the totally antisymmetric quantity $R_{[ijk]}$ is denoted by R. Further commutators are calculated to be

$$\begin{aligned}
[M_i, Q_{jk}] &= 0,\\
[M_i, Q_{ij}] &= 4\{M_i, M_j\} + 16J_{ij},\\
[M_i, R] &= 4\{M_k, J_{ij}\} - 4\{M_j, J_{ik}\},\\
[J_{ij}, Q_{ij}] &= 4\{M_i, J_{ij}\} - 4\{M_j, J_{ij}\} - 8(k_j^2 - 1)M_i + 8(k_i^2 - 1)M_j,\\
[J_{ij}, Q_{ik}] &= 4\{M_i, J_{jk}\} - 4\{M_j, J_{ik}\},\\
[J_{ij}, R] &= 4\{J_{ij}, J_{jk}\} - 4\{J_{ij}, J_{ik}\} - 8(k_i^2 - 1)J_{jk} + 8(k_j^2 - 1)J_{ik},
\end{aligned}$$

where $\{A, B\} = AB + BA$. The expression for the commutators of the Q and R are

$$[Q_{ij}, Q_{ik}] = 4\{M_i, Q_{jk}\}, \quad [Q_{ij}, R] = -4\{J_{ij}, Q_{ik}\} - 4\{J_{ij}, Q_{jk}\}.$$

All the commutators of the operators M_i, J_{mn}, Q_{pq} and R can be expressed in terms of quadratic symmetric products of themselves. The algebra therefore is closed quadratically. There are relations between the symmetric products of the generators of this algebra. The exhaustive list of these is as follows.

$$\begin{aligned}
Q_{ij}^2 &= \frac{8}{3}\{J_{ij}, M_i, M_j\} + \frac{64}{3}\{M_i, M_j\} + 16\omega^2 J_{ij}^2 - 16(1 - k_j^2)M_i^2\\
&\quad - 16(1 - k_i^2)M_j^2 - \frac{128}{3}\omega^2 J_{ij} - 64\omega^2(1 - k_i^2)(1 - k_j^2)\\
\{Q_{ij}, Q_{ik}\} &= \frac{8}{3}\{J_{ij}, M_i, M_k\} + \frac{8}{3}\{J_{ik}, M_i, M_j\} - \frac{8}{3}\{J_{jk}, M_i, M_i\}\\
&\quad + 32\omega^2(1 - k_i^2)\{J_{ij}, J_{ik}\} - 32(1 - k_i^2)M_j M_k - 64\omega^2(1 - k_i^2)J_{jk}\\
\{Q_{ij}, R\} &= \frac{8}{3}\{J_{ij}, J_{ij}, M_k\} - \frac{8}{3}\{J_{ij}, J_{ik}, M_j\} - \frac{8}{3}\{J_{ij}, J_{jk}, M_i\}\\
&\quad - \frac{64}{3}\{J_{ij}, M_k\} - \frac{64}{3}\{J_{ik}, M_j\} - \frac{64}{3}\{J_{jk}, M_i\}\\
&\quad + 16(1 - k_i^2)\{J_{jk}, M_j\} + 16(1 - k_j^2)\{J_{ik}, M_i\} - 64(1 - k_i^2)(1 - k_j^2)M_k
\end{aligned}$$

$$
\begin{aligned}
R^2 = &-\frac{4}{3}\{J_{ij}, J_{ik}, J_{jk}\} + \frac{64}{3}\{J_{ij}, J_{ik}\} + \frac{64}{3}\{J_{ij}, J_{jk}\} + \frac{64}{3}\{J_{ik}, J_{jk}\} \\
&- 16(1-k_k^2)J_{ij}^2 - 16(1-k_j^2)J_{ik}^2 - 16(1-k_i^2)J_{jk}^2 + \frac{128}{3}(1-k_k^2)J_{ij} \\
&+ \frac{128}{3}(1-k_j^2)J_{ik} + \frac{128}{3}(1-k_i^2)J_{jk} + 64(1-k_i^2)(1-k_j^2)(1-k_k^2)
\end{aligned}
$$

where $\{A, B, C\} = ABC + CAB + BCA$. Note that only five operators from (5.8)–(5.9) are functionally independent [**37**] and for all the coordinate systems that provide separable solutions for the Schrödinger equation the operators characterizing the separation are always combinations of the M_i and J_{ij}.

In the limiting case $k_i = \frac{1}{2}$, we obtain a quadratic algebra too. In this case

$$Q_{ij} = 2(L_{ij}D_{ij} + D_{ij}L_{ij}), \qquad R = \{L_{ik}, \{L_{ij}, L_{kj}\}\},$$

and instead of operators $\{M_i, J_{ij}, Q_{ij}, R\}$ we can consider as a basis for the symmetries the Demkov tensor—D_{ij} and the components of orbital momentum—L_{ij}. In this regard we arrive at the Lie algebra corresponding to the symmetries of the isotropic oscillator [**7**].

6. Some Questions and a New Approach

The preceding examples motivate some questions:

(1) Is there a uniform procedure to find superintegrable potentials on a constant curvature space?
(2) Is there a way of proving completeness, i.e., of demonstrating that all superintegrable potentials for a space have been found?
(3) Is superintegrability always associated with multivariable separation?
(4) What is the relationship between superintegrability and the existence of quadratic algebras?

First, let us consider the issue of constructing the superintegrable potentials on a given space. We will look at three approaches to the problem.

(1) Use what is known from separation of variables theory.

This is the most common approach to finding superintegrable potentials. One classifies the separable systems for the zero potential, determines a general expression for a potential separating in each one of these systems, and then tries to identify potentials that simultaneously separate in several systems. Although the approach is useful for finding examples, there are serious problems in obtaining exhaustive lists. Separable coordinates occur in equivalence classes: two coordinate systems related by a symmetry in the motion group of the space are consided as equivalent. A potential may separate in an ellipsoidal coordinate system, but the ellipsoidal coordinates may not be centered at the origin or aligned with the Cartesian coordinate axes. Such systems are difficult to uncover and many of the lists of superintegrable potentials contain gaps, even for 2-dimensional Minkowski space and hyperboloids. Furthermore, for $n > 2$ there is no proof that superintegrability necessarily implies multiseparability.

(2) Derive superintegrable potentials on a space as restrictions of free particle systems on higher dimensional spaces. (Split off ignorable coordinates.)

A very simple example will give the basic idea. Consider the free particle Hamiltonian on Euclidean four-space:

$$\mathcal{H}_4 = p_1^2 + p_2^2 + p_3^2 + p_4^4, \quad ds^2 = dx_1^2 + dx_2^2 + dx_3^2 + dx_4^2.$$

There are 4 separable coordinate systems for the Hamilton-Jacobi equation that have two angular variables θ_1, θ_2. The simplest of these is

$$x_1 = r_1 \cos\theta_1, \quad x_2 = r_1 \sin\theta_1, \quad x_3 = r_2 \cos\theta_2, \quad x_4 = r_2 \sin\theta_2.$$

The other 3 systems are of the form $u_1, u_2, \theta_1, \theta_2$ where $r_j = f_j(u_1, u_2)$, $j = 1, 2$. In the new coordinates we have

$$\mathcal{H}_4 = p_{r_1}^2 + p_{r_2}^2 + \frac{p_{\theta_1}^2}{r_1^2} + \frac{p_{\theta_2}^2}{r_2^2}, \quad ds^2 = dr_1^2 + dr_2^2 + r_1^2 d\theta_1^2 + r_2^2 d\theta_2^2.$$

Note that the θ_j are *ignorable coordinates*, i.e., the metric tensor in the new coordinates is independent of the θ_j. Now restrict to the symplectic submanifold such that $p_{\theta_j} = \alpha_j$, $d\theta_j = 0$, where the α_j are constants. Then the Hamiltonian becomes

$$\mathcal{H}_4 = p_{r_1}^2 + p_{r_2}^2 + \frac{\alpha_1^2}{r_1^2} + \frac{\alpha_2^2}{r_2^2}, \quad ds^2 = dr_1^2 + dr_2^2.$$

Locally, we can regard the Hamiltonian as defined in Euclidean 2-space and with the potential $V = \alpha_1^2/r_1^2 + \alpha_2^2/r_2^2$. The restricted Hamiltonian-Jacobi equation remains separable in 4 coordinate systems $\{u_1, u_2\}$, hence this is a superintegrable system. Boyer, Kalnins and Winternitz have applied this idea to the restriction of a free Hamiltonian on Hermitian hyperbolic spaces to hyperboloids in order to derive previously unknown superintegrable systems, [**3, 4**]. This is a powerful method, but there is no evident way to prove by this approach that all superintegrable systems have been obtained.

(3) New way: use integrabilty conditions. This last approach is the one that we shall advocate in the remainder of this paper.

7. Complex Euclidean 2-Space Treated in Detail

We consider, in detail, the case

$$\mathcal{H} = p_1^2 + p_2^2 + V(x, y) \tag{7.1}$$

where $p_1 = p_x$, $p_2 = p_y$ and all variables are complex. Let us assume that in addition to the classical Hamiltonian we have two quadratic constants of the motion

$$\mathcal{L}_h = \sum_{k,j=1}^{2} a_{(h)}^{kj}(x, y) p_k p_j + W_{(h)}(x, y) \equiv \ell_h + W_{(h)}, \quad h = 1, 2 \tag{7.2}$$

which must satisfy

$$\{\mathcal{H}, \mathcal{L}_h\} = 0$$

with $\{\}$ the usual Poisson bracket. We require that the set $\{d\mathcal{H}, d\mathcal{L}_1, d\mathcal{L}_2\}$ is linearly independent, so that $\mathcal{H}$, $\mathcal{L}_1$, $\mathcal{L}_2$ is a maximal set of functionally independent constants of the motion. It is clear that $\mathcal{R} = \{\mathcal{L}_1, \mathcal{L}_2\}$ is a constant of the motion, so it and $\mathcal{R}^2$ must be expressible as an analytic function of $\mathcal{H}$, $\mathcal{L}_1$, $\mathcal{L}_2$:

$$\mathcal{R}^2 = \mathcal{F}(\mathcal{L}_0, \mathcal{L}_1, \mathcal{L}_2), \quad \mathcal{H} \equiv \mathcal{L}_0. \tag{7.3}$$

Note that $\mathcal{R}$ has the form

$$\mathcal{R} = \sum_{k,l,m=1}^{2} c^{klm} p_k p_l p_m + \sum_{k=1}^{2} d^k p_k, \tag{7.4}$$

but that it doesn't follow that $\mathcal{R}^2$ is necessarily a polynomial as a function of $\mathcal{L}_0$, $\mathcal{L}_1$, $\mathcal{L}_2$. We will find conditions that guarantee that $\mathcal{F}$ is a third-order polynomial in its arguments.

Using the identity

$$\{\mathcal{K}, \mathcal{G}\} = \sum_{h=0}^{2} \{\mathcal{K}, \mathcal{L}_h\} \frac{\partial \mathcal{G}}{\partial \mathcal{L}_h} \tag{7.5}$$

for a continuously differentiable function $\mathcal{G}(\mathcal{L}_h)$, we find the relations

$$\{\mathcal{L}_1, \mathcal{R}\} = \frac{1}{2}\frac{\partial \mathcal{F}}{\partial \mathcal{L}_2}, \quad \{\mathcal{L}_2, \mathcal{R}\} = -\frac{1}{2}\frac{\partial \mathcal{F}}{\partial \mathcal{L}_1}. \tag{7.6}$$

Thus, the constants of the motion $\{\mathcal{L}_1, \mathcal{R}\}, \{\mathcal{L}_2, \mathcal{R}\}$ are easily computed once $\mathcal{F}$ is known. Further, if $\mathcal{F}$ is a polynomial in the invariants, then so are $\{\mathcal{L}_1, \mathcal{R}\}$, and $\{\mathcal{L}_2, \mathcal{R}\}$.

We first determine the conditions that the function

$$\mathcal{L} = \sum_{j,k=1}^{2} a^{jk}(x,y) p_k p_j + W(x,y), \quad a^{jk} = a^{kj}, \tag{7.7}$$

must satisfy to be a constant of the motion. The requirement is $\{\mathcal{H}, \mathcal{L}\} = 0$ where

$$\{f, g\} = \sum_{j=1}^{2} \left(\frac{\partial f}{\partial p_j}\frac{\partial g}{\partial x_j} - \frac{\partial f}{\partial x_j}\frac{\partial g}{\partial p_j} \right), \quad (x_1, x_2) = (x, y), \tag{7.8}$$

and

$$\mathcal{H} = p_1^2 + p_2^2 + V(x,y). \tag{7.9}$$

The conditions are thus

$$\begin{aligned} \frac{\partial a^{11}}{\partial x} &= 0 \quad 2\frac{\partial a^{12}}{\partial x} + \frac{\partial a^{11}}{\partial y} = 0 \\ \frac{\partial a^{22}}{\partial y} &= 0 \quad \frac{\partial a^{22}}{\partial x} + 2\frac{\partial a^{12}}{\partial y} = 0, \end{aligned} \tag{7.10}$$

and

$$\frac{\partial W}{\partial x} - a^{11}\frac{\partial V}{\partial x} - a^{12}\frac{\partial V}{\partial y} = 0, \quad \frac{\partial W}{\partial y} - a^{12}\frac{\partial V}{\partial x} - a^{22}\frac{\partial V}{\partial y} = 0. \tag{7.11}$$

The solution for the terms quadratic in the p_j is

$$a^{11} = \alpha_1 y^2 + \alpha_2 y + \alpha_3' \tag{7.12}$$

$$a^{12} = -\alpha_1 xy - \frac{1}{2}\alpha_2 x - \frac{1}{2}\alpha_4 y + \frac{1}{2}\alpha_5 \tag{7.13}$$

$$a^{22} = \alpha_1 x^2 + \alpha_4 x + \alpha_3'', \tag{7.14}$$

where the α_k are constants. The requirement that $\partial_x W_y = \partial_y W_x$ leads from (7.11) to the second order partial differential equation for the potential

$$\begin{aligned}(7.15)\quad &\frac{1}{2}(2\alpha_1 xy + \alpha_2 x + \alpha_4 y - \alpha_5)(V_{xx} - V_{yy}) + (\alpha_1[y^2 - x^2] + \alpha_2 y - \alpha_4 x + \alpha_3)V_{xy} \\ &\qquad = (-3\alpha_1 y - \frac{3}{2}\alpha_2)V_x + (3\alpha_1 x + \frac{3}{2}\alpha_4)V_y,\end{aligned}$$

where $\alpha_3 = \alpha_3' - \alpha_3''$. This is a fundamental equation in our approach. We denote the solution space of this equation by

$$(7.16)\qquad [\alpha_1, \ldots, \alpha_5].$$

Let us now return to our assumption that the Hamilton-Jacobi equation admits two constants of the motion:

$$\mathcal{L}_h = \sum_{j,k=1}^{2} a^{jk}_{(h)} p_k p_j + W_{(h)}, \quad h = 1, 2.$$

These two operators together with $\mathcal{H}$ are assumed functionally independent. The constant of the motion $\mathcal{L}_1$ leads to the condition (7.16) on the potential V; whereas $\mathcal{L}_2$ leads to the second condition

$$(7.17)\qquad [\beta_1, \ldots, \beta_5].$$

The potential must lie in the intersection of the solution spaces (7.16), (7.17) for these two conditions. It follows that the equations

$$(7.18)\qquad V_{xx} - V_{yy} = AV_x + BV_y, \quad V_{xy} = CV_x + DV_y$$

must hold, where

$$(7.19)\qquad \begin{aligned} A\mathcal{E} &= \frac{3}{2}H_{12}(x^2 + y^2) - 3H_{14}xy + 3H_{13}y - \frac{3}{2}H_{24}x + \frac{3}{2}H_{23} \\ B\mathcal{E} &= \frac{3}{2}H_{14}(x^2 + y^2) - 3H_{12}xy - 3H_{13}x + \frac{3}{2}H_{24}y + \frac{3}{2}H_{34} \\ 2C\mathcal{E} &= -3H_{14}y^2 + \left(-\frac{3}{2}H_{24} + 3H_{15}\right)y + \frac{3}{2}H_{25} \\ 2D\mathcal{E} &= 3H_{12}x^2 + \left(-\frac{3}{2}H_{24} - 3H_{15}\right)x - \frac{3}{2}H_{45} \\ 2\mathcal{E} &= -H_{12}xy^2 + H_{14}x^2y - H_{12}x^3 + H_{14}y^3 - 2H_{13}xy + H_{24}(x^2 + y^2) \\ &\quad + H_{15}(x^2 - y^2) + (H_{34} - H_{25})y + (H_{45} - H_{23})x - H_{35}, \end{aligned}$$

and $H_{k\ell} = -H_{\ell k} = \alpha_k\beta_\ell - \alpha_\ell\beta_k$.

From the fundamental equations (7.18) we can compute all of the third partial derivatives of V. Indeed

$$(7.20)\qquad \begin{aligned} V_{xxx} &= (A_x + BC + C_y + C^2 + A^2)V_x + (B_x + DB + D_y + CD + AB)V_y \\ &\quad + (A + D)V_{yy} \\ V_{xxy} &= (C_x + DC + AC)V_x + (D_x + D^2 + BC)V_y + CV_{yy} \\ V_{xyy} &= (C_y + C^2)V_x + (D_y + CD)V_y + DV_{yy} \\ V_{yyy} &= (-A_y + C_x + DC)V_x + (-B_y - AD + D_x + D^2 + BC)V_y \\ &\quad + (C - B)V_{yy}. \end{aligned}$$

Thus if the potential V is subject to the two conditions (7.16), (7.17), then V can depend on at most 3 parameters, in addition to a trivial additive constant. We can choose these parameters to be $V_x(x_0, y_0)$, $V_y(x_0, y_0)$, $V_{yy}(x_0, y_0)$ for any fixed regular point (x_0, y_0). Then $V_{xx}(x_0, y_0)$ and all higher derivatives can be computed by successive differentiation of relations (7.20). We require that our potential be *nondegenerate*, i.e., that it depend on 3 arbitrary parameters.

Then, the conditions $\partial_x V_{xxy} = \partial_y V_{xxx}$, $\partial_y V_{xxy} = \partial_x V_{xyy}$, $\partial_y V_{xyy} = \partial_x V_{yyy}$ for the fourth partial derivatives lead to the *integrability conditions*

$$\partial_x(2C - B) = \partial_y(2D + A) \qquad \text{(satisfied identically)} \tag{7.21}$$

$$\begin{aligned} C_{xx} - C_{yy} - A_{xy} &= 2CC_y - DA_y - 2CD_x + AA_y - AC_x \\ &\quad + CB_y + BC_y \end{aligned} \tag{7.22}$$

$$\begin{aligned} D_{xx} - D_{yy} - B_{xy} &= -2DD_x - CB_x + 2DC_y - BB_x \\ &\quad - BD_y + DA_x + AD_x. \end{aligned} \tag{7.23}$$

Note that if we have another constant of the motion $\mathcal{L}_3$ associated with a nondegenerate potential, then $\mathcal{L}_3$ must be a linear combination of $\mathcal{H}$, $\mathcal{L}_1$, $\mathcal{L}_2$. Indeed, if $\mathcal{L}_3$ is not a linear combination of the basis functions, then the potential V must satisfy an equation (7.15) that is linearly independent of the equations associated with $\mathcal{L}_1, \mathcal{L}_2$. This means an additional constraint on the solution space and that V can depend on at most two parameters, which is a contradiction.

We will use the conditions (7.22), (7.23) to classify the possible potentials V and the corresponding constants of the motion $\mathcal{L}_1, \mathcal{L}_2$. For this we note that it is only the three-dimensional subspace spanned by $\mathcal{H}, \mathcal{L}_1, \mathcal{L}_2$ that matters; we can choose any basis for this subspace. Hence we can replace the conditions (7.16), (7.17) by linear combinations of themselves without changing the potential. Moreover, to simplify the results we note that we can always subject the coordinates (x, y), and $\mathcal{L}_1, \mathcal{L}_2$ to a simultaneous Euclidean motion, i.e., we regard all translated and rotated potentials as members of the same equivalence class.

Multiplying both sides of (7.22) and (7.23) by $\mathcal{E}^3$ we obtain polynomial identities in x and y. Equating the coefficients of the various powers $x^n y^m$ we obtain conditions on the parameters H_{jk}. The simplest non-trivial condition, which is associated with the coefficient of a fifth order power in either of the equations, is

$$2H_{15}(H_{14}^2 - H_{12}^2) + H_{24}(H_{14}^2 + H_{12}^2) - 4H_{14}H_{12}H_{13} = 0. \tag{7.24}$$

We exploit these and the remaining conditions, and Euclidean motions to classify the possibilities for the $\mathcal{L}_j$. The full conditions (7.22) and (7.23), expressed in terms of the parameter H_{ij}, take several pages to list and are complicated to solve directly. (Indeed a symbol manipulation program was an important aid to our computations.) However, by dividing the problem up into special cases and using Euclidean motions, we can simplify the conditions and obtain a full solution. In the listing that follows we use the fact that the constants of the motion can each be expressed as a quadratic element in the enveloping algebra of the Euclidean group in the plane with basis elements

$$p_x, \quad p_y, \quad M = xp_y - yp_x,$$

to which a potential term $W(x, y)$ is added. (Strictly speaking, conditions (7.22) and (7.23) are only necessary conditions for existence of nondegenerate potentials.

CASE (3a). $H_{15}^2 + H_{13}^2 \neq 0$

(7.33) $$[1,0,0,0,0], \quad [0,0,1,0,0].$$

Here,

(7.34) $$L_1 = M^2 + W^{(1)}, \quad L_2 = p_x^2 + W^{(2)}$$

(7.35) $$V(x) = \alpha(x^2+y^2) + \frac{\beta}{x^2} + \frac{\gamma}{y^2}.$$

This potential permits separation in polar, elliptic and Cartesian coordinates.

CASE (3b). $H_{15}^2 + H_{13}^2 = 0$

(7.36) $$[1,0,0,0,0], \quad [0,0,2,0,\pm 2i].$$

Here,

$$L_1 = M^2 + W^{(1)}, \quad L_2 = p_+^2 + W^{(2)}$$

(7.37) $$V(x) = \alpha\frac{x^2+y^2}{(x+iy)^4} + \frac{\beta}{(x+iy)^2} + \gamma(x^2+y^2).$$

(There is a similar solution where the term p_+^2 in L_2 is replaced by p_-^2.) The potential permits separation in hyperbolic and polar coordinates.

CASE (3c). $H_{15}^2 + H_{13}^2 = 0$

(7.38) $$[1,0,c^2,0,0], \quad [0,0,2,0,\pm 2i].$$

Here,

$$L_1 = M^2 + c^2p_x^2 + W^{(1)}, \quad L_2 = p_+^2 + W^{(2)}$$

(7.39) $$V(x) = \frac{\alpha z}{\sqrt{z^2-c^2}} + \frac{\beta\bar{z}}{\sqrt{z^2-c^2}(z+\sqrt{z^2-c^2})^2} + \gamma z\bar{z}.$$

The potential permits separation in hyperbolic and elliptic coordinates.

CASE (4). $H_{12} = H_{13} = H_{14} = H_{15} = 0$, $\alpha_2 \neq 0$, $H_{24} \neq 0$

(7.40) $$[0,1,0,0,0], \quad [0,0,0,1,0].$$

Here,

(7.41) $$L_1 = -2Mp_x + W^{(1)}, \quad L_2 = -2Mp_y + W^{(2)}$$

(7.42) $$V(x) = \frac{\alpha}{\sqrt{x^2+y^2}} + \beta\frac{(\sqrt{x^2+y^2}+x)^{1/2}}{\sqrt{x^2+y^2}} + \gamma\frac{(\sqrt{x^2+y^2}-x)^{1/2}}{\sqrt{x^2+y^2}}.$$

Separation of variables is possible in two types of parabolic coordinates, the usual parabolic coordinates and the interchanged parabolic coordinates $x = \mu\nu, y = \frac{1}{2}(\mu^2-\nu^2)$.

CASE (5). $(H_{34}-H_{25})^2 + (H_{45}-H_{23})^2 \neq 0$

(7.43) $$[0,1,\alpha_3,\pm i,0], \quad [0,0,\pm i,0,1].$$

Here we choose the typical case

(7.44) $$L_1 = 4iMp_- + p_+^2 + W^{(1)}, \quad L_2 = p_-^2 + W^{(2)}$$

(7.45) $$V(x) = \alpha(x-iy) + \beta\Big(x+iy-\frac{3}{2}(x-iy)^2\Big) + \gamma\Big(x^2+y^2-\frac{1}{2}(x-iy)^3\Big).$$

where $\mathcal{L}_h = \ell_h + W^{(h)}$. Let $\mathcal{R} = \{\ell_1, \ell_2\}$. Then for each of the cases listed above it is straightforward to check that $\mathcal{R}^2 = \mathcal{P}_3(\ell_0, \ell_1, \ell_2)$ where $\mathcal{P}_3$ is a homogeneous third order polynomial in its arguments.[1] It follows that

$$\mathcal{R}^2 = \mathcal{F}(\mathcal{L}_0, \mathcal{L}_1, \mathcal{L}_2) = \mathcal{P}_3(\mathcal{L}_0, \mathcal{L}_1, \mathcal{L}_2) + \mathcal{F}_4(\mathbf{s}, \mathcal{L}_0, \mathcal{L}_1, \mathcal{L}_2), \tag{7.52}$$

where $\mathcal{F}_4$ is a fourth, second and zeroth order polynomial in the momenta p_x, p_y, and $\mathcal{F}_4(\mathbf{0}, \mathcal{L}_0, \mathcal{L}_1, \mathcal{L}_2) = 0$. Here, the parameters in the potential are denoted by $\mathbf{s} = (V_x^0, V_y^0, V_{yy}^0)$, evaluated at some fixed point (x_0, y_0) and $\mathcal{F}_4$ is a polynomial function of these parameters.

From (7.6) we have

$$\{\ell_1, \mathcal{R}\} = \frac{1}{2}\frac{\partial \mathcal{P}_3}{\partial \ell_2}(\ell_0, \ell_1, \ell_2),$$
$$\{\ell_2, \mathcal{R}\} = -\frac{1}{2}\frac{\partial \mathcal{P}_3}{\partial \ell_1}(\ell_0, \ell_1, \ell_2),$$

hence

$$\{\mathcal{L}_1, \mathcal{R}\} = \frac{1}{2}\frac{\partial \mathcal{P}_3}{\partial \mathcal{L}_2}(\mathcal{L}_0, \mathcal{L}_1, \mathcal{L}_2) + \frac{1}{2}\frac{\partial \mathcal{F}_4}{\partial \mathcal{L}_2}(\mathbf{s}),$$
$$\{\mathcal{L}_2, \mathcal{R}\} = -\frac{1}{2}\frac{\partial \mathcal{P}_3}{\partial \mathcal{L}_1}(\mathcal{L}_0, \mathcal{L}_1, \mathcal{L}_2) - \frac{1}{2}\frac{\partial \mathcal{F}_4}{\partial \mathcal{L}_1}(\mathbf{s}),$$

where the $\partial \mathcal{F}_4 / \partial \mathcal{L}_h(\mathbf{s})$ have only terms of orders two and zero in the momenta. It follows that the $\partial \mathcal{F}_4 / \partial \mathcal{L}_h(\mathbf{s})$ must be expressible as linear combinations of the $\mathcal{L}_h$. This shows that the commutators $\{\mathcal{L}_h, \mathcal{R}\}$ can be expressed as polynomials in $\mathcal{L}_0, \mathcal{L}_1, \mathcal{L}_2$. It is then a simple matter to verify that $\mathcal{F}$ itself is a polynomial in $\mathcal{L}_0, \mathcal{L}_1, \mathcal{L}_2$.

A complete list of the quadratic algebra relations for each of the cases studied above is given in [**23**]. In view of relations (7.6) it is sufficient to give the relation $R^2 = F(L_0, L_1, L_2)$ for each case. An example is

CASE (1). $[1,0,0,0,0]$, $[0,0,0,1,0]$

$$\mathcal{R}^2 = 16\mathcal{L}_1^2\mathcal{H} - 16\mathcal{L}_2^2\mathcal{L}_1 - 32(\beta+\gamma)\mathcal{L}_2^2 + 64\alpha(\beta-\gamma)\mathcal{L}_2 + 16\alpha^2\mathcal{L}_1 - 256\beta\gamma\mathcal{H} - 32\alpha^2(\beta+\gamma).$$

There are analogous quantum algebras for superintegrable systems arising from the potentials we have already computed. The only difference is that the Poisson bracket is now replaced by the commutator bracket $[A, B] = AB - BA$ and the operators H, L_1 and L_2 are the obvious (formally self-adjoint) symmetry partial differential operators.

$$H - \partial_x^2 + \partial_y^2 + V(x,y), \quad L_h = \sum_{k,j=1}^{2} \partial_k (a_{(h)}^{kj}) \partial_j + W_{(h)}(x,y), \quad h = 1, 2. \tag{7.53}$$

Just as for the Hamilton-Jacobi case, if we have another constant of the motion L associated with a maximal potential, then L must be a linear combination of H, L_1, L_2. Indeed, if L is in self-adjoint form, then the conditions that $[H, L] = 0$ are identical with (7.10), (7.11). Thus, if L is not a linear combination of the basis functions, then the potential V must satisfy an equation (7.15) that is linearly

[1] Moreover, it is straightforward to verify that the cases corresponding to nondegenerate potentials are the *only* cases where $\mathcal{P}_3$ is a homogeneous third order polynomial in its arguments. Thus the possible quadratic algebras generated by second order elements in the Euclidean Lie algebra correspond one-to-one with nondegenerate potentials.

independent of the equations associated with L_1, L_2. This means an additional constraint on the solution space and that V can depend on at most two parameters, which is a contradiction.

Furthermore the proof of the existence of quadratic algebra relations goes through almost unchanged for the operator case: $[L_1, L_2]^2 = R^2$ and $[L_1, R], [L_2, R]$ can be expressed as (symmetric) polynomials in the operators H, L_1, L_2. To make the prior construction go through, one need only note that since R^2 is a formally self-adjoint 6th order differential symmetry operator, the 5th order terms are fixed linear functions of the 6th order terms. The expressions $\{A, B\} = AB + BA$ and $\{A, B, C\} = ABC + CAB + BCA$ are operator symmetrizers. The detailed results are contained in [**23**]. An example is

CASE (1).

$$[L_2, R] = 8L_2^2 + 8HL_2 + 8\alpha^2,$$

$$[L_1, R] = 8\{L_2, L_1\} + 16(1 + 2\beta + 2\gamma)L_2 + 32\alpha(\gamma - \beta),$$

$$R^2 = 16L_1^2 H - \frac{8}{3}\{L_2, L_2, L_1\} - 16\left(2\beta + 2\gamma + \frac{11}{3}\right)L_2^2 - \frac{176}{3}HL_1 + 64\alpha(\beta - \gamma)L_2 + 16\alpha^2 L_1 + \left(-\frac{32}{3} + 96\gamma + 96\beta + 256\beta\gamma\right)H - \frac{32}{3}\alpha^2(3\beta + 3\gamma - 1).$$

8. Conclusions

We have used the concept of a "nondegenerate potential" to add structure to the study of superintegrable classical and quantum mechanical systems in $E_{2,C}$. We have shown how to classify all such systems in a straightforward manner. Furthermore:

(1) Each system is associated with a pair of constants of the motion in the classical case, and a pair of symmetry operators in the quantum case, that generate a quadratic algebra.
(2) There is a one-to-one correspondence between superintegrable systems and free-field symmetry operators that generate quadratic algebras.
(3) Superintegrability implies multiseparability.

We have analogous results for superintegrable systems on the complex 2-sphere, [**24**]. The real sphere and the real hyperboloid are special cases. Again we find and fill gaps in the known list of superintegrable systems. The next major challenge is to extend this analysis to higher dimensional systems.

References

1. V. I. Arnold, *Mathematical methods of classical mechanics* (translated by K. Vogtmann and A. Weinstein), Graduate Texts in Math., vol. 60, Springer-Verlag, New York, 1978.
2. D. Bonatos, C. Daskaloyannis, and K. Kokkotas., *Deformed oscillator algebras for two-dimensional quantum superintegrable systems*, Phys. Rev. A **50** (1994), no. 5, 3700–3709.
3. C. P. Boyer, E. G. Kalnins, and P. Winternitz, *Completely integrable relativistic Hamiltonian systems and separation of variables in Hermitian hyperbolic spaces*, J. Math. Phys. **24** (1983), no. 8, 2022–2034.
4. C. P. Boyer, E. G. Kalnins, and P. Winternitz, *Separation of variables for the Hamilton-Jacobi equation on complex projective spaces*, SIAM J. Math. Anal. **16** (1985), 93–109.

5. F. Calogero, *Solution of a three-body problem in one dimension*, J. Math. Phys. **10** (1969), no. 12, 2191–2196.
6. Lewis Carroll, *Alice's adventures in Wonderland*, Macmillan, London, 1865; St. Martins Press, New York, 1991.
7. Yu. N. Demkov, *Symmetry group of the isotropic oscillator*, Soviet Phys. JETP **26** (1954), 757; **36** (1959), 63–66.
8. L. P. Eisenhart, *Enumeration of potentials for which one-particle Schrödinger equations are separable*, Phys. Rev. **74** (1948), 87–89.
9. A. Erdélyi et al., *Higher transcendental functions*, Vol. II, McGraw-Hill, New York, 1953.
10. N. W. Evans, *Superintegrability of the Calogero-Moser system*, Phys. Lett. A **95** (1983), 279–281.
11. N. W. Evans, *Superintegrability in classical mechanics*, Phys. Rev. A **41** (10) (1990), 5666–5676; *Group theory of the Smorodinsky-Winternitz system*, J. Math. Phys. **32** (1991), 3369–3375.
12. J. Friš, V. Mandrosov, Ya. A. Smorodinsky, M. Uhlir, and P. Winternitz, *On higher symmetries in quantum mechanics*, Phys. Lett. **16** (1965), 354–356.
13. J. Friš, Ya. A. Smorodinskii, M. Uhlír, and P. Winternitz, *Symmetry groups in classical and quantum mechanics*, Soviet J. Nuclear Phys. **4** (1967), 444–450.
14. Ya. A. Granovsky, A. S. Zhedanov, and I. M. Lutzenko, *Quadratic algebra as a 'hidden' symmetry of the Hartmann potential*, J. Phys. A **24** (16) (1991), 3887–3894.
15. C. Grosche, G. S. Pogosyan, and A. N. Sissakian, *Path integral discussion for Smorodinsky-Winternitz potentials. I. Two- and three-dimensional Euclidean space*, Fortschr. Phys. **43** (1995), 453–521.
16. E. G. Kalnins, *Separation of variables for Riemannian spaces of constant curvature*, Pitman Monographs Surveys Pure Appl. Math., vol. 28, Longman, Harlow, England, 1986.
17. E. G. Kalnins and W. Miller, Jr., *The theory of orthogonal R-separation for Helmholtz equations*, Adv. Math. **51** (1984), 91–106.
18. E. G. Kalnins and W. Miller, Jr., *Separation of variables on n-dimensional Riemannian manifolds. I. The n-sphere S_n and Euclidean n-space R_n*, J. Math. Phys. **27** (7) (1986), 1721–1736.
19. E. G. Kalnins and W. Miller, *Hypergeometric expansions of Heun polynomials*, SIAM J. Math. Anal. **22** (1991), 1450–1459.
20. E. G. Kalnins and W. Miller, Jr., *q-Algebra representations of the Euclidean, pseudo-Euclidean and oscillator algebras, and their tensor products*, Symmetries and Integrability of Difference Equations (D. Levi et al., eds.), CRM Proc. Lecture Notes, vol. 9, Amer. Math. Soc., Providence, RI, 1996.
21. E. G. Kalnins, W. Miller Jr., and G. S. Pogosyan, *Superintegrability and associated polynomial solutions. Euclidean space and the sphere in two dimensions*, J. Math. Phys. **37** (12) (1996), 6439–6467.
22. E. G. Kalnins, W. Miller, and G. S. Pogosyan, *Superintegrability in three dimensional Euclidean space*, J. Math. Phys. **40** (1999), 708–725.
23. E. G. Kalnins, W. Miller, and G. S. Pogosyan, *Completeness of multiseparable superintegrability in $E_{2,C}$*, IMA preprint 1678 (submitted for publication).
24. E. G. Kalnins, W. Miller, and G. S. Pogosyan, *Completeness of multiseparable superintegrability on the complex 2-sphere*, IMA preprint 1686 (submitted for publication).
25. E. G. Kalnins, W. Miller, and M. V. Tratnik, *Families of orthogonal and biorthogonal polynomials on the n-sphere*, SIAM J. Math. Anal. **22** (1991), 272–294.
26. L. Landau and E. Lifshitz, *The classical theory of fields*, Addison-Wesley, Providence, RI, 1959 (translated from Russian).
27. P. Létourneau and L. Vinet, *Superintegrable systems: polynomial algebras and quasi-exactly solvable Hamiltonians*, Ann. Phys. **243** (1995), 144–168.
28. A. A. Makarov, Ya. A. Smorodinsky, Kh. Valiev, and P. Winternitz, *A systematic search for nonrelativistic systems with dynamical symmetries*, Nuovo Cimento A **52** (1967), 1061–1084.
29. L. G. Mardoyan, G. S. Pogosyan, A. N. Sissakian, and V. M. Ter-Antonyan, *Elliptic basis for a circular oscillator*, Nuovo Cimento B **88** (1985), 43–56;
Two-dimensional hydrogen atoms. I. Elliptic bases, Theor. Math. Phys. **61** (1984), 1021–1034; *Hidden symmetry, separation of variables and interbasis expansions in the two-dimensional hydrogen atom*, J. Phys. A **18** (1985), 455–466.

30. W. Miller, Jr., *Lie theory and special functions*, Academic Press, New York, 1968.
31. W. Miller, Jr., *Symmetry and separation of variables*, Addison-Wesley, Providence, RI, 1977.
32. W. Miller Jr., *Mechanisms for variable separation in partial differential equations and their relationship to group theory*, Symmetries and Non-Linear Phenomena (D. Levi and P. Winternitz, eds.), World Scientific, 1988, pp. 188–221.
33. M. F. Rañada, *Superintegrable $n = 2$ systems, quadratic constants of motion, and potentials of Drach*, J. Math. Phys. **38** (8) (1997), 4165–4178.
34. V. N. Shapovalov, *Stäckel spaces*, Siberian Math. J. **20** (1980), 790–800.
35. N. Vilenkin, *Special functions and the theory of group representations*, Transl. Math. Monogr., vol. 22, Amer. Math. Soc., Providence, RI, 1968.
36. P. Winternitz, I. Lukac, and Ya. A. Smorodinskii, *Quantum numbers of the little group of the Poincaré group*, Soviet J. Nuclear Phys. **7** (1968), 139–145.
37. S. Wojciechowski, *Super-integrability of the Winternitz system*, Phys. Lett. A **147** (1990), 483–486.

INSTITUTE FOR MATHEMATICS AND ITS APPLICATIONS, UNIVERSITY OF MINNESOTA, MINNEAPOLIS, MN 55455, USA

E-mail address: miller@ima.umn.edu

Centre de Recherches Mathématiques
CRM Proceedings and Lecture Notes
Volume **26**, 2000

Integrability and Symmetry of the XXZ Model

Tetsuji Miwa

ABSTRACT. I discuss the integrability and symmetry of the XXZ Hamiltonian by using the representation theory of quantum affine sl_2 algebra, and give the particle and local picture for the space of its eigenvectors.

1. Particle Structure of Integrable Hamiltonian

I will discuss the integrability of models in statistical mechanics from the viewpoint of representation theory, or I would say, symmetries. A symmetry of Hamiltonian is an operator which commutes with it. The system is called integrable if the Hamiltonian has a lot of symmetries. To make this statement precise, I consider a specific example: the XXZ Hamiltonian

$$H = -\frac{1}{2}\sum_n (\sigma_n^x\sigma_{n+1}^x + \sigma_n^y\sigma_{n+1}^y + \Delta\sigma_n^z\sigma_{n+1}^z).$$

Here, Δ is a real parameter. I restrict to the region

$$\Delta < -1.$$

It is useful to set

$$\Delta = \frac{q+q^{-1}}{2}.$$

We have $-1 < q < 0$ in the above region. The operators σ_n^x, σ_n^y and σ_n^z are the Pauli matrices acting on the n-th component of the tensor product

$$F = \otimes_n V_n, \quad V_n \simeq \mathbf{C}^2.$$

The main problem is the diagonalization of H. Namely, we want to know the eigenvalues (energy spectrum) and the corresponding eigenvectors. We are most interested in the infinite volume limit. For a finite N, we may set the cyclic boundary condition such that $V_n = V_{n+N}$. In the limit $N \to \infty$, the space F becomes infinite-dimensional. Mathematically, it is not even a priori defined. For finite N, we can label the eigenvectors by their eigenvalues; we order eigenvectors by their eigenvalues from the smallest to the largest. Letting N large, we trace the behavior of eigenvalues and eigenvectors, those which have zero or finite (i.e., not growing with N) gaps above the smallest eigenvalue.

1991 *Mathematics Subject Classification.* 05E05, 15A52, 17B37, 58F07, 70F10.
This is the final form of the paper.

We say the system has a particle structure if the following statement holds. I consider the case of a massive particle. The eigenvector corresponding to the smallest eigenvalue is called the ground state. There is a gap between the smallest and the next smallest eigenvalue. This gap m is called the mass. In the infinite volume limit, the continuous spectrum starts at this point. There exists a parameter p called quasi momentum, and the energy spectrum (normalized at the smallest) is given by some function $\epsilon(p)$ $(p \in \mathbf{R}/2\pi\mathbf{Z})$. The minimum of $\epsilon(p)$ is attained at $p = 0$ and the value is m. The set of eigenvalues parametrized in this way is called the single particle spectrum. In general, we have multi-particle spectrum. For example, 2-particle spectrum are parametrized by a pair of quasi momentums (p_1, p_2), and the energy is given by $\epsilon(p_1) + \epsilon(p_2)$.

The particle structure is a very strong statement about a system of infinite degrees of freedom. I do not know under which assumptions a particle structure exists. However, we know a lot of examples for which a particle structure exists, in particular, among integrable systems. It is an open problem if a particle structure exists in non-integrable (or near-integrable) systems.

Eigenvalues may have degeneracy. If the degeneracy is finite (for finite N it is certainly so) and not large, we must be able to talk about the corresponding eigenvectors. However, if (in the infinite volume limit) infinite degeneracy occurs, we need some other mean to specify a particular eigenvector from the infinite set. The particle structure is enough for this. For example, for a two particle eigenvector, we have not only the energy $\epsilon(p_1)+\epsilon(p_2)$ but also the number of particle, i.e., 2, and the two quasi momentums p_1 and p_2, which specify the finite-dimensional subspace.

A question is how to single out the set of momentums out of the system. Integrability does the job as I will now explain. Suppose that we have a set of symmetries X_i of the Hamiltonian. Assume that they are not only commuting with H but also commuting each other. We diagonalize Hamiltonian along with these commuting set of operators. Then, simultaneous eigenspaces give a decomposition of an eigenspace of H; if there are enough number of symmetries, the decomposed subspaces become finite-dimensional even in the infinite volume limit. In this case, we call the system integrable.

In the case of the XXZ Hamiltonian, we can construct commuting set of operators in the following way. We have an operator $R(\zeta)$ acting on $\mathbf{C}^2 \otimes \mathbf{C}^2$. This operator is dependent not only on ζ but also q. Since we fix q in our discussion, we do not write the q-dependence explicitly unless it is important. Consider the cyclic boundary condition of size N, and set

$$T(\zeta) = \mathrm{trace}_a\, R_{a1}(\zeta) R_{a2}(\zeta) \cdots R_{aN}(\zeta).$$

The index a indicates an auxiliary space $V_a \simeq \mathbf{C}^2$. We take the trace on this space, and the operator $T(\zeta)$ acts on the tensor product $\otimes_n V_n$. For an appropriate choice of thc matrix R, we have

$$[H, T(\zeta)] = 0.$$

In fact, H is contained in $T(\zeta)$:

$$H = \mathrm{const.}\ \zeta \frac{d}{d\zeta} T(\zeta).$$

The smallest eigenvalue of H corresponds to the largest eigenvalue of $T(\zeta)$. We normalize $R(\zeta)$ in such a way that the largest eigenvalue of $T(\zeta)$ is 1. The n-particle spectrum is characterized by the equality of the eigenvalue

$$\log T(\zeta) = \sum_{i=1}^{n} t(\zeta, p_i). \tag{1.1}$$

Since ζ is an arbitrary parameter, this single equality determines the set of momentums $p_1, \ldots, p_n$.

For finite N, the values of p_i are not arbitrary (because the total space is finite-dimensional). They take only finitely many possibilities that are determined by the so-called Bethe equations. I will not go into this discussion but explain the particle structure, i.e., why the transfer matrix has such spectrum by means of representation theory.

2. Non-Abelian Symmetries

Let us discuss more about the symmetry and degeneracy of Hamiltonian. I defined the integrability by the existence of commuting (i.e., abelian) symmetries so that the degeneracy of the Hamiltonian resolves to finite degeneracy by the simultaneous diagonalization of these commuting operators. If the degeneracy is still not trivial (i.e., the simultaneous eigenspaces are not one-dimensional), then we can expect more symmetries of the Hamiltonian, which explains the degeneracy. This time we have further symmetries which form a non-abelian family. For example, if $\Delta = 1$, the Hamiltonian (now it is called the XXX Hamiltonian) has the sl_2 symmetry: we consider V_n as the two dimensional representation of the Lie algebra sl_2; there is an action of sl_2 on tensor product $\otimes_n V_n$; this action commutes with the Hamiltonian. In fact, this action commutes with the transfer matrix $T(\zeta)$. Therefore, the simultaneous eigenspaces are closed under the sl_2 action. If this action is irreducible, the eigenvectors are completely labeled by the sl_2 action. However, this is not true.

In the infinite volume limit, the degeneracy may increase because a fine difference of the spectrum may disappear in the limit. A singlet and a triplet, that have different eigenvalues for finite N may degenerate to a quartet. Even if it happens, as for the sl_2 action, the four dimensional space is still a direct sum of one and three dimensional irreducible spaces.

This difficulty is resolved if we have larger symmetries in the infinite volume limit. For $\Delta < -1$, such larger symmetries really exist; they form the quantum affine algebra $U_q(\widehat{\mathrm{sl}}_2)$. We consider V_n as the two dimensional representation of $U_q(\widehat{\mathrm{sl}}_2)$. For finite N, the tensor product $V_1 \otimes \cdots \otimes V_N$ has an $U_q(\widehat{\mathrm{sl}}_2)$ action given by the standard coproduct. However, this action does not commute with the finite transfer matrix. The transfer matrix is defined by using cyclic boundary condition (i.e., no boundary), and does not fit to the tensor action with boundary at $n = 1$ and $n = N$ (or rather, $n = -N/2$ and $n = N/2$). Note that $U_q(\widehat{\mathrm{sl}}_2)$ contains two copies of $U_q(\mathrm{sl}_2)$. It is possible to modify the Hamiltonian to fit to the $U_q(\mathrm{sl}_2)$ action for one of the two [**12**]. However, it is not possible to adjust to both. On the other hand, in the infinite volume limit both of the boundaries go to infinity; thus we have $U_q(\widehat{\mathrm{sl}}_2)$ symmetry in this limit. This is only a heuristic argument. In the next section, I will explain how to construct a mathematical theory out of this idea. Among others, we must set appropriate boundary conditions for $n \to \pm\infty$. Take

$i, j = 0, 1$. In the region $\Delta < -1$, the proper boundary condition is such that the vector in V_n is frozen to v_{i+n} for $n \to \infty$, and v_{j+n} for $n \to -\infty$. Here, $v_{0 \bmod 2}$ and $v_{1 \bmod 2}$ are the weight vectors of the two dimensional representation. We denote the corresponding space by $F^{i,j}$.

Before passing, let me discuss the following natural question: What is the space F in the infinite volume limit as a representation space of $U_q(\widehat{\mathrm{sl}}_2)$? We interprete the particle structure in the language of representation theory. The action of $U_q(\widehat{\mathrm{sl}}_2)$ on $\mathbf{C}^2$ admits one parameter, say

$$\zeta = e^{\sqrt{-1}p}.$$

We denote this representation by V_ζ. For $\zeta = 1$, we denote it simply by V.

The particle structure in this language reads as follows. The ground state is the one-dimensional subspace of $F^{0,0}$ or $F^{1,1}$. The single particle state corresponds to the two dimensional representation V_ζ for some p, and it belongs to $F^{0,1}$ or $F^{1,0}$. In general, an n-particle state corresponds to

$$V_{\zeta_1} \otimes \cdots \otimes V_{\zeta_n},$$

and belongs to $F^{i,i}$ (if n is even) or $F^{i,1-i}$ (if n is odd).

3. Intertwiners and Half-Infinite Tensor Product

The particle picture given in the last section has a defect. The structure of tensor product, which is the main structure of the original system, has been completely forgotten. For example, the spin operator σ_1^z certainly acts on F in the original picture, but its action is not given in the above particle picture.

I present a different picture, having clear meaning in the representation theory, that suits for description of local operators. We call it the local picture.

We start from a half infinite tensor product

$$F_{\mathrm{left}} = \cdots \otimes V \otimes V \otimes V.$$

We want to identify this space with a representation space of $U_q(\widehat{\mathrm{sl}}_2)$. The action is very complicated on the tensor product. However, in the crystal limit $q \to 0$, the action of the Chevalley generators e_i, f_i $(i = 0, 1)$ drastically simplifies. For example, the action of f_1 is as follows: Suppose we have a vector

$$\cdots \otimes v_{i_3} \otimes v_{i_2} \otimes v_{i_1}.$$

We consider any neighboring pair $v_0 \otimes v_1$ contained in this sequence to be a singlet. We remove all such pairs. Because of the boundary condition, (we denote by F^i_{left} the space with the boundary condition $i_n = i + n \bmod 2$ for $i = 0, 1$) for large n, $i_n = 0, 1$ alternate. Therfore, we are left with a finite subsequence. We repeat the same procedure to this subsequence until the subsequnce reduces to

$$\underbrace{v_1 \otimes \cdots \otimes v_1}_{m} \otimes \underbrace{v_0 \otimes \cdots \otimes v_0}_{n}.$$

Then, the f_1 action on the original sequence is such that this subsequence changes to

$$\underbrace{v_1 \otimes \cdots \otimes v_1}_{m+1} \otimes \underbrace{v_0 \otimes \cdots \otimes v_0}_{n-1}.$$

The crystal actions of other Chevalley generators are similarly given.

Using the crystal action we can calculate the character of the half-infinite tensor product. The result suggests the following identification:

$$F^i_{\text{left}} \simeq V(\Lambda_i). \tag{3.1}$$

Here, by $V(\lambda)$ I mean the irreducible highest weight representation of $U_q(\widehat{\mathrm{sl}}_2)$ with highest weight λ.

Historically, the computation of such characters was initiated by Baxter in his calculation of the one point correlation functions [**1**]. He invented the method of corner transfer matrix for this purpose. The connection to the representation theory in the character level was extensively studied in Kyoto [**3, 4, 5**]. It led to Kashiwara's theory of crystals [**2, 10, 11**], and finally in [**7, 6, 8, 9**], the above identification was established and used in the diagonalization of the Hamiltonian and the computation of the general correlation functions.

Now, I will explain how to identify F^i_{left} with $V(\Lambda_i)$. First, let me give a general definition of intertwiners. Suppose we have representations M_1 and M_2 of an algebra A. A mapping $\phi : M_1 \to M_2$ is called an intertwiner if the following diagram is commutative:

$$\begin{array}{ccc} M_1 & \xrightarrow{\phi} & M_2 \\ \downarrow & & \downarrow \\ M_1 & \xrightarrow{\phi} & M_2 \end{array}$$

Here, the arrow $M_i \to M_i$ means the A-action.

There exists a unique intertwiner of the form

$$\Phi(\zeta) : V(\Lambda_i) \xrightarrow{\simeq} V(\Lambda_{1-i}) \otimes V_\zeta.$$

We define the components $\Phi_j(\zeta) : V(\Lambda_i) \to V(\Lambda_{1-i})$ by

$$\Phi(\zeta)|v\rangle = \sum_{j=0,1} \Phi_j(\zeta)|v\rangle \otimes v_j.$$

The intertwiner $\Phi(\zeta)$ is an isomorphism in the sense we have an equality

$$\sum_{j=0,1} \Phi_j(-q^{-1}\zeta)\Phi_{1-j}(\zeta) = \mathrm{id}. \tag{3.2}$$

This is called the inversion relation. I will use this equality in the next section.

By using $\underbrace{\Phi(1) \circ \cdots \circ \Phi(1)}_{n}$ we can identify $V(\Lambda_i)$ with

$$V(\Lambda_{i+n \bmod 2}) \otimes \underbrace{V \otimes \cdots \otimes V}_{n}.$$

In this way, we can see the local tensor components of F^i_{left} .

The right half of $F^{i,j}$,

$$V \otimes V \otimes V \otimes \cdots ,$$

which we denote by F^j_{right}, is identified with the dual space of $V(\Lambda_j)$. Therefore, the total space is identified with the space

$$F^{i,j}_{\text{local}} = V(\Lambda_i) \otimes V(\Lambda_j)^*.$$

This is the local picture for the space of eigenvectors of the Hamiltonian.

4. Local and Particle Picture

In this section, I will connect the local picture with the particle picture. As I have explained, the ground state is the trivial representation of $U_q(\widehat{\mathrm{sl}}_2)$. It is the maximal eigenvector of $T(\zeta)$, i.e., the one corresponding to the largest eigenvalue. How is it contained in $F^{i,i}_{\mathrm{local}}$?

In order to characterize this vector, we follow Baxter's argument on the corner transfer matrix (CTM). In our present discussion, the basic fact which connects to the CTM method is that the matrix elements of $R(\zeta)$ are nothing but the Boltzmann weight of the six-vertex model. This is a model in statistical mechanics on the two dimensional lattice. The statistical sum is taken over the fluctuation variables sitting on bonds of the lattice. The transfer matrix $T(\zeta)$ corresponds to a column of the total lattice, with sums taken over the variablces on the vertical bonds on this column. A CTM corresponds to a quadrant of the total lattice, with sums taken over all the internal bonds of this quadrant. In the limit $N \to \infty$, it acts on the half-infinite tensor product, F^i_{left}. Baxter has shown that the product of two CTMs, say the NW and SW ones, has no dependence on the parameter ζ (except for a divergent scalar factor), and if we write it as

$$(-q)^D,$$

the spectrum of D belongs to $\mathbf{Z}_{\geq 0}$. Under the identification (3.1), the operator D is identified with the scaling operator which counts the degree of weight vectors in the highest weight representation.

Following Baxter [**1**] (page 375) we see that the maximal eigenvector is obtained as

$$T(\zeta)^N|\text{boundary}\rangle$$

in the limit $N \to \infty$, where $|\text{boundary}\rangle$ is a vector of the form $\otimes_n v_{i_n}$ with $i_n = i + n \bmod 2$. In short, the maximal eigenvector is given by the statistical sum over one half of the two-dimesional lattice. This is equivalent to the product of two CTMs, i.e., $(-q)^D$ discussed above. In conclusion, using the canonical identification,

$$F^{i,j}_{\mathrm{local}} = \mathrm{Hom}_{\mathbf{C}}\big(V(\Lambda_j), V(\Lambda_i)\big),$$

the ground state in the sector $F^{i,i}_{\mathrm{local}}$ is identified with

$$(-q)^D \in \mathrm{End}_{\mathbf{C}}\big(V(\Lambda_i)\big).$$

Let us derive the spectrum of the transfer matrix (1.1). We fully exploit the relations satisfied by the intertwiners $\Phi(\zeta)$ and $\Psi(\zeta)$. They are summarized as follows.

$$\xi^{-D} \circ \Phi_j(\zeta) \circ \xi^D = \Phi_j(\zeta/\xi), \tag{4.1}$$

$$\xi^{-D} \circ \Psi_j(\zeta) \circ \xi^D = \Psi_j(\zeta/\xi), \tag{4.2}$$

$$\Phi_{j_2}(\zeta_2)\Phi_{j_1}(\zeta_1) = \sum_{j_1',j_2'} R(\zeta_1/\zeta_2)^{j_1',j_2'}_{j_1,j_2} \Phi_{j_1'}(\zeta_1)\Phi_{j_2'}(\zeta_2), \tag{4.3}$$

$$\Psi_{j_1}(\zeta_1)\Psi_{j_2}(\zeta_2) = \sum_{j_1',j_2'} R(\zeta_1/\zeta_2)^{j_1,j_2}_{j_1',j_2'} \Psi_{j_2'}(\zeta_2)\Psi_{j_1'}(\zeta_1), \tag{4.4}$$

$$\Phi_{j_1}(\zeta_1)\Psi_{j_2}(\zeta_2) = \tau(\zeta_1/\zeta_2)\Psi_{j_2}(\zeta_2)\Phi_{j_1}(\zeta_1). \tag{4.5}$$

The action of the transfer matrix $T(\zeta)$ on a vector $f \in \mathrm{End}_{\mathbf{C}}\big(V(\Lambda_0) \oplus V(\Lambda_1)\big)$ is given by

$$T(\zeta)f = \sum_j \Phi_j(\zeta) \circ f \circ \Phi_{1-j}(\zeta).$$

Consider the vector

$$|\xi_1, \ldots, \xi_n\rangle_{j_1,\ldots,j_n} = \Psi_{j_1}(\xi_1) \circ \cdots \circ \Psi_{j_n}(\xi_n) \circ (-q)^D.$$

Using (4.1), (4.5) and (3.2), we obtain

$$\begin{aligned} T(\zeta)|\xi_1, \ldots, \xi_n\rangle_{j_1,\ldots,j_n} &= \sum_j \Phi_j(\zeta) \circ \Psi_{j_1}(\xi_1) \circ \cdots \circ \Psi_{j_n}(\xi_n) \circ (-q)^D \circ \Phi_{1-j}(\zeta) \\ &= \prod_m \tau(\zeta/\xi_m)|\xi_1, \ldots, \xi_n\rangle_{j_1,\ldots,j_n}. \end{aligned}$$

Setting $t(\zeta, p) = \log \tau(\zeta/e^{\sqrt{-1}p})$, we obtain (1.1). The equations (4.3) and (4.4) are important because they characterize these operators and connect the correlation functions and the form factors to the quantum Knizhnik-Zamolodchikov equation. I will not go into the details.

References

1. R. J. Baxter, *Exactly solved models in statistical mechanics*, Academic Press, London, 1982.
2. E. Date, M. Jimbo, and T. Miwa, *Representations of $U_q\big(\mathrm{gl}(n, \mathbf{C})\big)$ at $q = 0$ and the Robinson-Shensted correspondence*, Physics and Mathematics of Strings, World Scientific, Singapore, 1990, pp. 185–211.
3. E. Date, M. Jimbo, A. Kuniba, T. Miwa, and M. Okado, *Exactly solvable SOS models. I. Local height probabilities and theta function identities*, Nuclear Phys. B **290** (1987), 231–273.
4. E. Date, M. Jimbo, A. Kuniba, T. Miwa, and M. Okado, *Exactly solvable SOS models. II. Proof of the star-triangle relation and combinatorial identities*, Conformal Field Theory and Solvable Lattice Models (Kyoto, 1986), Adv. Stud. Pure Math., vol. 16, 1988, p. 17–122.
5. E. Date, M. Jimbo, A. Kuniba, T. Miwa, and M. Okado, *One dimensional configuration sums in vertex models and affine Lie algebra characters*, Lett. Math. Phys. **17** (1989), 69–77.
6. B. Davies, O. Foda, M. Jimbo, T. Miwa, and A. Nakayashiki, *Diagonalization of the XXZ Hamiltonian by vertex operators*, Comm. Math. Phys. **151** (1993), 89–153.
7. O. Foda and T. Miwa, *Corner transfer matrices and quantum affine algebras*, Internat. J. Modern Phys. A **7**, Suppl. 1 (1993), 89.
8. M. Jimbo, K. Miki, T. Miwa, and A. Nakayashiki, *Correlation functions of the XXZ model for $\Delta < -1$*, Phys. Lett. A **163** (1992), 256–263.
9. M. Jimbo and T. Miwa, *Algebraic analysis of solvable lattice models*, CBMS Regional Conf. Ser. in Math., vol. 85, Amer. Math. Soc., Providence, RI, 1995.
10. M. Kashiwara, *Crystallizing the q-analogue of universal enveloping algebras*, Comm. Math. Phys. **133** (1990), 249–260.
11. K. Misra and T. Miwa, *Crystal bases for basic representation of $U_q(\widehat{\mathrm{sl}}_n)$*, Comm. Math. Phys. **134** (1990), 79–88.
12. V. Pasquier and H. Saleur, *Common structure between finite systems and conformal field theory through quantum groups*, Nuclear Phys. B **330** (1990), 523–556.

DEPARTMENT OF MATHEMATICS, FACULTY OF SCIENCE, KYOTO UNIVERSITY, KYOTO 606, JAPAN.

E-mail address: tetsuji@kusm.kyoto-u.ac.jp

Centre de Recherches Mathématiques
CRM Proceedings and Lecture Notes
Volume **26**, 2000

Characteristic Systems on Poisson Lie Groups and Their Quantization

Nicolai Reshetikhin

ABSTRACT. These lectures are an overview of the approach to integrable systems which is based on Poisson Lie groups, and which has been developed over the years as a Hamiltonian interpretation of Lax operators. They also contain a review of some relatively recent results about such systems related to simple Lie groups.

1. Introduction

In these lectures we will focus on a specific but very rich class of integrable systems. We will discuss characteristic Hamiltonian systems on Poisson Lie groups. The phase space of such a system is a symplectic leaf of a factorizable Poisson Lie group. Poisson commuting Hamiltonians are restrictions of adjoint-invariant functions on the group to the symplectic leaf.

Most of the known integrable systems are of this kind, and most of them are related to Poisson Lie structures on loop groups. In these lectures the attention will be given to characteristic systems on simple Lie groups with standard Poisson Lie structures and to their quantization.

The first examples of such systems (on a Lie algebra, rather than on a Poisson Lie group) was given by Kostant [**Ko**] when he gave a Lie theoretical interpretation of Toda systems. This approach was developed for Lie algebras in works of Adler [**A**], van Moerbeke, Reyman, Semenov-Tian-Shansky, and others. In its present form (see the lectures) it was formulated in [**STS**]. One should emphasize that the theory of integrable systems has many aspects. It involves a lot of interesting analysis with various applications, in particular to the theory of soliton equations, a lot of interesting algebraic geometry (algebraically integrable systems etc.), interesting applications to mechanics and to statistical mechanics. Here we will focus on the role of Lie theory in integrable systems.

There are many important features of characteristic systems in the classical case. For example, those related to algebraic Poisson Lie groups are most likely algebraically integrable (see [**RSTS**]). But, perhaps, the most interesting feature of these systems is that their quantization is very closely related to the representation

2000 *Mathematics Subject Classification.* 53D17, 37J35, 17B62.
This is the final form of the paper.

theory of the corresponding quantum groups. Quantization will not be discussed in in these lectures, but it should be kept in mind that it is one of the strongest motivations for considering characteristc integrable systems.

In the first sections we present some general notions about integrable systems. Then there is an overview of Poisson Lie groups [**Dr, STS**], of the structure of symplectic leaves in Poisson Lie groups. Next we define characteristic systems and describe their integrability. Finally we discuss in detail the structure of simplectic leaves in simple Poisson Lie groups and describe Coxeter-Toda systems as examples.

2. Integrable Systems in Hamiltonian Mechanics

2.1. Recall that a symplectic manifold is a pair $(\mathcal{M}, \omega)$, where $\mathcal{M}$ is a manifold and ω is a nondegenerate closed 2-form on $\mathcal{M}$. In these lectures we will consider two types of symplectic manifolds: real smooth and complex algebraic. Accordingly, we will consider either smooth functions on $\mathcal{M}$ or algebraic functions. When $\mathcal{M}$ is real smooth the form will be assumed smooth, when $\mathcal{M}$ is complex algebraic the symplectic form will be assumed complex algebraic as well. We will not specify the type when it will not cause confusion.

A Hamiltonian dynamics on a symplectic manifold is determined by the choice of a function H on $\mathcal{M}$ which is called the Hamiltonian of the system. The trajectories in such a system are flow lines of the Hamiltonian vector field $v_H = \omega^{-1}(dH)$ generated by H.

The symplectic manifold on which the Hamiltonian dynamics system "lives" is the phase space of the system. The algebra of smooth functions $C(\mathcal{M})$ on the phase space is the algebra of classical observables. The symplectic structure on $\mathcal{M}$ equips the algebra of functions on $\mathcal{M}$ with the structure of a Poisson algebra. Recall that a Poisson algebra is a vector space with

- the structure of acommutative associative algebra;
- the structure of a Lie algebra (with the bracket $\{,\}$, called the Poisson bracket);
- such that the Lie algebra acts by derivations on the commutative algebra $(\{ab, c\} = a\{b, c\} + b\{a, c\})$.

On $C(\mathcal{M})$ the commutative associative algebra structure is given by the pointwise product, and the Lie algebra structure is given by the Poisson bracket

$$\{f, g\} = \omega^{-1}(df \wedge dg). \tag{2.1}$$

Algebraically, the evolution is the (local) action of $\mathbb{R}$ on the classical algebra of observables $t \in \mathbb{R} : f \to f_t$, where f_t is the solution to the Hamilton equations

$$\frac{df_t}{dt} = \{H, f_t\}, \quad f_0 = f. \tag{2.2}$$

This is a group action of the additive group $\mathbb{R}$ $f_{t+s} = (f_t)_s$ by automorphisms of the Poisson algebra $C(\mathcal{M})$: $(fg)_t = f_t g_t$, $\{f, g\}_t = \{f_t, g_t\}$.

2.2. In Hamiltonian mechanics the classical algebra of observables is the algebra of functions on a symplectic manifold. A more general example of a Poisson algebra is the algebra of functions on Poisson manifolds. A Poisson manifold is a pair $(\mathcal{M}, p)$, where $\mathcal{M}$ is a manifold (again we will consider either real smooth or complex algebraic manifolds), and $p \in \wedge^2 T\mathcal{M}$ is a bivector field such that the bracket

$$\{f, g\} = p(df \wedge dg) \tag{2.3}$$

is a Lie bracket on $C(\mathcal{M})$. When p is non-degenerate, the form $\omega \in \wedge^2 T^*\mathcal{M}$ is closed and non-degenerate and therefore determines a symplectic structure on $\mathcal{M}$. In general p can be degenerate.

A symplectic submanifold spanned by all piecewise Hamiltonian vector fields passing through $x \in \mathcal{M}$ is called a symplectic leaf passing through x. Locally $\mathcal{M}$ is fibered by symplectic leaves.

A function on a Poisson manifold obviously determines a Hamiltonian system on each symplectic leaf of this manifold. In some cases it is convenient to analyze these systems simultaneously for all symplectic leaves.

2.3. Liouville Integrable Hamiltonian Systems. The Hamiltonian system generated by $H \in C(\mathcal{M})$ is called *integrable* if there exist n functionally independent[1] $F_1, \dots, F_n \in C^\infty(\mathcal{M})$ such that

$$\{F_i, F_j\} = 0 = \{F_i, H\}.$$

Algebraically this means that the functions $F_1, F_2, \dots, F_n$ generate a maximal Poisson commutative subalgebra $I(\mathcal{M})$ in the algebra of functions on $\mathcal{M}$, and the Hamiltonian H belongs to this subalgebra.

The functions $F_1, \dots, F_n$ are called the *first integrals* of the system. The subalgebra $I(\mathcal{M})$ is called the *subalgebra of first integrals.*

Geometrically, n independent functions in involution (Poisson commutative) determine the Lagrangian projection $F : \mathcal{M} \to \Lambda \subset \mathbb{R}^n$, $F(x) = \big(F_1(x), \dots, F_n(x)\big)$ where Λ (the image of F) is an n-dimensional open subvariety in $\mathbb{R}^n$. The Hamiltonian H generates integrable dynamics if the flow lines of v_H are parallel to the fibers (equivalent to commutativity of H with F_i). Clearly the fibers of F are level surfaces of the functions $F_1, \dots, F_n$:

$$M_c = \{F_i(x) = c_i | x \in M\}.$$

We will also denote the level surfaces by $F^{-1}(c)$.

We will say that M_c is non-degenerate if the form $dF_1 \wedge \cdots \wedge dF_n$ does not vanish on M_c.

REMARK 2.1. More generally, an integrable system consists of a symplectic manifold $\mathcal{M}$, a Lagrangian $F : \mathcal{M} \to \Lambda$, and a Hamiltonian which is a pull-back of a function on Λ.

THEOREM 2.2 (Arnold-Liouville). *For an integrable system on a symplectic manifold the following holds*:

(1) *The Hamiltonian vector field generated by H is parallel to level surfaces of integrals (fibers of F).*
(2) *Each connected component of a generic level surface has a canonical affine structure: i.e., an atlas on M_c such that transition functions are affine transformations of $\mathbb{R}^n$ (compositions of translations and linear transformations), such that the flow lines of the Hamiltonian are linear in these coordinates.*
(3) *The compact connected components of level surfaces are tori.*

[1]This means that $dF_1 \wedge \cdots \wedge dF_n \neq 0$ on an open dense subset of $\mathcal{M}$.

The affine structure in the second statement is determined by the flow lines of the Hamiltonians $F_1, \ldots, F_n$. The affine coordinates on level surfaces are usually called the *angle variables.* The coordinates parametrizing level surfaces are called *action variables.*

If $(\phi_1, \ldots, \phi_n)$ is an such affine coordinate system then flow lines of H are linear:

$$\phi_i(t) = \omega_i(H, c_1, \ldots, c_n)t + \phi_i. \tag{2.4}$$

The "frequencies" ω_i are important characteristics of the system.

REMARK 2.3. It is worth emphasizing that integrable Hamiltonian systems have more "functorial" ingredients than Hamiltonian systems. A "functorial", geometrical, ingredient of a Hamiltonian system is a symplectic structure on a manifold (on a phase space of a system). In the integrable case we have not only this but also a Lagrangian projection given by the level surfaces of integrals.

Thus, from a geometrical point of view, integrable systems are Lagrangian projections of symplectic manifolds. This structure has "intrinsic" geometrical meaning.

2.4. It is possible that a Hamiltonian system on a symplectic manifold of dimension $2n$ admits only $r < n$ independent commuting integrals and is still integrable. In this case it should have $2n - r$ integrals which form a Poisson subalgebra with the Poisson center generated by r Poisson commuting integrals. If such a situation arises, an analog of the Liouville theorem holds.

We will call such systems *degenerate integrable systems.*

Here we recall the generalization of the Liouville integrability which was introduced in [**N**]. Let

$$\tilde{\pi} : \mathcal{P} \to B_k$$

be a Poisson projection where B_k has trivial Poisson structure, and let $b \in B_k$ be a nonsingular point.

DEFINITION 2.4. (1) The projection $\tilde{\pi}$ is *regular* at the point $b \in B_k$ if connected components of fibers are Poisson subvarieties which have single open dense symplectic leaf.

(2) The projection $\tilde{\pi}$ is *regular* if it is regular on an open dense subset of B_k.

(3) The Poisson manifold $\mathcal{P}_n$ is *regular of rank* k if it admits a regular projection to a k-dimensional base B_k.

Let $\mathcal{M}_{2n}$ be a symplectic manifold of dimension $2n$. Assume that we have a Poisson projection π to a regular Poisson manifold $\mathcal{P}_{2n-k}$ of dimension $2n - k$ and of rank k, i.e., we have:

$$\mathcal{M}_{2n} \xrightarrow{\pi} \mathcal{P}_{2n-k} \xrightarrow{\tilde{\pi}} B_k.$$

We will say that $c \in \mathcal{P}_{2n-k}$ is *generic* if $\tilde{\pi}(c)$ is regular and c belongs to the symplectic leaf which is open dense in the connected component of $\tilde{\pi}^{-1}(\tilde{\pi}(c))$ which contains c.

THEOREM 2.5. (1) *Any k smooth functions on B_k with differentials linearly independent at the regular point $b \in B_k$ generate k Hamiltonian flows on $(\tilde{\pi} \circ \pi)^{-1}(b) \subset \mathcal{M}_{2n}$ whose flow lines give an affine coordinate system on the k-dimensional submanifold $(\tilde{\pi} \circ \pi)^{-1}(b) \subset \mathcal{M}_{2n}$.*

(2) *The Hamiltonian flow on $\mathcal{M}_{2n}$ generated by the pull-back of a function on B_k is linear in this affine coordinate system.*

This theorem is a generalization of Liouville's theorem to the case when invariant tori may have dimension less than n.

Let D be a neighborhood of a regular point $b \in B_k$. Choose a generic $c \in \tilde{\pi}^{-1}(b)$ and let $\pi^{-1}(b)$, $U \ni c$ be its neighborhood. Choose the trivialization of $\tilde{\pi}$ over D:

$$\tilde{\pi}^{-1}(D) \simeq \tilde{\pi}^{-1}(b) \times D.$$

Let $W \subset \tilde{\pi}^{-1}(D)$ be such that with respect to this trivialization,

$$W \simeq U \times D.$$

Together with the choice of the trivialization of π over W this gives the isomorphism

$$f : \pi^{-1}(W) \simeq \pi^{-1}(c) \times U \times D.$$

Any k smooth functions on B_k with independent differentials at $b \in B_k$ give local coordinates in a neighborhood of b. Denote these coordinates by $(I_1, \ldots, I_k)$. The corresponding affine coordinates on $(\tilde{\pi} \circ \pi)^{-1}(b)$ are denoted by $(\varphi_i, \ldots, \varphi_k)$.

THEOREM 2.6. *Assume that $\pi^{-1}(c)$ is compact. Then there exists a trivialization $f : \pi^{-1}(W) \simeq \pi^{-1}(c) \times U \times D$ such that the symplectic form ω on $\mathcal{M}$ has the form*

$$\omega = f^*\left(\sum_{i=1}^{k} dI_i \wedge d\varphi_i + \pi^*(\omega_{\pi^{-1}(b)})\right)$$

where $\omega_{\pi^{-1}(b)}$ is the symplectic form on the open dense symplectic leaf of the connected component of $\pi^{-1}(b)$ which contains c.

One can replace real smooth manifolds by complex manifolds (complex algebraic) and Poisson structures by complex holomorphic (complex algebraic) structures. In the complex algebraic category there is a natural notion of algebraicaly integrable systems. The phase space of an algebraically integrable system is a complex algebraic manifold with holomorphic algebraic symplectic structure, which is nonsingular on a Zariski open subset of the phase space. In such a system fibers over Zariski open subsets of the base of the Lagrangian projection given by integrals are abelian varieties.

This is a very important class of integrable systems. One should expect that characteristic integrable systems related to algebraic Poisson Lie groups are algebraically integrable. Also, one may expect that they all admit quantization, and that formal quantizations can be classified similarly to regular integrable systems in the smooth real category.

3. Poisson Lie Groups

DEFINITION 3.1. A Lie group G (either finite dimensional or topological infinite-dimensional) is called a *Poisson-Lie group* if

(1) G is a Poisson manifold;
(2) the group multiplication $G \times G \longrightarrow G$ is a Poisson map.

Let $p \in TG \simeq \mathfrak{g} \times G$ be the Poisson tensor which defines a Poisson Lie structure on G. Here and below we trivialize tangent and cotangent bundles on a group via left translations. The Poisson tensor p determines the Poisson bracket on the algebra of smooth functions on G:

$$\{f, g\} = \langle p, d_l f \wedge d_l g\rangle.$$

Here, $d_l f \in T^*G \simeq \mathfrak{g}^* \times G$ is the left differential of f, i.e.,

$$\langle d_l f(x), \xi \rangle = \frac{d}{dt} f(e^{t\xi} x) \Big|_{t=0} \quad \forall \xi \in \mathfrak{g}.$$

Similarly, for the right differential we will use the notation $d_r f$:

$$\langle d_r f(x), \xi \rangle = \frac{d}{dt} f(x e^{t\xi}) \Big|_{t=0} \quad \forall \xi \in \mathfrak{g}.$$

The compatibility of the Poisson structure with the group multiplication is equivalent to the following cocycle property of p:

$$\mathrm{Ad}_x^{\otimes 2}(p(y)) - p(xy) + p(x) = 0.$$

Here $p(x) \in \wedge^2 T_x G \simeq \wedge^2 \mathfrak{g}$. It is clear from the cocycle property that $p(e) = 0$. Therefore a Poisson-Lie group cannot be a symplectic manifold.

REMARK 3.2. If G is algebraic and $C(G)$ is the space of algebraic functions on G, $C(G)$ is a Hopf algebra with pointwise multiplication and the coproduct

$$(\Delta f)(x, y) = f(xy)$$

as structure maps. If, in addition, G is a Poisson-Lie group, $C(G)$ is a *Poisson-Hopf algebra*, i.e.,

- $C(G)$ is both a Poisson and a Hopf algebra;
- $\Delta(\{f, g\}) = \{\Delta f, \Delta g\}$, where the Poisson bracket on $C(G) \otimes C(G)$ is defined by $\{a \otimes b, c \otimes d\} = \{a, c\} \otimes bd + ac \otimes \{b, d\}$.

4. Lie Bialgebras

4.1. Basic Notions.

DEFINITION 4.1. A pair consisting of the Lie algebra $\mathfrak{g}$ and the linear map $\delta : \mathfrak{g} \longrightarrow \mathfrak{g} \wedge \mathfrak{g}$ is called a *Lie bialgebra*, if

(1) δ is a 1-cocycle;
(2) $(\delta \wedge id) \circ \delta = 0$.

Let $\mathfrak{g}^*$ be the dual vector space to $\mathfrak{g}$. If the Lie algebra is finite dimensional it is the dual vector space of all linear functions on $\mathfrak{g}$. If it is a graded Lie algebra with finite dimensional graded components, then by $\mathfrak{g}^*$ we mean the graded vector space with graded components dual to those of $\mathfrak{g}$. More generally, by a *dual space* we understand the pair consisting of a vector space $\mathfrak{g}^*$ and the bilinear non degenerate map $\langle \cdot, \cdot \rangle : \mathfrak{g}^* \otimes \mathfrak{g} \to \mathbb{C}$.

THEOREM 4.2. *If* $(\mathfrak{g}, \delta)$ *is a Lie bialgebra, then* $(\mathfrak{g}^*, \delta_*)$ *with*

$$\langle [l_1, l_2]_*, a \rangle = \langle l_1 \wedge l_2, \delta(a) \rangle, \quad \langle \delta_*(l), a \wedge b \rangle = \langle l, [a, b] \rangle$$

is a Lie bialgebra as well.

The proof of this is an elementary exercise.

Thus, Lie bialgebras always appear in dual pairs $\big((\mathfrak{g}, \delta), (\mathfrak{g}^*, \delta_*)\big)$. This is why sometimes we will denote Lie bialgebras by pairs $(\mathfrak{g}, \mathfrak{g}^*)$ of Lie algebras on $\mathfrak{g}$ and on $\mathfrak{g}^*$.

DEFINITION 4.3. $(\mathfrak{g}^*, \delta_*)$ is called the *Lie bialgebra dual to* $(\mathfrak{g}, \delta)$.

For a Poisson-Lie group (G,p) with the Lie algebra $\mathfrak{g} = \mathrm{Lie}(G)$ define the bracket $[\cdot,\cdot]_* : \mathfrak{g}^* \otimes \mathfrak{g}^* \longrightarrow \mathfrak{g}^*$ as

$$[l_1, l_2]_* = \langle dp(e), l_1 \wedge l_2 \rangle.$$

The equality makes sense, since $dp(e) \in \mathfrak{g}^* \otimes \mathfrak{g} \wedge \mathfrak{g}$. The following theorem establishes the correspondence between Lie bialgebras and Poisson Lie groups.

THEOREM 4.4. (1) *The bracket* $[\cdot,\cdot]_*$ *determines a Lie algebra structure on* $\mathfrak{g}^*$.

(2) *The linear map* $\delta : \mathfrak{g} \longrightarrow \mathfrak{g} \wedge \mathfrak{g}$ *which is dual to* $[\cdot,\cdot]_* : \mathfrak{g}^* \wedge \mathfrak{g}^* \longrightarrow \mathfrak{g}^*$ *is a* 1*-cocycle on* $\mathfrak{g}$, *i.e.,*

$$\delta([a,b]) = [a, \delta(b)] + [\delta(a), b],$$

where $[x, y \wedge z] = [x,y] \wedge z + y \wedge [x,z]$ *for* $x, y, z \in \mathfrak{g}$, *and therefore determines the structure of a Lie bialgebra on* $\mathfrak{g}$.

DEFINITION 4.5. The Lie bialgebra described above is called the *tangent Lie bialgebra* to a Poisson Lie group.

It is easy to see that if $f : (G,p) \to (G',p')$ is a homomorphism of Poisson Lie groups (a morphism of groups and of Poisson manifolds), then the corresponding map $f' : (\mathfrak{g},\delta) \to (\mathfrak{g}',\delta')$ obtained from f by linearization in the neighborhood of the identity is a homomorphism of the Lie bialgebras (both f' and its dual map are homomorphisms of Lie algebras). This means that the theorem above describes the functor from the category of Poisson Lie groups to the category of Lie bialgebras whose action on the objects was described above. There is also a functor in the opposite direction described by the following theorem.

THEOREM 4.6. *For any finite dimensional Lie bialgebra* $(\mathfrak{g},\delta)$ *there exists a unique connected simply connected Poisson-Lie group* (G,p), *such that* $(\mathfrak{g},\delta)$ *is its tangent Lie bialgebra.*

Such a correspondence between Poisson Lie groups and Lie bialgebras gives a natural notion of a dual Poisson Lie group. We will say the Poisson Lie group (G^*, p_*) is dual to the Poisson Lie group (G,p). Notice that among all duals there exists a unique connected simply connected Poisson Lie group. The duality between Poisson Lie groups can be illustrated by the following diagram:

$$\begin{array}{ccc} (\mathfrak{g},\delta) & \xrightarrow{\text{Lie bialgebra duality}} & (\mathfrak{g}^*,\delta_*) \\ \uparrow & & \downarrow \\ (G,p) & \xrightarrow{\text{Poisson duality}} & (G^*,p_*) \end{array}$$

Here, $(\mathfrak{g},\delta)$ is the tangent Lie bialgebra of (G,p), and (G^*,p_*) is a Poisson Lie group with the tangent Lie bialgebra $(\mathfrak{g}^*,\delta_*)$.

4.2. Quasitriangular Lie Bialgebras.

DEFINITION 4.7. Two Lie bialgebra structures are *related by a twist*, (i.e., they are equivalent), if the corresponding 1-cocycles are cohomologous one to the other, i.e., when there exists $f \in \mathfrak{g} \wedge \mathfrak{g}$ such that

$$\delta_1(a) - \delta_2(a) = [a, f] \quad \forall a \in \mathfrak{g}.$$

In this case, the 1-cocycle δ_1 is cohomologous to the 1-cocycle δ_2 in the Chevalley complex for the Lie algebra $\mathfrak{g}$ with coefficients in $\mathfrak{g} \wedge \mathfrak{g}$.

If the, 1-cocycle δ for the Lie bialgebra $(\mathfrak{g}, \delta)$ is cohomologous to 0, i.e., if there exists an element $\rho \in \mathfrak{g} \wedge \mathfrak{g}$, such that

$$\delta(x) = [x, \rho], \tag{4.1}$$

the Lie bialgebra $(\mathfrak{g}, \delta)$ is called a *coboundary Lie bialgebra.* An element ρ determines a coboundary Lie bialgebra with cobracket (4.1) if and only if

$$[\rho_{12}, \rho_{13}] + [\rho_{12}, \rho_{23}] + [\rho_{13}, \rho_{23}] \in (\wedge^3 \mathfrak{g})_{\text{inv}},$$

where $(\wedge^3 \mathfrak{g})_{\text{inv}}$ is the subspace in $\wedge^3 \mathfrak{g}$ which is invariant with respect to the adjoint action of $\mathfrak{g}$.

The coboundary Lie bialgebra is a *quasitriangular* Lie bialgebra if there exists an element $r^{(0)} \in S^2(\mathfrak{g})_{\text{inv}}$ such that $r = r^{(0)} + \rho \in \mathfrak{g} \otimes \mathfrak{g}$ (here we assume that $\mathfrak{g} \wedge \mathfrak{g} \subset \mathfrak{g} \otimes \mathfrak{g}$ as skew symmetric tensors) satisfying the classical Yang-Baxter equation:

$$[r_{12}, r_{13}] + [r_{12}, r_{23}] + [r_{13}, r_{23}] = 0.$$

A quasitriangular Lie bialgebra is called *triangular* if $r^{(0)} = 0$, and it is called *factorizable* if $r^{(0)}$ determines a non-degenerate symmetric bilinear form on $\mathfrak{g}^*$.

5. Factorizable Lie Bialgebras and Poisson-Lie Groups

5.1. Factorizable Lie Bialgebras. Let $\mathfrak{g}$ be a factorizable Lie bialgebra with classical r-matrix $r \in \mathfrak{g} \otimes \mathfrak{g}$.

LEMMA 5.1. *The linear maps* $r_\pm : \mathfrak{g}^* \longrightarrow \mathfrak{g}$

$$r_+(l) := \langle r, l \otimes id \rangle, \quad r_-(l) := -\langle r, id \otimes l \rangle \quad \forall l \in \mathfrak{g}^*;$$

are Lie bialgebra homomorphisms.

We will leave the proof of this lemma as an exercise. It follows from the classical Yang-Baxter equation for r.

The linear map $I : \mathfrak{g}^* \to \mathfrak{g}$, $l \mapsto r_+(l) - r_-(l)$ is a linear isomorphism. It is called the *factorization map.*

COROLLARY 5.2. *The subspaces* $\mathfrak{g}_+ = \mathrm{Im}(r_+)$ *and* $\mathfrak{g}_- = \mathrm{Im}(r_-)$ *are Lie subbialgebras in the Lie bialgebra* $\mathfrak{g}$.

LEMMA 5.3.
(1) *The subspaces* $n_\pm = r_\pm(\ker(r_\mp))$ *are Lie ideals in* $\mathfrak{g}_\pm$, *respectively.*
(2) *The map* $\theta : \mathfrak{g}_+/n_+ \to \mathfrak{g}_-/n_-$ *which sends the residue class of* $r_+(e)$ mod n_+ *to* $r_-(e)$ mod n_- *is well-defined and is an isomorphism of Lie algebras.*

PROOF. The first statement follows from the facts that $\ker(r_\pm)$ are Lie ideals in $\mathfrak{g}^*$ and that $r_\pm$ are Lie algebra homomorphisms.

Let $\ell \in \mathfrak{g}^*$, $n \in \ker(r_+)$, $m \in \ker(r_-)$. Consider a representative $r_+(\ell) + r_+(m)$ of $r_+(\ell)$ mod (n_+). By linearity $r_+(\ell) + r_+(m) = r_+(\ell + m) = r_+(\ell + n + m)$. Now, $r_-(n + \ell + m) = r_-(\ell + n) = r_-(\ell) + r_-(n)$ represents the equivalence class $r_-(\ell)$ mod (n_-). Therefore the map $\theta : \mathfrak{g}_+/n_+ \to \mathfrak{g}_-/n_-$ is well-defined and is a linear isomorphism.

We leave it as an exercise to prove that it is a Lie algebra homomorphism. $\square$

THEOREM 5.4. (1) *Every element* $x \in \mathfrak{g}$ *admits a unique factorization*

$$x = x_+ - x_-,$$

where $x_\pm \in \mathfrak{g}_\pm$, *and* $\theta(x_+ \bmod n_+) = x_- \bmod n_-$.

(2) *The Lie algebra* $\mathfrak{g}^*$ *is isomorphic to the following Lie subalgebra of* $\mathfrak{g}_+ \oplus \mathfrak{g}_-$:

$$\{(x_+, x_-) \in \mathfrak{g}_+ \oplus \mathfrak{g}_- \mid \theta(x_+ \bmod n_+) = x_- \bmod n_-\}.$$

(3) *If we model* $\mathfrak{g}^*$ *as in* (2) *the factorization map* $I : x \mapsto r_+(x) - r_-(x)$ *acts as* $(x_+, x_-) \mapsto x_+ - x_-$, *where on the right hand side we consider* $\mathfrak{g}_\pm$ *as Lie subalgebras of* $\mathfrak{g}$.

5.2. Factorizable Poisson Lie Groups. Because we have a bijection between Poisson Lie groups and Lie bialgebras we have Poisson Lie counterparts of the Lie bialgebras introduced above: coboundary, quasitriangular, triangular, factorizable.

If (G, p) is a quasitriangular Poisson Lie group, the Poisson tensor has the following explicit description:

$$p(x) = \mathrm{Ad}_x(r) - r \in \wedge^2 TG \simeq \wedge^2 g.$$

Here we trivialized the tangent bundle by left translations. For the Poisson brackets on a quasitriangular Poisson Lie group we have:

$$\{f_1, f_2\} = \langle r, d_l f_1 \wedge d_l f_2 \rangle - \langle r, d_r f_1 \wedge d_r f_2 \rangle,$$

where d_l and d_r are, respectively, left and right differentials on G.

For factorizable Poisson Lie groups we have

- Maps $r_\pm$ lift to Lie group homomorphisms $r_\pm : G^* \to G$;
- $G_\pm = \mathrm{Im}(r_\pm) \subset G$ are Poisson Lie subgroups (connected simply connected);
- $N_\pm = \mathrm{Im}(\ker(r_\pm)) \subset G_\pm$ are normal Lie subgroups;
- Lie algebra isomorphism $\theta : \mathfrak{g}_+/n_+ \xrightarrow{\sim} \mathfrak{g}_-/n_-$ lifts to Lie group isomorphism $\theta : G_+/N_+ \xrightarrow{\sim} G_-/N_-$;
- Lie group G^* can be modeled as:

$$G^* = \{(g_+, g_-) \in G_+ \times G_- \mid \theta(g_+ \bmod N_+) = g_- \bmod N_-\};$$

- There exist open dense subsets $G', G'' \subset G$ such that for each $g \in G'$ there exists a unique factorization $g = g_+ g_-^{-1}$, $\theta(g_+ \bmod N_+) = g_- \bmod N_-$, and for each $g \in G''$ there exists a unique factorization $g = g_-^{-1} g_+$ with the same conditions on $g_\pm$;
- Left, respectively right, *factorization maps* $G^* \to G$ map (g_+, g_-) to $(g_+ g_-^{-1})$, respectively to $g_-^{-1} g_+$.

Here we assume that G^* is represented as a subgroup of $G_+ \times G_-$.

6. The Double of a Lie Bialgebra

6.1. The Double of a Lie Bialgebra. The construction of the double of a Lie bialgebra makes a factorizable Lie bialgebra out of any dual pair of Lie bialgebras. Below we will work with finite dimensional Lie bialgebras. All constructions work in an obvious way for dual pairs of infinite dimensional Lie bialgebras as well.

Let $(\mathfrak{g}, \delta)$ be a Lie bialgebra and let $(\mathfrak{g}^*, \delta_*)$ be its dual. Let $\mathrm{ad}^* : \mathfrak{g} \otimes \mathfrak{g}^* \longrightarrow \mathfrak{g}^*$ be the coadjoint action of $\mathfrak{g}$ on $\mathfrak{g}^*$:

$$\mathrm{ad}^*_x(l) = l([x, \cdot]) \quad \forall x \in \mathfrak{g}, l \in \mathfrak{g}^*,$$

and let $\widetilde{\mathrm{ad}}^* : \mathfrak{g}^* \otimes \mathfrak{g} \longrightarrow \mathfrak{g}$ be the coadjoint action of $\mathfrak{g}^*$ on $\mathfrak{g}$ (assuming that we have identified $\mathfrak{g}^{**}$ with $\mathfrak{g}$).

THEOREM 6.1. (1) *There exists a unique Lie bialgebra structure on* $\mathcal{D}(\mathfrak{g}) = \mathfrak{g} \oplus \mathfrak{g}^*$, *such that*

(a) *the form* $\langle (x, l), (y, m) \rangle = l(y) + m(x)$ *is* $\mathcal{D}(\mathfrak{g})$*-invariant;*

(b) $\mathfrak{g}, \mathfrak{g}^* \hookrightarrow \mathcal{D}(\mathfrak{g})$ *are Lie sub-bialgebras.*

(2) *This Lie bialgebra has the following structure:*

(a) $[(x, l), (y, m)] = \big([x, y] + \widetilde{\mathrm{ad}}^*_l(y) - \widetilde{\mathrm{ad}}^*_m(x), [l, m]_* + \mathrm{ad}^*_x(m) - \mathrm{ad}^*_y(l)\big)$,

(b) $\mathcal{D}(\mathfrak{g}) = \mathfrak{g} \oplus \mathfrak{g}^*$ *as a co-bialgebra, i.e.,* $\delta_{\mathcal{D}(\mathfrak{g})}(x + l) = \delta(x) - \delta^*(l)$ *for* $x \in \mathfrak{g}$ *and* $l \in \mathfrak{g}^*$.

Observe that the Lie bialgebras $(\mathfrak{g}, \delta)$ and $(\mathfrak{g}^*, -\delta_*)$ are Lie sub-bialgebras in $(\mathcal{D}(\mathfrak{g}), \delta_{\mathcal{D}(\mathfrak{g})})$.

DEFINITION 6.2. The Lie bialgebra $(d(\mathfrak{g}), \delta_{\mathcal{D}(\mathfrak{g})})$ is called the *double* of the Lie bialgebra $(\mathfrak{g}, \delta)$.

Let $\{e_i\}$ be a linear basis in the Lie algebra $\mathfrak{g}$, and let $\{e^*\}$ be a dual linear basis in $\mathfrak{g}^*$. Denote by $c^k_{i,j}$ the structural constants of the Lie algebra $\mathfrak{g}$ in this basis, and by f^{ij}_k the structural constants of the Lie algebra $\mathfrak{g}^*$ in the dual basis. Then for $\mathcal{D}(\mathfrak{g})$ we have in terms of the structure constants of $\mathfrak{g}$ and $\mathfrak{g}^*$:

$$[e_i, e_j] = c^k_{i,j} e_k, \tag{6.1}$$

$$[e^i, e^j] = f^{i,j}_k e^k, \tag{6.2}$$

$$[e_i, e^j] = c^j_{l,i} e^l + f^{j,k}_i e_k, \tag{6.3}$$

and

$$\delta_{\mathcal{D}}(e_i) = \sum_{jk} e_j \wedge e_k, \quad \delta_{\mathcal{D}}(e^i) = -\sum_{kl} e^k \wedge e^l.$$

PROPOSITION 6.3. *The double* $\mathcal{D}(\mathfrak{g})$ *is a factorizable Lie bialgebra with* $r = \sum_i e^i \otimes e_i \in \mathfrak{g}^* \otimes \mathfrak{g} \hookrightarrow \mathcal{D}(\mathfrak{g}) \otimes \mathcal{D}(\mathfrak{g})$, *where* $\{e_i\}$ *denotes a basis of* $\mathfrak{g}$ *and* $\{e^i\}$ *the corresponding dual basis of* $\mathfrak{g}^*$.

As a factorizable Lie bialgebra the double of a Lie bialgebra $(\mathfrak{g}, \delta)$ has the following properties:

- $\mathcal{D}(\mathfrak{g})_+ = \mathfrak{g}$, $\mathcal{D}(\mathfrak{g}^*)_- = \mathfrak{g}^{*\mathrm{opp}}$,
- $\mathfrak{n}_+ = \mathfrak{g}, \mathfrak{n}_- = \mathfrak{g}^{*\mathrm{opp}}$,
- for any element $x \in \mathcal{D}(\mathfrak{g})$ there exists a unique factorization $x = x_+ - x_-$ where $x_\pm \in \mathcal{D}(G)_\pm$.

Let $(\mathfrak{g}, \delta; r)$ be a factorizable Lie bialgebra.

THEOREM 6.4. *For a factorizable Lie bialgebra* $(\mathfrak{g}, \delta; r)$ *the following holds.*

(1) $\mathcal{D}(\mathfrak{g}) \simeq \mathfrak{g} \oplus \mathfrak{g}$ *as a Lie algebra. The isomorphism is given by the linear map*

$$\varphi : \mathfrak{g} \oplus \mathfrak{g}^* \ni (x, l) \longmapsto \big(x + r_+(l), x + r_-(l)\big) \in \mathfrak{g} \oplus \mathfrak{g}.$$

(2) *The r-matrix of the double* $r_{d(\mathfrak{g})}$ *can be expressed through the r-matrix for* $\mathfrak{g}$ *as follows:*

$$(\varphi \otimes \varphi)\, r_{\mathcal{D}(\mathfrak{g})} = r_{13} + r_{14} - r_{32} - r_{42},$$

where $r_{ij} \in (\mathfrak{g} \oplus \mathfrak{g}) \otimes (\mathfrak{g} \oplus \mathfrak{g})$ *and the indices indicate the copies of* $\mathfrak{g}$ *in this product which the r-matrix belongs to (assuming the enumeration from left to right).*

This theorem shows, in particular, that the iteration of the double construction does not produce "new" Lie algebras.

If $x = x_+ - x_-$ is the factorization in $\mathfrak{g}$, then in $\mathcal{D}(\mathfrak{g})$ the factorization is $(x_1, x_2) = (x + \xi_+, x + \xi_-)$ where $\xi_\pm \in \mathfrak{g}_\pm$ and $\theta(\xi_+ \bmod \mathfrak{n}_+) = \xi_- \bmod \mathfrak{n}_-$.

7. The Double of a Poisson Lie Group

7.1. The Double of a Poisson Lie Group.

DEFINITION 7.1. The *double* of a Poisson Lie group (G, p) is the connected simply connected Poisson Lie group whose tangent Lie bialgebra is the double of the tangent Lie bialgebra of (G, p).

The following diagram is commutative

$$\begin{array}{ccccc}
(\mathfrak{g}, \delta) & \hookrightarrow & \mathcal{D}(\mathfrak{g}) & \hookleftarrow & (\mathfrak{g}^*, -\delta_*) \\
\downarrow & & \downarrow & & \downarrow \\
(G, p) & \hookrightarrow & (\mathcal{D}, p_{\mathcal{D}}) & \hookleftarrow & (G^*, -p_*) = G^{*\mathrm{opp}}
\end{array}$$

The double is a factorizable Poisson Lie group,

$$\mathcal{D}(G)_+ = G \hookrightarrow \mathcal{D}(G)$$
$$\mathcal{D}(G)_- = G^{*\mathrm{opp}} \hookrightarrow \mathcal{D}(G)$$

and if $\mathcal{D}(G)^*$ is the Poisson Lie group dual to $\mathcal{D}(G)$, we have the following group isomorphism

$$\mathcal{D}(G)^* \cong G' \times G^*,$$

where G' is a Lie group with the multiplication opposite to G.

REMARK 7.2. On an open dense subset of $\mathcal{D}(G) \times \mathcal{D}(G)$ where the appropriate factorization exists, we have the following formula for the multiplication in $\mathcal{D}(G)$:

$$xy = x_+ {x_-}^{-1} y_+ {y_-}^{-1} = x_+ y_+^{x_-} {x_-^{y_+}}^{-1} {y_-}^{-1}.$$

Here we used notations x^{y_+} and $y_+^{x_-}$ for the dressing actions of y_+ on x_- and of x_- on y_+, respectively.

7.2. The Double of a Factorizable Poisson Lie Group. If G is a factorizable Poisson Lie group, the correspondence between Lie bialgebras and Poisson-Lie groups gives the following group isomorphism:

$$\mathcal{D} \simeq G \times G.$$

For the Poisson Lie subgroups $\mathcal{D}(G)_\pm$ we have:

$$\mathcal{D}(G)_+ = G \subset G \times G, \; g \mapsto (g, g),$$
$$\mathcal{D}(G)_- = G^* \subset G \times G, \; (\xi_+, \xi_-) \mapsto (\xi_+, \xi_-).$$

The left and right factorization maps $R_{l,r} : \mathcal{D}^* \longrightarrow \mathcal{D}$ for the double of a factorizable Poisson-Lie group are

$$R_l : G \times G^* \ni \big(g, (\xi_+, \xi_-)\big) \longmapsto (g\xi_+, g\xi_-) \in G \times G,$$
$$R_r : G \times G^* \ni \big(g, (\xi_+, \xi_-)\big) \longmapsto (\xi_+ g, \xi_- g) \in G \times G.$$

8. Finite Dimensional Examples

8.1. Lie Bialgebra With Trivial Lie Coalgebra Structure. Let $\mathfrak{g}$ be a trivial finite dimensional Lie bialgebra $(\mathfrak{g}, \delta \equiv 0)$. Then its dual Lie bialgebra is the Abelian Lie algebra on the vector space $\mathfrak{g}^*$ with the Lie cobracket dual to the Lie bracket on $\mathfrak{g}$.

The double of the trivial Lie bialgebra is the vector space $d(\mathfrak{g}) = \mathfrak{g} \oplus \mathfrak{g}^*$ with the Lie bracket

$$[(x,l),(y,m)] = \big([x,y], \mathrm{ad}_x^*(l) - \mathrm{ad}_y^*(m)\big).$$

8.2. Poisson Lie Groups Corresponding Lie Bialgebra With Trivial Lie Bialgebra Structure. Consider $\mathfrak{g}$ endowed with the trivial Lie bialgebra structure:

- The dual Lie bialgebra $\mathfrak{g}^*$:
  - the Lie structure is trivial, i.e., $[\cdot,\cdot]_* = 0$.
  - the bialgebra structure is given as follows:
    Let $\{e_i\}$ denote a basis of $\mathfrak{g}$ and let c_{ij}^k be the corresponding structure constants, i.e.,
    $$[e_i, e_j] = \sum_k c_{ij}^k e_k.$$
    Then we have with respect to the dual basis $\{e^i\}$ of $\mathfrak{g}^*$:
    $$\delta_*(e^i) = \sum_{k,l} c_{kl}^i e^k \wedge e^l.$$
- The Poisson-Lie groups G and $\mathfrak{g}^*$ associated with $\mathfrak{g}$ and $\mathfrak{g}^*$, respectively:
  - $(\mathfrak{g}, \delta)$ induces the trivial Poisson structure on G.
  - The Lie group associated with $(\mathfrak{g}^*, [\cdot,\cdot])$ is $\mathfrak{g}^*$ viewed as an Abelian group. The Poisson structure induced by $(\mathfrak{g}^*, \delta_*)$ on $\mathfrak{g}^*$ can be described by means of the Kostant-Kirillov construction, i.e., for $f, g \in C^\infty(\mathfrak{g}^*)$,
    $$\{f,g\}(x) = \langle x, [\nabla f(x), \nabla g(x)]\rangle \quad \forall x \in \mathfrak{g}^*.$$
- The double of $\mathfrak{g}$:
  $$d(\mathfrak{g}) = \mathfrak{g} \oplus \mathfrak{g}^*,$$
  where the Lie algebra structure is given by
  $$[(x,l),(y,m)] = \big([x,y], \mathrm{ad}_y^*(l) - \mathrm{ad}_x^*(m)\big).$$
- The Poisson-Lie group corresponding to $d(\mathfrak{g})$:
  $$\mathcal{D} = T^*G \simeq G \times \mathfrak{g}^* \quad \text{as a manifold.}$$
  - The multiplication:
    $$(g,l)(h,m) = (gh, \mathrm{Ad}_h^*(l) + m).$$

– The Poisson structure:

trivial along G, Lie-Kostant-Kirillov along $\mathfrak{g}^*$.

– The factorization of $\mathcal{D}$:

$$(g,l) = (g,0)(1,l) = \big(1, \mathrm{Ad}_g^*(l)\big)(g,0).$$

– The dressing map:

$$G \times \mathfrak{g}^* \longrightarrow \mathfrak{g}^*, (g,l) \longmapsto \mathrm{Ad}_g^*(l).$$

- The symplectic leaves of $\mathfrak{g}^*$ are the orbits of the Ad^*-action of G on $\mathfrak{g}^*$.

Thus, in this case, the dressing action of G on $\mathfrak{g}^*$ is just the coadjoint action. The action of $\mathfrak{g}^*$ on G is trivial.

8.3. Standard Lie Bialgebra Structure on a Simple Lie Algebra. Let $\mathfrak{g}$ be a simple complex Lie algebra. Fix a Borel subalgebra $\mathfrak{b}$. Let (H_i, e_i, f_i) $i = 1, \dots, r = \mathrm{rank}(\mathfrak{g})$, be elements of the Chevalley basis of $\mathfrak{g}$ for this choice of the Borel subalgebra which correspond to simple roots. It is well known that $\mathfrak{g}$ is freely generated by (H_i, e_i, f_i) modulo the determining relations

$$[H_i, H_j] = 0, \tag{8.1}$$

$$[H_i, e_j] = a_{ij} e_j, \tag{8.2}$$

$$[H_i, f_j] = -a_{ij} f_j, \tag{8.3}$$

$$[e_i, f_j] = \delta_{ij} H_i, \tag{8.4}$$

$$(\mathrm{ad}_{e_i})^{1-a_{ij}}(e_j) = 0, \quad i \neq j, \tag{8.5}$$

$$(\mathrm{ad}_{f_i})^{1-a_{ij}}(f_j) = 0, \quad i \neq j, \tag{8.6}$$

where (a_{ij}) denotes the Cartan matrix of $\mathfrak{g}$.

Consider a linear map $\delta : \mathfrak{g} \to \mathfrak{g} \wedge \mathfrak{g}$ acting on the generators as

$$\delta(H_i) = 0, \tag{8.7}$$

$$\delta(e_i) = d_i H_i \wedge e_i, \tag{8.8}$$

$$\delta(f_i) = d_i H_i \wedge f_i, \tag{8.9}$$

where d_i is the length of the i-th root, in particular,

$$d_i a_{ij} = a_{ij} d_j.$$

THEOREM 8.1. *There exists unique such linear map $\delta : \mathfrak{g} \to \mathfrak{g} \wedge \mathfrak{g}$ which has the following properties:*

(1) *δ is a 1-cocycle.*
(2) *$(\delta \wedge id) \circ \delta = 0$.*

REMARK 8.2. The definition of the standard Lie bialgebra structure on $\mathfrak{g}$ requires the choice of a Borel subalgebra $\mathfrak{b} \subset \mathfrak{g}$.

THEOREM 8.3. (1) *$(\mathfrak{g}, \delta)$ is factorizable with*

$$r = \frac{1}{2} \sum_{ij} (B^{-1})_{ij} H_i \otimes H_j + \sum_{a>0} e_\alpha \otimes f_\alpha,$$

where $B_{ij} = d_i a_{ij}$ is the symmetrized Cartan matrix.

(2) *The Lie subalgebras* $\mathfrak{g}_+$ *and* $\mathfrak{g}_-$ *are* $\mathfrak{b}$ *and* $\mathfrak{b}_-$ *respectively, where* $\mathfrak{b}$ *is the upper Borel subalgebra and* $\mathfrak{b}_-$ *is the lower Borel subalgebra. The Lie subalgebras* $\mathfrak{n}_\pm$ *are upper and lower Borel subalgebras, respectively. The quotient Lie algebras* $\mathfrak{b}_\pm/\mathfrak{n}_\pm$ *can be naturally identified with* $\mathfrak{h}$. *After this identification the isomorphism* θ *is the identity automorphism of* $\mathfrak{h}$.

The Borel subalgebra $\mathfrak{b}$ (generated by H_i, e_i) is a Lie sub-bialgebra in $(\mathfrak{g}, \delta)$.

THEOREM 8.4. *The double of the Borel subalgebra has the following structure:* $\mathcal{D}(\mathfrak{b}) = \mathfrak{b} \oplus \mathfrak{b}^*$ *is isomorphic to the Lie algebra with generators* H_i, H_i^*, e_i, f_i , $i = 1, \ldots, \mathrm{rank}(\mathfrak{g})$, *and with determining relations*

$$\begin{aligned}
&[H_i, H_j] = 0, && [H_i^*, H_j] = 0,\\
&[H_i, e_j] = a_{ij} e_j, && [H_i^*, e_j] = a_{ij} e_j,\\
&[H_i, f_j] = -a_{ij} f_j, && [H_i^*, f_j] = -a_{ij} f_j,\\
&[e_i, f_j] = \frac{1}{2}\delta_{ij}(H_i + H_i^*), && [H_i^*, H_j^*] = 0,\\
&(\mathrm{ad}_{e_i})^{1-a_{ij}}(e_j) = 0, && (\mathrm{ad}_{f_i})^{1-a_{ij}}(f_j) = 0, \quad i \neq j,
\end{aligned}$$

and the Lie bialgebra structure is given by

$$\delta(H_i) = 0, \quad \delta(e_i) = d_i H_i \wedge e_i,$$
$$\delta(H_i^*) = 0, \quad \delta(f_i) = d_i H_i^* \wedge f_i.$$

We leave the proof of this theorem as an exercise.

COROLLARY 8.5. (1) *The Lie subalgebra* $\mathfrak{h}$ *generated by* $H_i - H_i^*$ *is a Lie bialgebra ideal, i.e., a Lie algebra ideal and a Lie sub-coalgebra* ($\delta(\mathfrak{h}) \subset \mathfrak{h} \wedge \mathfrak{h}$).

(2) $d(\mathfrak{b})/\mathfrak{h} \simeq \mathfrak{g}$ *as a Lie bialgebra.*

For a complete classification of factorizable Lie bialgebra structures on simple Lie algebras see [**KS**].

8.4. Standard Poisson Lie Structure on Simple Lie Groups. The standard Poisson Lie structure on a simple Lie group is determined by the standard Lie bialgebra structure on the corresponding simple Lie algebra. The Poisson Lie subgroups $G_\pm$ are Borel subgroups $B^\pm$ fixed by the choice of the classical r-matrix. The normal subgroups $N_\pm$ in $G_\pm$ are nilpotent Lie subgroups $N^\pm \subset B^\pm$. The quotient Lie groups $B^\pm/N^\pm$ are naturally isomorphic to the Cartan subgroup H and the isomorphism θ is the identity map of the Cartan subgroup into itself.

9. Symplectic Leaves in Poisson Lie Groups

9.1. Poisson Actions of Poisson Lie Groups. Let G be a Poisson Lie group and M a Poisson manifold.

DEFINITION 9.1. A map $G \times M \to M$ (resp. $M \times G \to M$) is called a *left* (resp. *right*) *Poisson action* of a Poisson Lie group G if it is a group action and if it is a Poisson map. Such manifolds are called *Poisson* G*-manifolds.*

DEFINITION 9.2. A Poisson G-manifold M is called a *Poisson homogeneous manifold* if G acts transitively on M.

PROPOSITION 9.3. *If* M *is a Poisson* G*-manifold and* M/G *is a manifold, then* M/G *inherits a Poisson structure from* M.

9.2. The Dressing Action for Poisson Lie Groups. Let (G,p) be a Poisson-Lie group and let $\mathcal{D} = (\mathcal{D}(G), p_{\mathcal{D}})$ be its double.

The double is a factorizable Poisson-Lie group. Therefore, there exist dense open subsets $\mathcal{D}(G)'$ and $\mathcal{D}(G)''$ in $\mathcal{D}(G)$ (images of left and right factorization maps, respectively) which are neighborhoods of $1 \in \mathcal{D}(G)$, such that for all $x \in \mathcal{D}(G)'$:

$$x = g\xi^{-1},$$

and for all $x \in \mathcal{D}(G)''$

$$x = (\bar{\xi})^{-1}\bar{g}.$$

Here $g, \bar{g} \in G$ and $\xi, \bar{\xi} \in G^*$. We will keep these notations.

The factorization in $\mathcal{D}(G)$ gives maps

(1) $G \times G^{*\mathrm{opp}} \longrightarrow G^{*\mathrm{opp}}$, $(g, \xi) \longmapsto \bar{g}$,
(2) $G^{*\mathrm{opp}} \times G \longrightarrow G^{*\mathrm{opp}}$, $(\bar{\xi}, \bar{g}) \longmapsto \xi$,

defined for $g\xi^{-1} = \bar{\xi}^{-1}\bar{g} \in \mathcal{D}(G)' \cap \mathcal{D}(G)''$. Here $G^{*\mathrm{opp}}$ denotes the Poisson Lie group $(G^*, -p_*)$.

THEOREM 9.4. *If the factorization is defined globally then the maps* (1) *and* (2) *are, respectively, a left and right Poisson action of G on G^*.*

PROOF. Consider only (1). The map (2) can be treated in a completely similar way.

We have

$$g_1g_2\xi^{-1} = (\xi^{g_1g_2})^{-1}(g_1g_2)^{\xi}.$$

On the other hand,

$$g_1g_2\xi^{-1} = g_1(\xi^{g_2})^{-1}g_2^{\xi} = \left((\xi^{g_2})^{g_1}\right)^{-1}g_1^{\xi^{g_2}}g_2^{\xi}$$

and therefore

$$\xi^{g_1g_2} = (\xi^{g_2})^{g_1}.$$

The map (1) is the composition of the group multiplication and the projection $\pi_1 : D(G) \to D(G)/G \simeq G^{*\mathrm{opp}}$. Both of these maps are Poisson, therefore the map (1) is Poisson too. □

REMARK 9.5. The group multiplication on $\mathcal{D}(G)$ induces left and right actions of G^* on $\mathcal{D}(G)$ that are defined globally. These actions are Poisson since G is a Poisson Lie subgroup in a Poisson Lie group $\mathcal{D}(G)$. The left action of G^* commutes with the right action of G^*, therefore it descends to a Poisson action on the right cosets $\mathcal{D}(G)/G^*$. We will call this action the *dressing action* on the right cosets.

THEOREM 9.6. *If the factorization* (*and therefore the maps* (1) *and* (2)) *is defined globally, then the left dressing G-orbit passing through G^* coincides with the right orbit passing through this point.*

PROOF. Let $\xi \in G^*$ and $g \in G$. Let ξ^g be the image of ξ with respect to the left dressing action of g:

$$g\xi^{-1} = (\xi^g)^{-1}g^{\xi}.$$

Let ${}^g\xi \in G^*$ be the image of ξ with respect to the right action of $g \in G$:

$$\xi^{-1}g = {}^{\xi}({}^g\xi)^{-1}.$$

Inverting this relation we have

$${}^g\xi = \left((\xi^{-1})^{g^{-1}}\right)^{-1}.$$

On the other hand, $((\xi^{-1})^g)^{-1} = \xi^{\tilde{g}}$ where $\tilde{g} = g^{\xi^{-1}}$. Therefore ${}^g\xi$ lies on the orbit of left dressing action. Then the theorem follows.

In general the maps (1) and (2) do not determine the group action of G on G^* (since they are not globally defined). However, since they are defined in a neighborhood of $1 \in G$, the local actions are defined. They are Lie algebra homomorphisms: $\rho_{e,r} : \mathfrak{g} \to TG^*$ which are called *local dressing actions* (respectively left and right).

Let us describe the local dressing actions explicitly. For $X \in \mathfrak{g}$ define left and right invariant 1-forms on G^* such that $X_e(e) = X_r(e) = X \in \mathfrak{g} = T_e^*G^*$. Here $e \in G^*$ is the identity element. The Poisson tensor $p_* \in \wedge^2 TG^*$ determines the linear map $\check{p}_* : T^*G^* \to TG^*$.

LEMMA 9.7. *The left and right local dressing actions of G on G^* are given by the following maps*

$$\rho_e(X) = \check{p}_*(X_e),$$
$$\rho_r(X) = \check{p}_*(X_r).$$

The proof will be left as an exercise.

An orbit of a local dressing action of G on $x \in G^*$ is a neighborhood of x that is spanned by vector fields of local dressing action.

THEOREM 9.8. *The symplectic leaves of G^* locally coincide with the orbits of the local dressing action of G^*. If the dressing action is defined globally, then for any $x \in G^*$ the symplectic leaf passing through x coincides with the orbit of the dressing action of G passing through this point.*

It is clear that symplectic leaves locally coincide with orbits: the corresponding vector fields are images of 1-forms on G^* with respect to the map given by the Poisson tensor. So locally they span Hamiltonian vector fields passing through x. Thus, the orbit is a Poisson submanifold which locally coincides with the corresponding symplectic leaf; it is therefore a symplectic leaf.

The map $\phi : G \hookrightarrow \mathcal{D}(G) \to \mathcal{D}(G)/G^*$, which is the composition of the inclusion map and the projection map, is a Poisson map.

THEOREM 9.9. *The symplectic leaves of G are connected components of preimages of orbits of the action of G^* on $\mathcal{D}(G)/G^*$ by left multiplication with respect to the map $\phi : G \to \mathcal{D}(G)/G^*$ (see Remark 9.5).*

In other words symplectic leaves of G are preimages of points in $G^*\backslash\mathcal{D}(G)/G^*$ with respect to the natural projection $G \hookrightarrow \mathcal{D}(G) \to G^*\backslash\mathcal{D}(G)/G^*$.

10. Symplectic Leaves in Simple Lie Groups With Standard Poisson Lie Structure

10.1. Bruhat Decomposition of the Double of G. A simple Lie group G with fixed Borel subgroup B admits a Bruhat decomposition with respect to B:

$$G = \bigcup_{w \in W} BwB.$$

Here $BwB \overset{\text{def}}{=} B\dot{w}B$, where $\dot{w}$ is a representative of $w \in N_G(H)/H$ in $N_G(H)$ (clearly $B\dot{w}B$ depends only on the class $w \in N_G(H)/H$).

There is also a Bruhat decomposition of G with respect to B^-:

$$G = \bigcup_{w} B^- w B^-.$$

Recall [**KS**] that the double $D(G)$ is, as a group, isomorphic to $G \times G$. The cell decompositions of G therefore give the Bruhat decomposition of $D(G)$ with respect to $D^- = B^- \times B$:

$$D(G) = \bigcup_{(w_1,w_2)\in W\times W} D^-(w_1, w_2)D^-,$$

$D^-(w_1, w_2)D^- = B^- w_1 B^- \times B w_2 B$, where $W \times W = N_{D(G)}(H \times H)/H \times H$ is the Weyl group of $D(G)$. We can also represent $D^- \subset D(G)$ as

$$D^- = (H \times H)(N^- \times N^+) = (N^- \times N^+)(H \times H).$$

Then for the Bruhat cell $D^-(w_1, w_2)D^-$ we can write

$$D^-(w_1, w_2)D^- = (N^-_{w_1} \times N^+_{w_2})(H \times H)(\dot{w}_1, \dot{w}_2)D^-, \tag{10.1}$$

where $N^{\pm}_w = \{n \in N^{\pm} | \dot{w}^{-1} n \dot{w} \in N^{\mp}\}$ (clearly this definition of $N^{\pm}_w$ does not depend on the choice of $\dot{w}$).

10.2. Left Cosets $D(G)/j(G^-)$. Let $G^- = G^{*\mathrm{opp}}$ which may be identified with $\{(b^-, b) \in B^- \times B | \theta^-(b^-) = \theta(b)^{-1}\}$, a subgroup of $B^- \times B$, [**KS**]. We write $j : G^- \hookrightarrow B^- \times B$ for this identification. There is a natural isomorphism:

$$D^-/j(G^-) \simeq H. \tag{10.2}$$

The group $H \times H$ acts on the cosets $(\dot{w}_1, \dot{w}_2)D^-/j(G^-)$ by left multiplication:

$$\begin{aligned}(h, h')(\dot{w}_1, \dot{w}_2)(b^-, b)j(G^-) &= (\dot{w}_1, \dot{w}_2)(h_{w_1} b^-, h'_{w_2} b)j(G^-)\\ &= (\dot{w}_1, \dot{w}_2)(\mathrm{Ad}_{h_{w_1}} b^-, h'_{w_2} h_{w_1}(\mathrm{Ad}_{h^{-1}_{w_1}} b)j(G^-)\\ &= (\dot{w}_1, \dot{w}_2)\big(\theta^-(b^-), h'_{w_2} h_{w_1} \theta(b)\big)j(G^-).\end{aligned}$$

Here $(b^-, b) \in B^- \times B$, and we write $h_w = \dot{w}^{-1} h \dot{w}$. Using also (10.2) we conclude that this action has the stationary subgroup

$$H^{w_1,w_2} = \{(h, h') \in H \times H \mid h_{w_1} = {h'_{w_2}}^{-1}\}. \tag{10.3}$$

Thus, we have an isomorphism $D^-(w_1, w_2)D^-/j(G^-) \cong N^-_{w_1} \times N^+_{w_2} \times H$ and, in particular, $\dim\big(D^-(w_1, w_2)D^-/j(G^-)\big) = l(w_1) + l(w_2) + r$.

10.3. Double Cosets $j(G^-)\backslash D(G)/j(G^-)$. For double cosets, we have

$$j(G^-)(n^-_{w_1}, n^+_{w_2})(h_1, h_2)(\dot{w}_1, \dot{w}_2)(b^-, b)j(G^-) = j(G^-)(\tilde{h}_1, \tilde{h}_2)(\dot{w}_1, \dot{w}_2)j(G^-),$$

where $\tilde{h}_1 = h_1 \theta^-(b^-)_{w_1^{-1}}, \tilde{h}_2 = h_2 \theta^-(b^-)_{w_2^{-1}}$. According to (10.3), the set of such double cosets is isomorphic to

$$j(H)\backslash H \times H / j_{w_1 w_2}(H), \tag{10.4}$$

where $j(H) \subset H \times H$ is the subgroup that consists of the elements (h, h^{-1}), $h \in H$, and $j_{w_1,w_2}(h) = (h_{w_1}, h^{-1}_{w_2})$. The coset of $(h_1, h_2) \in H \times H$ in (10.4) is the set $\{(h, h^{-1})(1, h_1 h_2 h'' {h''}^{-1}_{w_2^{-1} w_1}) | h, h'' \in H\}$. Thus (10.4) is isomorphic to $H_{w_2^{-1} w_1}$, where H_w is the space of H-orbits on H with respect to the action

$$h : h' \to h' h h_w^{-1}. \tag{10.5}$$

All orbits are naturally isomorphic and we denote the one through 1 by H^w. Furthermore, H_w is isomorphic to $\ker(w_2^{-1}w_1 - \mathrm{id}) = \{h \in H \mid h_{w_2^{-1}w_1} = h\}$. Thus, we have proved:

PROPOSITION 10.1. *We have an isomorphism*

$$j(G^-)\backslash D^-(w_1, w_2)D^-/j(G^-) \simeq H_{w_2^{-1}w_1}. \tag{10.6}$$

Each $j(G^-)$-orbit corresponding to an element of this set is isomorphic to

$$N_{w_1}^- \times N_{w_2}^+ \times H^{w_2^{-1}w_1}. \tag{10.7}$$

In particular, each such orbit has the dimension

$$\ell(w_1) + \ell(w_2) + \dim\bigl(\mathrm{coker}(w_2^{-1}w_1 - 1)\bigr).$$

Notice that the isomorphism (10.6) and the isomorphisms between the $j(G^-)$-orbits and the sets (10.7) are not canonical but depend on the choice of representatives $\dot{w}_1, \dot{w}_2$. What we really have here is a fiber bundle

$$D^-(w_1, w_2)D^-/j(G^-) \to j(G^-)\backslash D^-(w_1, w_2)D^-/j(G^-) \tag{10.8}$$

over the torus $H_{w_2^{-1}w_1}$ whose fibers are $j(G^-)$-orbits.

10.4. Symplectic Leaves of G and Double Bruhat Cells. Double Bruhat cells are defined as intersections of B-Bruhat cells and B^--Bruhat cells:

$$G^{w_1,w_2} = B^-w_1B^- \cap Bw_2B.$$

It is known that $\dim(G^{w_1,w_2}) = l(w_1) + l(w_2) + r$ (for example [**FZ**]).

Let $\varphi : G \overset{i}{\hookrightarrow} D \to D/j(G^-)$ be the composition of the diagonal embedding with the natural projection. According to (10.1) we have

$$D^-(w_1, w_2)D^-/j(G^-) \cong (N_{w_1}^- \times N_{w_2}^+)(\dot{w}_1, \dot{w}_2)i(H).$$

Define

$$\Gamma := \{\varepsilon \in H | \varepsilon^2 = 1\}.$$

THEOREM 10.2. *The restriction of φ to $G^{w_1,w_2} \to D^-(w_1, w_2)D^-/j(G^-)$ is a cover map with group of deck transformations Γ. Its image is a dense open subset.*

Here is an outline of the proof. Let $g = n_{w_1}^-\dot{w}_1b^- = n_{w_2}^+\dot{w}_2b^+ \in B^-w_1B^- \cap Bw_2B$, where $n_{w_1}^- \in N_{w_1}^-$, $n_{w_2}^+ \in N_{w_2}^+$ and $b^\pm \in B^\pm$. Then we have

$$\varphi(g) = (g, g)j(G^-) = (n_{w_1}^-\dot{w}_1b^-, n_{w_2}^+\dot{w}_2b^+)j(G^-).$$

Therefore $\varphi(g)$ is an element of $D^-(w_1, w_2)D^-/j(G^-)$.

Conversely, assume $x_1 = n_{w_1}^-\dot{w}_1b^-$ and $x_2 = n_{w_2}^+\dot{w}_2b^+$; then $(x_1, x_2)j(G)$. This class has a representative of the form $(g, g)j(G)$ if and only if there exists $(\eta_+, \eta_-) \in G_-$ such that $n_{w_1}^-\dot{w}_1\eta_- = n_{w_2}^+\dot{w}_2\eta_+$. According to [**FZ**] such elements exist on a dense open subset of $N_{w_1}^- \times N_{w_2}^+$. Therefore the image of φ is open dense in $D^-(w_1, w_2)D^-/j(G^-)$. Furthermore,

$$\varphi(g\varepsilon) = (g\varepsilon, g\varepsilon^{-1})j(G^-) = \varphi(g)$$

for each $\varepsilon \in \Gamma$. This shows that Γ acts fixed point freely on the preimages of points. Since $i(\Gamma) = i(H) \cap j(H)$ is the kernel of φ, Γ is the group of deck transformations.

Since the symplectic leaves in G are connected components of preimages of $j(G^-)$-orbits in $D(G)/j(G^-)$ we obtain the following description of leaves.

COROLLARY 10.3. *The double Bruhat cell G^{w_1,w_2} is a collection of symplectic leaves of G each one being a connected component of the preimage of a double coset $j(G^-)\backslash D^-(w_1,w_2)D^-/j(G^-)$.*

10.5. Coordinates on Symplectic Leaves of G. Let B^+, $B^- \subset \mathrm{SL}_2$ be positive and negative Borel subgroups in SL_2 with Lie algebras generated by H, e and H, f, respectively. They are Poisson Lie subgroups for the corresponding standard Poisson Lie structure on SL_2.

There is a natural isomorphism $B^+ \simeq \mathbb{C}^\times \times \mathbb{C}$ given by the matrix elements of the 2-dimensional representation of B^+:

$$\begin{pmatrix} A & B \\ 0 & A^{-1} \end{pmatrix}.$$

Here A is a coordinate on $\mathbb{C}^\times$ and B is a coordinate on $\mathbb{C}$. The standard Poisson structure on SL_2 gives the following Poisson brackets between coordinate functions:

$$\{A, B\} = AB.$$

The symplectic leaves of B^+ consist of a 1-parameter family of 0-dimensional simplectic leaves which consists of elements with $B = 0$ and of a single 2-dimensional symplectic leaf S^+ which consists of elements with $B \neq 0$.

Similarly, the negative Borel subgroup B^- is isomorphic to $\mathbb{C}^\times \times \mathbb{C}$, the coordinates being matrix elements in the 2-dimensional representation:

$$\begin{pmatrix} D & 0 \\ C & D^{-1} \end{pmatrix}$$

and with the Poisson brackets

$$\{D, C\} = DC.$$

The simplectic leaves of B^- also consist of a 1-parametric family of 0-dimensional symplectic leaves consisting of diagonal matrices and a single 2-dimensional symplectic leaf which consits of elements with $C \neq 0$. The Weyl group of SL_2 consists of two elements 1 and w_0. The element w_0 acts by reflection on the Cartan subgroup $w_0(h) = h^{-1}$, $h \in H$. According to the classification of symplectic leaves of simple Lie groups given above this gives four types of symplectic leaves in SL_2:

- 1-parameter family of 0-dimensional symplectic leaves

$$\left\{\begin{pmatrix} t & 0 \\ 0 & t^{-1} \end{pmatrix}\right\}, \quad t \in \mathbb{C}^\times,$$

  which belong to $B^+ \cap B^-$;
- 2-dimensional symplectic leaf

$$\left\{\begin{pmatrix} A & B \\ 0 & A^{-1} \end{pmatrix}, A, B \in \mathbb{C}^\times\right\},$$

  which is $B^+ \cap B^- w_0 B^-$;
- 2-dimensional symplectic leaf

$$\left\{\begin{pmatrix} A & 0 \\ C & A^{-1} \end{pmatrix}, A, C \in \mathbb{C}^\times\right\},$$

  which is $B^+ w_0 B^+ \cap B^-$;

- 1-parameter family of 2-dimensional symplectic leaves

$$\left\{ \begin{pmatrix} A & B \\ tB & A^{-1} \end{pmatrix}, A, B \in \mathbb{C}^\times \right\}, \quad t \in \mathbb{C}^\times,$$

which belong to $B^+w_0B^+ \cap B^-w_0B^-$.

10.6. Here we will describe coordinate systems on open dense subsets of double Bruhat cells which were introduced in [**FZ**].

Let $\mathrm{SL}_2(i)$ be the subgroup in G generated by the Chevalley generators H_i, e_i, f_i corresponding to the i-th simple root. It is a Poisson Lie subgroup in G. Denote by $B(i)$ and by $B(i)^-$ the upper and lower Borel subgroups in $\mathrm{SL}_2(i)$ (determined by fixing a standard Poisson Lie structure on G).

Let $S_i^\pm$ be 2-dimensional symplectic leaves in $B(i)^\pm$ and let A_i, B_i and C_i, D_i be the coordinates on S_i^+ and S_i^-, respectively, given by matrix elements in 2-dimensional representation.

They have Poisson brackets

$$\{A_i, B_i\} = -d_iA_iB_i, \quad \{D_i, C_i\} = d_iD_iC_i \tag{10.9}$$

where d_i is the length of the i-th simple root.

The double Bruhat cell $G^{u,v}$ has a natural decomposition $G^{u,v} = G'^{u,v} \times H(u,v)$, where $H(u,v)$ is the subgroup of H generated by the elements corresponding to simple roots which do not belong to $|u| \cup |v|$. Here $|u|$ is the support of the element $u \in W$; it is a subset of $\{1, 2, \dots, r\}$ which consists of only those indices which appear in a reduced decomposition of u.

Fix a reduced decomposition of elements $u, v \in W$ $u = s_{i_1} \dots s_{i_k}$, $v = s_{j_1} \dots s_{j_l}$. Denote $r(u,v) = \mathrm{Card}\{|u| \cup |v|\}$. Then $\dim(G^{u,v}) = l + k + r(u,v)$.

Let $p_1, \dots, p_{k+l}$ be a shuffle of the sequence $\{i_1, \dots, i_k, -j_1, \dots, -j_l\}$.

Consider the image of

$$S_{i_1}^+ \times \dots \times S_{i_k}^+ \times S_{j_l}^- \times \dots \times S_{j_1}^-$$

under the multiplication, the shuffle and the inverse map:

$$(b_{i_1}, \dots, b_{i_k}, b_{j_l}^-, \dots, b_{j_1}^-) \mapsto b_{i_1} \dots b_{i_k} (b_{j_1}^-)^{-1} \dots (b_{j_l}^-)^{-1}. \tag{10.10}$$

Denote by $G_{i_1,\dots,i_k|j_1,\dots,j_l}$ the image of this map. It is clear that $G_{\{i\}|\{j\}}$ is isomorphic to $\mathbb{C}^{k+l+r(u,v)}$, and therefore is a complex torus which is open dense in $G^{u,v}$.

THEOREM 10.4 ([**FZ**]). *For each pair of reduced decompositions and shuffles $\{i\}$, $\{j\}$ and $\{i'\}$, $\{j'\}$, the subvarieties $G_{\{i\}|\{j\}}$ and $G_{\{i'\}|\{j'\}}$ are birationally isomorphic and they are birationally isomorphic to $G^{u,v}$.*

Let us use the following notations:

$$x_i(t) = \exp(te_i), \quad x_{-i}(t) = \exp(tf_i).$$

We will write s^{h_i} for the elements $\left(\begin{smallmatrix} s & 0 \\ 0 & s^{-1} \end{smallmatrix}\right)$ in $\mathrm{SL}_2(i) \subset G$.

The set $G_{\{i\}|\{j\}}$ can be described as

$$G_{\{i\}|\{j\}} = \left\{ \prod_{i \in |u| \cup |v|} s_i^{h_i} x_{p_1}(t_1) \dots x_{p_{k+l}}(t_{k+l}) \;\middle|\; s_i, t_i \in \mathbb{C} \right\},$$

where $p_1, \dots, p_k = i_1, \dots, i_k$ and $p_{k+1}, \dots, p_{k+l} = -j_1, \dots, -j_l$. The map (10.10) is Poisson and therefore $G_{\{i\}|\{j\}}$ is a Poisson subvariety. Poisson brackets between the coordinate functions $\{s_i, t_i\}$ can be computed from the brackets (10.9) and from the definition of the map (10.10):

$$\{s_i, s_j\} = 0, \tag{10.11}$$

$$\{s_i, t_\alpha\} = d_i s_i t_\alpha, \quad i = i_\alpha, \tag{10.12}$$

$$\{t_\alpha, t_\beta\} = s(\alpha) a_{i_\alpha, j_\beta} d_{i_\beta} t_\alpha t_\beta, \quad \alpha < \beta, \tag{10.13}$$

where $s(\alpha) = \operatorname{sign}(p_\alpha)$. One should emphasize that the tori $G_{\{i\}|\{j\}}$ do not cover the double Bruhat cell $G^{u,v}$.

10.7. Casimir Functions. Here we will describe the Casimir functions on the double Bruhat cells whose level surfaces are symplectic leaves.

According to [**FZ**] define generalized minors as the following functions on the group G. Let G_0 be the subset in G formed by the elements which have Gaussian factorization $x = [x]_-[x]_0[x]_+$ with $[x]_\pm \in N^\pm$ and $[x]_0 \in H$. For a weight λ define a function

$$\Delta_\lambda(x) = [x]_0^\lambda.$$

Let ω_i be a highest weight of the i-th fundamental representation of G, and let$\bar{\bar{u}}$ and $\bar{v}$ be special representatives of elements $u, v \in W$ in G. The generalized minors are the following functions:

$$\Delta_{u\omega_i, v\omega_i}(x) = \Delta_{\omega_i}(\bar{\bar{u}}^{-1} x \bar{v}).$$

PROPOSITION 10.5. *The symplectic leaves of the double Bruhat cell $G^{u,v}$ are connected components of common level surfaces of the functions*

$$c_{u,v,t}(x) = \prod_{i=1}^{r} \Delta_{v\omega_i,\omega_i}(x)^{t_i} \Delta_{\omega_i,u^{-1}\omega_i}(x)^{t_i},$$

where $t = \sum_{i=1}^{r} t_i \omega_i \in \ker(uv^{-1} - \mathrm{id}) \in \mathfrak{h}^*$.

Casimir functions are monomial on the tori described above, and they are multiplicative in the s, t coordinates on these tori [**Z**].

11. Characteristic Integrable Systems

11.1. Degenerate Integrability of Characteristic Hamiltonian System. Let (G, p) be a factorizable Poisson-Lie group. Let $I(G) \subset C^\infty(G)$ be the subspace of Ad_G-invariant functions on G.

THEOREM 11.1. (1) *$I(G)$ is a commutative Poisson algebra in $C^\infty(G)$.*

(2) *In a neighborhood of $t = 0$ the flow lines of the Hamiltonian flow induced by $H \in I(G)$ passing through $x \in G$ at $t = 0$ have the form*

$$x(t) = g_\pm(t)^{-1} x g_\pm(t),$$

where the mappings $g_\pm(t)$ are determined by

$$g_+(t) g_-(t)^{-1} = \exp\Big(tI\big(d_l H(x)\big)\Big),$$

and $I : \mathfrak{g}^* \longrightarrow \mathfrak{g}$ is the factorization isomorphism.

DEFINITION 11.2. A *characteristic Hamiltonian system* on a factorizable Poisson Lie group is a Hamiltonian system whose phase space is a symplectic leaf of a factorizable Poisson Lie group and whose Hamiltonian is an adjoint invariant function on G.

The above theorem implies that the equations of motion of a characteristic Hamiltonian system on a factorizable Poisson Lie group can be solved via factorization.

Characteristic Hamiltonian systems for simple Lie groups with standard Poisson Lie structure are degenerate integrable [**R**]. Let $S_{u,v} \subset G$ be a symplectic leaf which belongs to the Poisson subvariety $G^{u,v} \subset G$. Denote by $[S_{u,v}]$ the set of Ad_{G^*}-orbits in $G \times G$ intersecting $S_{u,v}$, and by $\mathrm{Ad}_G S_{u,v}$ the set of Ad_G-orbits in G intersecting $S_{u,v}$. We have natural Poisson projections

$$S_{u,v} \to [S_{u,v}] \to \mathrm{Ad}_G S_{u,v}. \tag{11.1}$$

Here the first map is the composition of the diagonal embedding $S_{u,v} \to S_{u,v} \times S_{u,v}$ and the natural projection $S_{u,v} \times S_{u,v} \to [S_{u,v}]$. The set $[S_{u,v}]$ naturally projects to $\mathrm{Ad}_G S_{u,v} \times \mathrm{Ad}_G S_{u,v}$, and the image belongs to the image of the diagonal embedding of $\mathrm{Ad}_G S_{u,v}$ into $\mathrm{Ad}_G S_{u,v} \times \mathrm{Ad}_G S_{u,v}$. This describes the second map in 11.1.

In other words, characteristic Hamiltonian systems are integrable and their Liouville tori considered as subsets in G are intersections of adjoint orbits of G and of orbits of the dressing action of G^*.

REMARK 11.3. It has been known that Liouville tori of characteristic integrable systems on factorizable Poisson Lie groups contain intersections of dressing orbits and adjoint orbits [**STS**]. However, the fact that these intersections are exactly Liouville tori is relatively new [**R**]. The first results of this type which go beyond Toda systems are due to [**DLNT**] and [**EFS**]. The most recent one is due to [**GS**], where the integrability of characteristic systems was proved for generic symplectic leaves on simple Lie algebras.

11.2. Coxeter-Toda Symplectic Leaves in *G*. Let w be a Coxeter element, i.e., in a reduced decomposition $w = s_{i_1} \dots s_{i_r}$ the sequence $\{i_1, \dots, i_r\}$ is a permutation of $\{1, 2, \dots, r\}$.

In the double Bruhat cell $G^{w,w}$ we have the complex torus $G_{\{i\},\{i\}}$:

$$G_{\{i\},\{i\}} = \left\{ \prod_{i=1}^{r} s_i^{h_i} x_{i_1}(t_1) \dots x_{i_r}(t_r) x_{-j_1}(\bar{t}_1) \dots x_{-j_r}(\bar{t}_r) \;\middle|\; s_i, t_i, \bar{t}_i \in \mathbb{C}^\times \right\}.$$

The map (10.10) sends the point with coordinates $A_{i_\alpha}, B_{i_\alpha}, C_{i_\alpha}, D_{i_\alpha}$ to the point in $G_{\{i\},\{j\}}$ with coordinates

$$t_\alpha = B_{i_\alpha} A_{i_\alpha} \prod_{\beta > \alpha} A_{i_\beta}^{-a_{i_\beta, i_\alpha}} \prod_{\beta} D_{i_\beta}^{a_{i_\beta, i_\alpha}}, \tag{11.2}$$

$$\bar{t}_\alpha = C_{i_\alpha} D_{i_\alpha}^{-1} \prod_{\beta > \alpha} D_{i_\beta}^{-a_{i_\beta, i_\alpha}}, \quad s_i = A_i / D_i. \tag{11.3}$$

Introduce new coordinates

$$\chi_{i_\alpha}^- = t_\alpha \bar{t}_\alpha, \tag{11.4}$$

$$\chi_i^+ = \chi_i^- \prod_{i=1}^{r} s_j^{a_{ji}}, \tag{11.5}$$

$$G_i = t_i / \bar{t}_i \prod_{j=1}^{r} s_j^{a_{ji}}. \tag{11.6}$$

PROPOSITION 11.4. *The functions* $\chi_i^{\pm}$, *and* G_i *on* $G_{\{i\},\{i\}}$ *have the following Poisson brackets*:

$$\begin{aligned}\{\chi_i^+,\chi_j^+\} &= \{\chi_i^-,\chi_j^-\} = 0,\\ \{\chi_i^+,\chi_j^-\} &= -2d_i a_{ij}\chi_i^+\chi_j^-,\\ \{\chi_i^{\pm},G_j\} &= 0.\end{aligned}$$

Using the definition of $\chi_i^{\pm}$ and elementary algebra we arrive at the following

PROPOSITION 11.5. **[HKKR]** *Let* V *be a finite-dimensional representation of* G *and* Ch_V *be its character. Then*

$$\begin{aligned}\mathrm{Ch}_V|_{G_{\{i\},\{i\}}} &= \mathrm{Ch}_V\left(\prod_{j=1}^{r}\left(\frac{\chi_j^+}{\chi_j^-}\right)^{h^j}\phi_{i_1}(g_{i_1})\dots\phi_{i_r}(g_{i_r})\right)\\ &= \mathrm{Ch}_V\left(\prod_{j=1}^{r}\left(\frac{\chi_j^+}{\chi_j^-}\right)^{h^j}\phi_{i_1}(\bar g_{i_1})\dots\phi_{i_r}(\bar g_{i_r})\right).\end{aligned}$$

Here $\phi_i : \mathrm{SL}_r(i) \hookrightarrow G$ is the embedding of SL_2 generated by e_i, h_i, f_i into G, g_i and $\bar g_i$ are elements of SL_2 whose image in 2-dimensional irreducible representation is given by the following weight basis of the 2-dimensional irreducible representation:

$$g_i = \begin{pmatrix}1-\chi_i^- & \chi_i^-\\ -1 & 1\end{pmatrix},\quad \bar g_i = \begin{pmatrix}1 & \chi_i^+\\ -1 & 1-\chi_i^+\end{pmatrix}.$$

The elements $\{h^j\}$ form the basis in $\mathfrak{h} \subset \mathfrak{g}$ corresponding to the fundamental weights: $h_j = \sum_i a_{ji}h^i$.

Thus, we have an integrable system on $\mathbb{C}^{2n}$ with Hamiltonians given by Proposition 11.5.

11.3. Here are a few examples of Coxeter-Toda Hamiltonians for classical Lie algebras. In all formulae below, the representation V is a vector representation of the corresponding Lie algebra. It is $(r+1)$-dimensional for A_r, $2r$-dimensional for C_r and D_r and $(2r+1)$-dimensional for B_r.

It is convenient to introduce the variable

$$\varphi_i = \prod_{j=1}^{r}\left(\frac{\chi_j^+}{\chi_j^-}\right)^{(a^{-1})_{ji}}.$$

Then we have

(1) A_r:

$$\mathrm{Ch}_V = \sum_{i=1}^{r+1}\frac{\varphi_i}{\varphi_{i-1}}(1-\chi_{i-1}^+).$$

(2) B_r:

$$\begin{aligned}\mathrm{Ch}_V = \sum_{i=1}^{r-1}\frac{\varphi_i}{\varphi_{i-1}}(1-\chi_{i-1}^+) + \frac{\varphi_r}{\varphi_{r-1}}(1-\chi_{r-1}^+) + (1-2\chi_r^+)\\ + \frac{\varphi_{r-1}}{\varphi_r}(1-\chi_r^+)^2 + \sum_{i=1}^{r-1}\frac{\varphi_{i-1}}{\varphi_i}(1-\chi_i^+).\end{aligned}$$

(3) C_r:

$$\mathrm{Ch}_V = \sum_{i=1}^{r-1} \frac{\varphi_i}{\varphi_{i-1}}(1-\chi_{i-1}^+) + \left(\frac{\varphi_r}{\varphi_{r-1}}\right)^2 (1-\chi_{r-1}^+) + \left(\frac{\varphi_{r-1}}{\varphi_r}\right)^2 (1-\chi_r^+) + \sum_{i=1}^{r-1} \frac{\varphi_{i-1}}{\varphi_i}(1-\chi_i^+).$$

(4) D_r:

$$\mathrm{Ch}_V = \sum_{i=1}^{r-2} \frac{\varphi_i}{\varphi_{i-1}}(1-\chi_{i-1}^+) + \varphi_r\varphi_{r-1}(\varphi_{r-2})^{-1}(1-\chi_{r-2}^+) + \varphi_r(\varphi_{r-1})^{-1}(1-\chi_{r-1}^+) + \varphi_{r-1}(\varphi_r)^{-1}(1-\chi_r^+)(\varphi_r)^{-1} \times (\varphi_{r-1})^{-1}\varphi_{r-2}(1-\chi_{r-1}^+)\sum_{i=1}^{r-2} \frac{\varphi_{i-1}}{\varphi_i}(1-\chi_i^+).$$

References

[A] M. Adler, *On a trace functional for formal pseudo-differential operators and the sympletic structure of the KdV-type equations*, Invent. Math. **50** (1979), 219–248.

[DLNT] P. Deift, L. C. Li, T. Nanda, and C. Tomei, *The Toda flow on a generic orbit is integrable*, Comm. Pure Appl. Math. **39** (1986), 183–232.

[Dr] V. G. Drinfeld, *Quantum groups*, Proc. Internat. Congress of Mathematicians (Berkeley, 1986), Amer. Math. Soc., Providence, RI, 1987, pp. 798–820.

[EFS] N. M. Ercolai, H. Flaschka, and S. F. Singer, *The geometry of the full Kostant-Toda lattice*, Prog. Math. **115** (1993), 181–225.

[FZ] S. Fomin and A. Zelevinsky, *Double Bruhat cells and total positivity*, J. Amer. Math. Soc. **12** (1999), 335–380.

[GS] M. I. Gekhtman and M. Z. Shapiro, *Non-commutative and commutative integrability of generic Toda flow in simple Lie algebras*, Comm. Pure Appl. Math. **52** (1999), 53–84.

[HKKR] T. Hoffmanh, J. Kellendonk, N. Kutz, and N. Reshetikhin, *Factorization dynamics and Coxeter-Toda lattices*, Comm. Math. Phys. (in print).

[KS] L. Korogodski and Y. Soibelman, *Algebras of functions on quantum groups*, Part I, Math. Surveys Monogr., vol. 56, Amer. Math. Soc., Providence, RI, 1998.

[Ko] B. Kostant, *The solution to a generalised Toda lattice and representation theory*, Adv. Math. **34** (1979), 195–338.

[N] N. N. Nekhoroshev, *Action-angle variables and their generalizations*, Trans. Moscow Math. Soc. **26** (1972), 180–197.

[R] N. Reshetikhin, *Integrability of characteristic Hamiltonian systems on simple Lie groups with standard Poisson Lie structure*, preprint.

[RSTS] A. G. Reyman and M. A. Semenov-Tian-Shansky, *Reduction of Hamiltonian systems, affine Lie algebras and Lax equations.* I, Invent. Math. **54** (1979), 81–100.

[STS] M. Semenov-Tian-Shansky, *Dressing transformations and Poisson group actions*, Publ. Res. Inst. Math. Sci. **21** (1985), 1237–1260.

[Z] A. Zelevinsky, *Connected components of real double Bruhat cells*, math.AG/0003231.

Department of Mathematics, University of California at Berkeley, Berkeley, CA 94720–3840, USA

E-mail address: reshetik@math.berkeley.edu

Centre de Recherches Mathématiques
CRM Proceedings and Lecture Notes
Volume **26**, 2000

Special Functions Associated with Calogero-Moser Type Quantum Systems

S. N. M. Ruijsenaars

ABSTRACT. These lecture notes concern various special functions playing a role in the diagonalization of (two-particle) relativistic Calogero-Moser quantum Hamiltonians. We first consider generalized gamma functions of hyperbolic, trigonometric and elliptic type (Euler's gamma function being of rational type), which arise in the various parameter regimes associated with the Hamiltonians. The pertinent hyperbolic Hamiltonian can be diagonalized via a 'relativistic' hypergeometric function R built from hyperbolic gamma functions. We discuss a number of features of this R-function, in particular its connection to the Askey-Wilson polynomials. Finally, we consider generalized Lamé functions, which solve the diagonalization problem in the elliptic regime. Their asymptotics involves the elliptic gamma function.

1. Preliminaries and Overview

The following is concerned with various special functions that play a role in the general program of constructing unitary eigenfunction transforms for Calogero-Moser type quantum dynamics. More specifically, we focus on the reduced two-particle Hamiltonian of the relativistic Calogero-Moser system and functions related to it. It is beyond the scope of these lecture notes to provide substantial background information on integrable particle systems of Calogero-Moser type. We refer to our lecture notes Ref. [**37**] for a rather detailed account and bibliographical information until 1995; some more recent references are mentioned below.

Here, we only include a quite concise sketch of the pertinent many-particle dynamics, which will also enable us to introduce various quantities and notation that we use as a starting point. The nonrelativistic Calogero-Moser system is defined by the partial differential operator

$$H_{\mathrm{nr}} \equiv -\frac{\hbar^2}{2m}\sum_{j=1}^{M}\partial_j^2 + \frac{g(g-\hbar)}{m}\sum_{\substack{,k=1\\ j<k}}^{M} V(x_j - x_k). \tag{1.1}$$

The most general potential function $V(x)$ allowed here is the Weierstrass $\wp$-function $\wp(x;\omega,\omega')$. To ensure at least *formal* self-adjointness of H_{nr} as an operator on

---

1991 *Mathematics Subject Classification.* 33–02, 33C99, 33D99, 33E99, 39A10.
This is the final form of the paper.

$L^2(\mathbb{R}^M, dx)$, we choose the coupling constant g real, and the periods ω and ω' in $(0,\infty)$ and $i(0,\infty)$, resp. (Of course, the particle mass m and Planck's constant $\hbar$ are chosen positive.)

Letting one or both periods of $\wp(x)$ go to ∞, one obtains special cases that are of independent interest and that are customarily denoted as follows:

$$(1.2)\ \text{(IV)} \qquad \wp(x; \frac{\pi}{2r}, \frac{ia}{2}), \qquad r, a > 0, \qquad \text{(elliptic)},$$

$$(1.3)\ \text{(III)} \qquad -\frac{r^2}{3} + \frac{r^2}{\sin^2 rx}, \qquad r > 0,\ a = \infty, \qquad \text{(trigonometric)},$$

$$(1.4)\ \text{(II)} \qquad \frac{\pi^2}{3a^2} + \frac{\pi^2}{a^2 \operatorname{sh}^2(\pi x/a)}, \qquad r = 0,\ a > 0, \qquad \text{(hyperbolic)},$$

$$(1.5)\ \text{(I)} \qquad 1/x^2, \qquad r = 0,\ a = \infty, \qquad \text{(rational)}.$$

(The parametrization of the periods is somewhat unconventional; it turns out to be convenient later on.)

These potentials give rise to quantum integrable M-particle systems, in the sense that there exist M independent PDOs of orders $k = 1, \ldots, M$ (with the $k = 2$ PDO given by (1.1)) that commute pairwise. The $k = 1$ PDO is the total momentum operator

$$P_{\mathrm{nr}} \equiv \frac{\hbar}{i} \sum_{j=1}^{M} \partial_j. \tag{1.6}$$

The Hamiltonian (1.1) is referred to as 'nonrelativistic', since it yields together with P_{nr} and the Galilean boost operator

$$B \equiv -m \sum_{j=1}^{M} x_j, \tag{1.7}$$

a representation of the nonrelativistic space-time symmetry Lie algebra (the associated symmetry group being the Galilei group). This is true for an *arbitrary* potential. By contrast, the relativistic Calogero-Moser Hamiltonian to be presented shortly gives rise to a representation of the Lie algebra of the Poincaré group that would be spoiled by a different choice of 'potential'.

The pertinent potential involves the Weierstrass σ-function $\sigma(x; \omega, \omega')$. We find it convenient, however, to work with the function

$$s(r, a; x) \equiv \sigma\left(x; \frac{\pi}{2r}, \frac{ia}{2}\right) \exp\left(-\frac{\eta r x^2}{\pi}\right). \tag{1.8}$$

The features of this function that will be crucial below are the following. First of all, $s(x)$ is an entire, odd, and π/r-antiperiodic function. Second, $s(x)$ satisfies the analytic difference equation (from now on AΔE)

$$\frac{s(x + ia/2)}{s(x - ia/2)} = -\exp(-2irx). \tag{1.9}$$

Third, the zeros of $s(x)$ for x in the lattice $\pi r^{-1}\mathbb{Z} + ia\mathbb{Z}$ (whose presence follows from the previous properties) are simple, and one has

$$s(x) = x + O(x^3), \quad x \to 0. \tag{1.10}$$

(Observe that by Liouville's theorem $s(x)$ is the unique function with all of the features listed thus far.) Finally, one has the limits

$$\lim_{a\uparrow\infty} s(r,a;x) = \frac{\sin rx}{r}, \quad \text{(uniformly on compacts)}, \tag{1.11}$$

$$\lim_{r\downarrow 0} s(r,a;x) = \frac{\operatorname{sh} \pi x/a}{\pi/a}, \quad \text{(uniformly on compacts)}. \tag{1.12}$$

With this infomation at our disposal, we are prepared to detail the many-particle relativistic Calogero-Moser Hamiltonian and total momentum operator. They are defined via the analytic difference operators (henceforth AΔOs)

$$S_{\pm k} \equiv \sum_{\substack{I\subset\{1,\ldots,M\} \\ |I|=k}} \prod_{\substack{i\in I \\ j\notin I}} f_{\mp}(x_i - x_j) \exp\Big(\mp i\frac{\hbar}{mc}\sum_{l\in I}\partial_l\Big) \prod_{\substack{i\in I \\ j\notin I}} f_{\pm}(x_i - x_j), \quad k = 1,\ldots,M, \tag{1.13}$$

where

$$f_{\pm}(x) \equiv \big(s(x \pm ig/mc)/s(x)\big)^{1/2}, \quad g \downarrow 0 \Rightarrow f_{\pm}(x) \to 1. \tag{1.14}$$

Specifically, one has

$$H_{\text{rel}} \equiv mc^2(S_1 + S_{-1}), \tag{1.15}$$

$$P_{\text{rel}} \equiv mc(S_1 - S_{-1}). \tag{1.16}$$

The elliptic system thus obtained is denoted IV_{rel}, whereas the trigonometric, hyperbolic and rational specializations are denoted by III_{rel}, II_{rel} and I_{rel}, resp.

The additional parameter $c > 0$ is physically interpreted as the speed of light. The nonrelativistic limit $c \to \infty$ yields

$$H_{\text{rel}} = Mmc^2 + H_{\text{nr}} + C + O(c^{-2}), \quad P_{\text{rel}} = P_{\text{nr}} + O(c^{-2}), \tag{1.17}$$

as is readily verified. It is true, but not easy to verify, that one has

$$[H_{\text{rel}}, P_{\text{rel}}] = 0. \tag{1.18}$$

It is at this point that the special choice of $s(x)$ in $f_{\pm}(x)$ is crucial. Indeed, replacing $s(x)$ by other functions spoils the commutativity of H_{rel} and P_{rel} (and hence their interpretation as time and space translation generators), whereas the choice of $s(x)$ is immaterial for the commutators

$$[H_{\text{rel}}, B] = i\hbar P_{\text{rel}}, \quad [P_{\text{rel}}, B] = i\hbar H_{\text{rel}}/c^2. \tag{1.19}$$

(That is, the operator B (1.7) now plays the role of the Lorentz boost generator.)

More generally, all of the AΔOs $S_{\pm k}$ (1.13) commute pairwise, which is why the quantum dynamics H_{rel} (or, equivalently, the slightly simpler 'light cone Hamiltonian' S_1) defines an integrable M-particle quantum system. Originally, the commutativity was reduced to a sequence of functional equations satisfied by the Weierstrass σ-function [**31**]. In recent years, some powerful technology in the area of integrable systems has been applied to the relativistic quantum systems [**8, 18, 13, 19, 23, 2, 1, 22**], yielding in particular alternative proofs of commutativity.

With quantum integrability known, the central problem is to construct/discover an explicit unitary (joint) eigenfunction transform, diagonalizing the above M-particle Hamiltonians H_{nr} and H_{rel} (and their commuting integrals) as real-valued multiplication operators. In this connection we recall that the 3-dimensional many-particle Coulomb Hamiltonian is widely believed to encompass most of chemistry and biology. Of course, searching for explicit eigenfunctions for more than two particles in the latter context seems (at least to date) a hopeless enterprise. But here we are dealing with 1-dimensional many-particle systems that are moreover integrable, so that the starting point is incomparably simpler.

In our lecture notes Ref. [**37**] we have summarized some literature concerning the many-particle problem just delineated, and we now continue by mentioning a few more recent relevant references. They include: a paper by Opdam [**29**], who proves in particular unitarity of the suitably normalized eigenfunction transform for the $\mathrm{II}_{\mathrm{nr}}$-Hamiltonian; work by Felder/Varchenko concerning the $\mathrm{II}_{\mathrm{nr}}$ and $\mathrm{IV}_{\mathrm{nr}}$ eigenfunctions [**11, 14**]; a paper by Hasegawa [**18**], who finds a finite-dimensional space of theta function type eigenfunctions for the $\mathrm{IV}_{\mathrm{rel}}$ case; an article by Chalykh [**7**], who announces a proof of the duality between the $\mathrm{II}_{\mathrm{nr}}$ and $\mathrm{I}_{\mathrm{rel}}$ systems (which we conjectured in Ref. [**32**]); a paper by van Diejen/Vinet [**9**] on the compactified trigonometric regime eigenfunction transform; work on quantum separation of variables by Kuznetsov/Sklyanin [**26**]; papers by Billey [**6**] and Hikami/Komori [**20**], who construct eigenfunctions of the $\mathrm{IV}_{\mathrm{rel}}$-Hamiltonian by an algebraic Bethe ansatz. (All of the elliptic results just mentioned concern the case $g/\hbar \in \mathbb{N}$, and do not adress Hilbert space properties.)

This list is by no means exhaustive. In particular, the ongoing activity in the area of q-KZ and q-KZB equations will probably turn out to be relevant to the eigenfunction program for the $\mathrm{II}_{\mathrm{rel}}$ and $\mathrm{IV}_{\mathrm{rel}}$ systems.

Here we will be dealing exclusively with the $M = 2$ Hamiltonians. Transforming to sum and difference variables, we can omit the center-of-mass part to obtain (setting from now on $\beta = 1/mc$ and $\hbar = m = 1$)

$$H_0 = -\frac{d^2}{dx^2} + g(g-1)\wp(x), \tag{1.20}$$

$$H_\beta = \left(\frac{s(r,a;x-i\beta g)}{s(r,a;x)}\right)^{1/2} T_{i\beta} \left(\frac{s(r,a;x+i\beta g)}{s(r,a;x)}\right)^{1/2} + (i \to -i), \quad \beta \equiv 1/mc > 0. \tag{1.21}$$

Here, the translations $T_{\pm i\beta}$ are defined by

$$(T_\alpha F)(x) \equiv F(x-\alpha), \quad \alpha \in \mathbb{C}. \tag{1.22}$$

The time-independent Schrödinger equation $H\Psi = E\Psi$, $E \in \mathbb{R}$, then reduces to a second-order ODE/AΔE in the nonrelativistic/relativistic case.

Now for second-order ordinary differential operators (and second-order *discrete* difference operators) there exists a well-developed eigenfunction expansion theory, the so-called Weyl-Titchmarsh-Kodaira theory. (See, e.g., Refs. [**44, 10, 5, 30, 16, 21**] for accounts of WTK theory from various complementary viewpoints.) By contrast, for second-order AΔOs of the type considered here it is not even clear that eigenfunctions do exist for arbitrary parameters. But whenever Ψ solves $H_\beta\Psi = E\Psi$, then any function $\hat{\Psi}$ of the form $m(x)\Psi(x)$, with $m(x)$ an $i\beta$-periodic meromorphic multiplier, satisfies $H_\beta\hat{\Psi} = E\hat{\Psi}$ as well.

The infinite-dimensional ambiguity in eigenfunctions that we have just described is a formidable obstacle in associating to H_β a well-defined self-adjoint operator $\hat{H}_\beta$ on $L^2(\mathbb{R}, dx)$ that is diagonalized as multiplication by E on a suitable

L^2-space via a unitary eigenfunction transform. In fact, from the few examples discussed in Section 4 one already concludes that expectations based on experience with WTK theory (or, alternatively, with the heuristics substituted for WTK theory in much of the physics literature) can be quite wrong for formally self-adjoint AΔOs.

Let us now sketch the organization of the remaining subject matter. It is divided into three sections. Each of these sections has a short introduction, outlining in particular a further subdivision into subsections. Briefly, Section 2 concerns generalized gamma functions that play an auxiliary role in the diagonalization problem delineated above. On the one hand, they can be exploited to transform H_β (1.21) (and all of the AΔOs $S_{\pm k}$ (1.13), for that matter) to an AΔO with meromorphic coefficients. On the other hand, suitable combinations encode the asymptotics for $x \to \pm\infty$ of the eigenfunctions studied in Sections 3 and 4. The generalized gamma functions are defined as solutions to first-order AΔEs with properties that render the solution unique. To be more specific, they satisfy a 'minimality' requirement and a suitable normalization condition.

Section 3 deals with a 'relativistic' hypergeometric function R that is defined in terms of an integral representation. The integrand consists of products of the 'hyperbolic gamma function' from Section 2. The R-function can be used in particular to solve the Schrödinger AΔE for the $\mathrm{II}_{\mathrm{rel}}$ case. This involves a specialization of four free parameters to one, in much the same way as two free parameters in the ${}_2F_1$-function can be specialized to one in order to solve the $\mathrm{II}_{\mathrm{nr}}$-ODE.

In Section 4 we consider eigenfunctions for the elliptic setting. We recall first what is known concerning the nonrelativistic case (1.20) and then study eigenfunctions for the relativistic case (1.21), especially with an eye on their Hilbert space properties.

A novel feature of the relativistic eigenfunctions (as compared to their nonrelativistic counterparts) is that they are explicitly known for a dense set in the natural parameter space $r, a, \beta > 0$, $g \in \mathbb{R}$. Moreover, the pertinent (discrete energy) eigenfunctions turn out to be pairwise orthogonal only when further restrictions are imposed, and they exhibit a quite unexpected invariance under the interchange of a and β.

2. Generalized Gamma Functions

Most of the results to be presented in this section can be found in our paper Ref. [**33**]. In that paper the results are obtained within the context of a general theory dealing with AΔEs of the form

$$\frac{F(z+ia/2)}{F(z-ia/2)} = \Phi(z), \quad a > 0. \tag{2.1}$$

Here, $\Phi(z)$ is a given meromorphic function, and a meromorphic function $F(z)$ obeying (2.1) is to be found. Though we will focus on rather special functions $\Phi(z)$, some insight in more general features is helpful. Therefore, we devote Subsection 2.1 to a sketch of the general theory.

In the applications of the theory to integrable particle systems of Calogero-Moser type one encounters right-hand side functions $\Phi(z)$ that are rational, hyperbolic, trigonometric or elliptic. Since such functions admit a factorization, and since AΔEs of the form (2.1) have an obvious group structure, one is led to the problem of constructing especially convenient 'building block functions'.

We refer to the latter as generalized gamma functions. This terminology is motivated by the rational case considered in Subsection 2.2, where the relevant gamma function amounts to Euler's gamma function. Subsequently, trigonometric, elliptic and hyperbolic gamma functions are introduced and studied in Subsections 2.3–2.5.

2.1. First Order AΔEs. A key point to be stressed first is that meromorphic solutions to (2.1) form an infinite-dimensional vector space, whenever one meromorphic solution exists. Indeed, when a meromorphic $F(z)$ solves (2.1), then $m(z)F(z)$ is another solution for any ia-periodic meromorphic function $m(z)$. Therefore it is crucial to single out special solutions by further requirements. (We need not address existence questions here, since we only consider AΔEs admitting quite explicit meromorphic solutions.)

We begin by imposing a rather weak requirement: we restrict attention to functions $\Phi(z)$ without poles and zeros in a strip $|\operatorname{Im} z| < s$, $s > 0$. (This can be often achieved by a suitable shift of z.) Then we can take logarithms to trade the multiplicative AΔE (2.1) for an additive one,

$$f(z + ia/2) - f(z - ia/2) = \phi(z), \quad |\operatorname{Im} z| < s. \tag{2.2}$$

Let us next require that $\phi(z)$ have at worst polynomial increase on the real axis. Then we may and will restrict attention to solutions $f(z)$ that are polynomially bounded in the strip $|\operatorname{Im} z| \leq a/2$ and that are moreover analytic for $|\operatorname{Im} z| < s + a/2$.

We call solutions $f(z)$ of the form just delineated *minimal* solutions: Their analytic behavior and asymptotics for $|\operatorname{Re} z| \to \infty$ are optimal. The polynomial boundedness is critical in proving that minimal solutions are unique up to an additive constant, whenever they exist. (Note that entire ia-periodic functions such as $\operatorname{ch}(2\pi z/a)$ are not polynomially bounded.)

It is readily verified that the resulting function $F(z) = \exp\big(f(z)\big)$ extends to a meromorphic solution of (2.1) that has no poles and zeros for $|\operatorname{Im} z| < s + a/2$, and whose logarithm is polynomially bounded for $|\operatorname{Im} z| \leq a/2$. Once more, solutions to (2.1) with the latter properties are termed *minimal*, and now they are unique up to a multiplier $c\exp(2\pi kz/a)$, $c \in \mathbb{C}^*$, $k \in \mathbb{Z}$. (The exponential ambiguity reflects the ambiguity in the branch choice for $\ln \Phi(z)$.)

A large class of right-hand sides arises as follows. Assume that $\phi(z)$ satisfies (in addition to the above)

$$\phi(x) \in L^1(\mathbb{R}), \quad \hat{\phi}(y) \in L^1(\mathbb{R}), \quad \hat{\phi}(y) = O(y), \quad y \to 0. \tag{2.3}$$

Here, $\hat{\phi}(y)$ denotes the Fourier transform, normalized by

$$\hat{\phi}(y) \equiv \frac{1}{2\pi} \int_{-\infty}^{\infty} dx\, \phi(x) e^{ixy}. \tag{2.4}$$

Then the function

$$f(z) = \int_{-\infty}^{\infty} dy\, \frac{\hat{\phi}(2y)}{\operatorname{sh} ay} e^{-2iyz}, \quad |\operatorname{Im} z| \leq a/2, \tag{2.5}$$

is clearly well defined, and analytic for $|\operatorname{Im} z| < a/2$. Proceeding formally, it is also obvious that $f(z)$ satisfies (2.2) (by virtue of the Fourier inversion formula).

As a matter of fact, it can be shown that the function $f(z)$ defined by (2.5) analytically continues to $|\operatorname{Im} z| < s + a/2$ and indeed fulfils the AΔE (2.2) (cf. Theorem II.2 in Ref. [**33**]). Moreover, $f(z)$ is bounded in the strip $|\operatorname{Im} z| \leq a/2$ and goes to 0 for $|\operatorname{Re} z| \to \infty$ in the latter strip. Thus $f(z)$ gives rise to a minimal solution $F(z)$ of the original AΔE (2.1).

Another extensive class of right-hand sides $\Phi(z)$ admitting minimal solutions arises by assuming that $\Phi(z)$ has a real period, in addition to the standing assumption of absence of zeros and poles in a strip $|\operatorname{Im} z| < s$. To anticipate our later needs, we denote this period by π/r, $r > 0$. Now $\phi(x) = \ln \Phi(x)$, $x \in \mathbb{R}$, is well defined up to a multiple of $2\pi i$. We assume first that $\Phi(x)$ has zero winding number around 0 in the period interval $[-\pi/2r, \pi/2r]$. Then $\phi(x)$ is a smooth π/r-periodic function too. Defining its Fourier coefficients by

$$\hat{\phi}_n \equiv \frac{r}{\pi} \int_{-\pi/2r}^{\pi/2r} dx\, \phi(x) e^{2inrx}, \quad n \in \mathbb{Z}, \tag{2.6}$$

we also assume at first $\hat{\phi}_0 = 0$.

With these assumptions in effect, it is clear that the function

$$f(z) = \frac{1}{2} \sum_{n \in \mathbb{Z}^*} \frac{\hat{\phi}_n e^{-2inrz}}{\operatorname{sh} nra}, \quad |\operatorname{Im} z| \le a/2, \tag{2.7}$$

is well defined, and analytic for $|\operatorname{Im} z| < a/2$. Again, it is formally obvious (by Fourier inversion) that $f(z)$ satisfies the AΔE (2.2). Once more, it can be shown that $f(z)$ (2.7) has an analytic continuation to $|\operatorname{Im} z| < s + a/2$ and solves (2.2) (cf. Theorem II.5 in Ref. [**33**]). Thus one obtains minimal solutions to the AΔE (2.1).

Next, we point out that the assumption $\hat{\phi}_0 = 0$ is not critical. Indeed, when $\hat{\phi}_0 \neq 0$, one need only add the function $\hat{\phi}_0 z/ia$ to $f(z)$ (2.7) to obtain a minimal solution. Of course, this entails that $f(z)$ is no longer π/r-periodic.

The obvious generalization detailed in the previous paragraph is relevant to the case in which $\Phi(z)$ has winding number $l \in \mathbb{Z}^*$ on $[-\pi/2r, \pi/2r]$. Then one needs to take the z-derivative of the AΔE (2.2) to obtain a rhs $\phi'(z)$ that is π/r-periodic, but for which the zeroth Fourier coefficient of $\phi'(x)$ equals $2irl$, cf. (2.6). Thus the solution $f'(z)$ obtained as just sketched has a term linear in z, and so $f(z)$ is the sum of a π/r-periodic function and a quadratic function. Clearly, this still gives rise to a minimal solution $F(z) = \exp f(z)$ to the AΔE (2.1).

The idea underlying this construction is easily generalized to functions $\Phi(z)$ for which a suitable derivative of $\ln \Phi(z)$ has period π/r. Likewise, the Fourier transform method yielding the minimal solution (2.5) can be generalized to $\Phi(z)$ for which $\phi(z) \equiv \partial_z^k \ln \Phi(z)$ has the three properties (2.3) for a suitable $k \in \mathbb{N}^*$. (See Theorems II.3 and II.6 in Ref. [**33**] for the details.)

The more general class of functions $\Phi(z)$ in the Fourier transform setting which we have just delineated is relevant for the rational and hyperbolic gamma functions of Subsections 2.2 and 2.5. The trigonometric and elliptic gamma functions of Subsections 2.3 and 2.4 can be viewed as minimal π/r-periodic solutions to AΔEs defined by a π/r-periodic function $\Phi(z)$ that has zero winding number on $[-\pi/2r, \pi/2r]$.

2.2. The Rational Gamma Function. We define the rational gamma function by

$$G_{\mathrm{rat}}(a; z) \equiv \Gamma\Big(-\frac{iz}{a} + \frac{1}{2}\Big). \tag{2.8}$$

The Γ-function AΔE $\Gamma(x+1) = x\Gamma(x)$ entails that $G_{\rm rat}$ solves the AΔE

$$\frac{G(z+ia/2)}{G(z-ia/2)} = -\frac{iz}{a}. \tag{2.9}$$

More generally, when $\Phi(z)$ is an arbitrary rational function, we can solve the AΔE (2.1) by a function $F(z)$ of the form

$$F(z) = \exp(c_0 + c_1 z)\frac{\prod_{j=1}^{M} G_{\rm rat}(a; z-\alpha_j)}{\prod_{k=1}^{N} G_{\rm rat}(a; z-\beta_k)}. \tag{2.10}$$

Indeed, when we let the integers N, M vary over $\mathbb{N}$ and c_1, α_j, β_k over $\mathbb{C}$, then we obtain *all* rational functions on the rhs of (2.1) (save for the zero function, of course).

We now explain the relation of $G_{\rm rat}$ to the notion of 'minimal solution'. To this end we first of all need a shift of z on the rhs of (2.9) in order to get a function $\Phi(z)$ that is analytic and free of zeros in a strip around the real axis. Taking

$$\Phi(z) \equiv -i(z+ia/2)/a, \tag{2.11}$$

a simple contour integration yields

$$\frac{1}{2\pi}\int_{-\infty}^{\infty} dx\, e^{ixy}\partial_x^2 \ln \Phi(x) = \begin{cases} -y\exp(ay/2), & y<0, \\ 0, & y \geq 0. \end{cases} \tag{2.12}$$

Therefore $\phi(z) \equiv \partial_z^2 \ln \Phi(z)$ has the three properties (2.3), and the corresponding minimal solution (2.5) is then given by

$$f(z) = -2\int_{-\infty}^{0} dy\, \frac{ye^{ay}}{\operatorname{sh} ay} e^{-2iyz}. \tag{2.13}$$

The point is now that the rhs of (2.13) equals $\partial_z^2 \ln G_{\rm rat}(a; z+ia/2)$, so that $G_{\rm rat}(a; z+ia/2)$ may be characterized as a *minimal* solution to the AΔE with rhs (2.11). It is not hard to check the asserted equality via Gauss' formula for the psi function (the logarithmic derivative of $\Gamma(x)$). But one may in fact reobtain various results on the gamma function (such as Gauss' formula) by taking the function (2.13) as a starting point. Put differently, if one would not have been familiar with the gamma function beforehand, one would have been led to it (and to a substantial part of its theory) via the 'minimal solution' approach sketched above.

We have worked out the details of this useful perspective on Euler's gamma function in Appendix A of our paper Ref. [**33**]. Here we only add a few more remarks on several features that are important with an eye on the generalized gamma functions introduced and studied in later subsections.

First, it should be observed that the shift of z by $ia/2$ is arbitrary; any shift $z \to z + is$, $s > 0$, would yield substantially the same conclusions.

Second, a shift $z \to z - is$, $s > 0$, on the rhs of (2.9) yields in the same way as sketched above solutions to (2.9) of the form $c\exp\big((2k+1)\pi z/a\big)/\Gamma(iz/a + 1/2)$, $c \in \mathbb{C}^*$, $k \in \mathbb{Z}$. Hence the quotient of such a solution and the previous one $\Gamma(-iz/a + 1/2)$ is ia-periodic. This is in agreement with the well-known reflection equation, which becomes here

$$\Gamma\Big(\frac{iz}{a} + \frac{1}{2}\Big)\Gamma\Big(-\frac{iz}{a} + \frac{1}{2}\Big) = \pi \operatorname{ch}\Big(\frac{\pi z}{a}\Big)^{-1}. \tag{2.14}$$

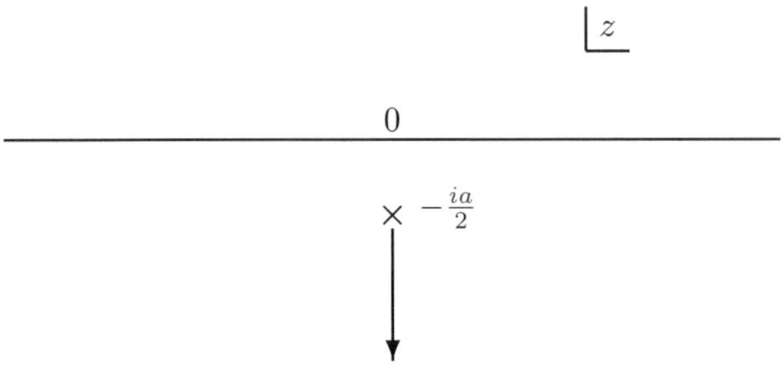

FIGURE 1. The pole sequence of the rational gamma function.

(This identity can also be derived from the AΔE-viewpoint, cf. Appendix A in Ref. [**33**].)

Third, we recall that the minimal solutions to the AΔE (2.1) with rhs (2.11) (obtained via twofold integration of $f(z)$ (2.13)) are analytic and zero-free for $|\operatorname{Im} z| < a$. From the AΔE one then sees that they are analytic and zero-free for all z not in $-ia\mathbb{N}^*$, whereas they have simple poles for $z = -iak$, $k \in \mathbb{N}^*$. (Similarly, the shift $z \to z - is, s > 0$, in (2.9) yields minimal solutions without poles and with a zero sequence in the upper half plane.) Of course, this is once again well known for the special function $G_{\text{rat}}(a; z)$ (2.8). To allow a graphic comparison of its pole sequence

$$z = -i(k + 1/2)a, \quad k \in \mathbb{N}, \quad \text{(poles)}, \tag{2.15}$$

to those of the generalized gamma functions introduced below, we symbolize it in Fig. 1.

2.3. The Trigonometric Gamma Function. In the same way as $G_{\text{rat}}(a; z)$ (2.8) serves as a building block to solve AΔEs with rational right-hand sides, the trigonometric gamma function can be used to handle trigonometric functions $\Phi(z)$ with period π/r. (More precisely, functions in the field $\mathbb{C}(\exp(2irz))$.) Specifically, letting

$$F(z) = \exp(c_0 + c_1 z + c_2 z^2) \frac{\prod_{j=1}^{M} G_{\text{trig}}(r, a; z - \alpha_j)}{\prod_{k=1}^{N} G_{\text{trig}}(r, a; z - \beta_k)}, \tag{2.16}$$

with $G_{\text{trig}}(r, a; z)$ solving

$$\frac{G(z + ia/2)}{G(z - ia/2)} = 1 - \exp(2irz), \tag{2.17}$$

one obtains all trigonometric functions by letting N, M vary over $\mathbb{N}$ and c_1, c_2, α_j, β_k over $\mathbb{C}$ in the quotient $F(z - ia/2)/F(z + ia/2)$.

Iterating the AΔE (2.17), one obtains

$$G(z + ila) = \prod_{k=1}^{l} (1 - \exp[2ir(z + (k - 1/2)ia)]) \cdot G(z) \tag{2.18}$$

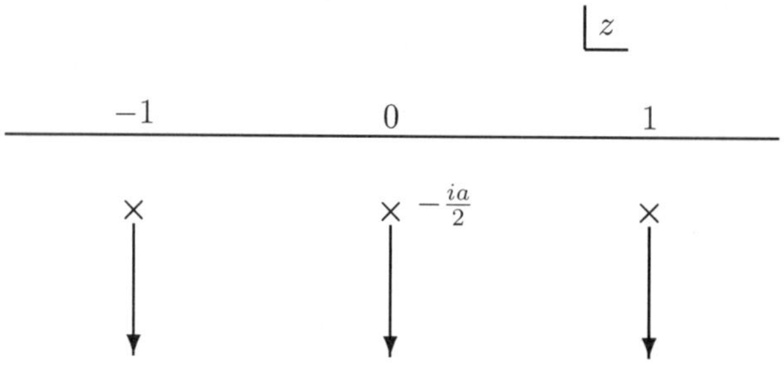

FIGURE 2. The pole sequences of the trigonometric gamma function (with $r = \pi$).

for all solutions. An obvious solution to (2.17) is, therefore,

$$G_{\text{trig}}(r, a; z) \equiv \prod_{k=1}^{\infty} (1 - q^{2k-1} e^{2irz})^{-1}, \quad q \equiv e^{-ar}. \tag{2.19}$$

Indeed, the infinite product clearly converges, yielding a meromorphic solution without zeros and with poles for

$$z = j\pi/r - i(k + 1/2)a, \quad j \in \mathbb{Z}, \ k \in \mathbb{N}, \quad \text{(poles)}, \tag{2.20}$$

as symbolized in Fig. 2.

Another representation for G_{trig} arises by writing

$$\prod_{k=1}^{\infty} (1 - q^{2k-1} e^{2irz})^{-1} = \exp\Big(- \sum_{k=1}^{\infty} \ln(1 - q^{2k-1} e^{2irz}) \Big), \tag{2.21}$$

and using the elementary Fourier series

$$\ln(1 - e^{2ir(z+is)}) = - \sum_{n=1}^{\infty} \frac{e^{2inr(z+is)}}{n}, \quad s > 0. \tag{2.22}$$

Indeed, this gives rise to the formula

$$G_{\text{trig}}(r, a; z) = \exp\Big(\sum_{n=1}^{\infty} \frac{e^{2inrz}}{2n \operatorname{sh} nra} \Big), \quad \operatorname{Im} z > -a/2, \tag{2.23}$$

as is easily checked.

Taking $z \to z + is$, $s > 0$, in the AΔE (2.17), it is of the form discussed in Subsection 2.1. Taking next logarithms and using (2.22), one deduces that (2.23) amounts to the special solution (2.7). That is, $G_{\text{trig}}(r, a; z)$ is once more a *minimal* solution to the (shifted) AΔE (2.17). As such, it is uniquely determined by the asymptotics

$$G_{\text{trig}}(r, a; z) \sim 1, \quad \operatorname{Im} z \to \infty, \tag{2.24}$$

which can be read off from (2.23).

Our trigonometric gamma function is closely related to Thomae's q-gamma function $\Gamma_q(x)$: One has

$$G_{\mathrm{trig}}(r,a;z)=\exp(c_0+c_1z)\Gamma_{\exp(-2ar)}\Big(-\frac{iz}{a}+\frac{1}{2}\Big), \tag{2.25}$$

for suitable constants c_0, c_1. (See, e.g., Ref. [**17**] for information on $\Gamma_q(x)$.) The AΔE-perspective from which G_{trig} arises leads to various other features that are detailed in Ref. [**33**]. Here we only point out that it is evident from the infinite product representation (2.19) that one can allow r to vary over the (open) right half plane, whereas one cannot take $r\in i\mathbb{R}$. This is why one needs another building block function to handle hyperbolic right-hand side functions $\Phi(z)$ in the AΔE (2.1), cf. Subsection 2.5.

2.4. The Elliptic Gamma Function. The elliptic gamma function is the minimal solution to the AΔE

$$\frac{G(z+ia/2)}{G(z-ia/2)}=\exp\Big(-\sum_{n=1}^{\infty}\frac{\cos(2nrz)}{n\,\mathrm{sh}(nrb)}\Big),\quad |\operatorname{Im} z|<b, \tag{2.26}$$

obtained via the formula (2.7):

$$G_{\mathrm{ell}}(r,a,b;z)\equiv\exp\Big(i\sum_{n=1}^{\infty}\frac{\sin(2nrz)}{2n\,\mathrm{sh}(nra)\,\mathrm{sh}(nrb)}\Big),\quad |\operatorname{Im} z|<(a+b)/2. \tag{2.27}$$

Indeed, our definition entails that the functions

$$F(z)=\exp(c_0+c_1z+c_2z^2+c_3z^3)\frac{\prod_{j=1}^{M}G_{\mathrm{ell}}(r,a,b;z-\alpha_j)}{\prod_{k=1}^{N}G_{\mathrm{ell}}(r,a,b;z-\beta_k)}, \tag{2.28}$$

give rise to all elliptic right-hand side functions $\Phi(z)$ with periods π/r and ib.

To explain why this is true, we recall first that any elliptic function with periods π/r and ib admits a representation as

$$c\prod_{j=1}^{N}\frac{\sigma(z-\gamma_j;\pi/2r,ib/2)}{\sigma(z-\delta_j;\pi/2r,ib/2)}, \tag{2.29}$$

where $c,\gamma_j,\delta_j\in\mathbb{C}$. (To readers who would like to read up on elementary properties of elliptic functions such as the one just mentioned, we recommend Ref. [**45**].) The crux is now that the function on the rhs of (2.26) is of the form

$$\exp(d_0+d_1z+d_2z^2)\sigma(z+ib/2;\pi/2r,ib/2). \tag{2.30}$$

Hence the functions $F(z+ia/2)/F(z-ia/2)$, with $F(z)$ given by (2.28), yield all functions (2.29) (with $c\neq 0$) by choosing

$$N=M,\quad \alpha_j=\gamma_j+ib/2,\quad \beta_j=\delta_j+ib/2,\quad j=1,\ldots,N, \tag{2.31}$$

and appropriate constants c_1, c_2, c_3 determined by the constants c, d_0, d_1 and d_2.

Next, we mention that the elliptic gamma function can also be written as an infinite product

$$G_{\mathrm{ell}}(r,a,b;z)=\prod_{m,n=1}^{\infty}\frac{1-q_a^{2m-1}q_b^{2n-1}e^{-2irz}}{1-q_a^{2m-1}q_b^{2n-1}e^{2irz}},\quad q_a\equiv e^{-ar},\quad q_b\equiv e^{-br}. \tag{2.32}$$

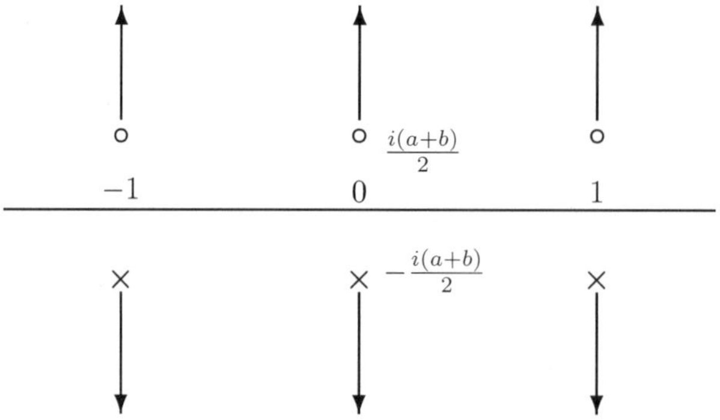

FIGURE 3. The zero and pole sequences of the elliptic gamma function (with $r = \pi$).

To check this, one need only proceed as in the trigonometric case: The infinite product can be written as the exponential of a series; using the Fourier series (2.22) and summing the resulting geometric series yields (2.27).

From (2.32) one can read off meromorphy in z and the locations of poles and zeros. Specifically, one obtains the doubly-infinite sequences

$$z = j\pi/r - i(k+1/2)a - i(l+1/2)b, \quad j \in \mathbb{Z},\ k,l \in \mathbb{N}, \quad \text{(poles)}, \tag{2.33}$$

$$z = j\pi/r + i(k+1/2)a + i(l+1/2)b, \quad j \in \mathbb{Z},\ k,l \in \mathbb{N}, \quad \text{(zeros)}. \tag{2.34}$$

They are depicted in Fig. 3.

It is immediate from the product representations (2.19) and (2.32) that one has

$$G_{\text{trig}}(r,a;z) = \lim_{b\uparrow\infty} G_{\text{ell}}(r,a,b;z-ib/2). \tag{2.35}$$

It is also not difficult to see that for the renormalized function

$$\tilde{G}_{\text{ell}}(r,a,b;z) \equiv G_{\text{ell}}(r,a,b;z)\exp\left(\frac{\pi^2 z}{6irab}\right), \tag{2.36}$$

the $r \downarrow 0$ limit exists. This yields the function

$$G_{\text{hyp}}(a,b;z) = \lim_{r\downarrow 0} \tilde{G}_{\text{ell}}(r,a,b;z), \tag{2.37}$$

studied in the next subsection. For further properties of the elliptic gamma function we refer to Subsection III.B in Ref. [**33**].

2.5. The Hyperbolic Gamma Function. The defining AΔE of the hyperbolic gamma function reads

$$\frac{G(z+ia/2)}{G(z-ia/2)} = 2\,\text{ch}(\pi z/b). \tag{2.38}$$

Clearly, any solution $G_{\text{hyp}}(a,b;z)$ of (2.38) can be used to solve equation (2.1) with $\Phi(z)$ an arbitrary hyperbolic function with period ib. Indeed, all functions $\Phi \in \mathbb{C}\big(\exp(2\pi z/b)\big)$ arise via functions of the form

$$F(z) = \exp(c_0 + c_1 z + c_2 z^2)\frac{\prod_{j=1}^{M} G_{\text{hyp}}(a,b;z-\alpha_j)}{\prod_{k=1}^{N} G_{\text{hyp}}(a,b;z-\beta_k)}. \tag{2.39}$$

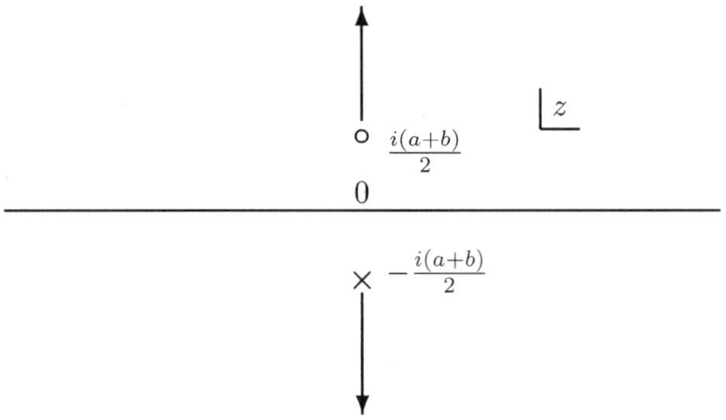

FIGURE 4. The zero and pole sequences of the hyperbolic gamma function.

As before, there is a certain arbitrariness in the choice of AΔE for the building block function. Our choice $2\,\mathrm{ch}(\pi z/b)$ together with the requirement that the solution be minimal will lead us to a function $G_{\mathrm{hyp}}(a, b; z)$ with various features that would be spoiled by any other choice, however. (In particular, the constant 2 cannot be changed without losing the remarkable $(a \leftrightarrow b)$-invariance of the hyperbolic gamma function (2.40).)

Following the method to construct minimal solutions sketched in Subsection 2.1, one readily verifies that $\phi(z) \equiv \partial_z^3 \ln\bigl(2\,\mathrm{ch}(\pi z/b)\bigr)$ has the three properties (2.3). The Fourier transform can be done explicitly, and integrating up three times then yields our hyperbolic gamma function,

$$G_{\mathrm{hyp}}(a, b; z) \equiv \exp\left(i \int_0^\infty \frac{dy}{y} \left(\frac{\sin(2yz)}{2\,\mathrm{sh}(ay)\,\mathrm{sh}(by)} - \frac{z}{aby} \right) \right), \quad |\operatorname{Im} z| < (a+b)/2. \tag{2.40}$$

Since $G_{\mathrm{hyp}}(a, b; z)$ has no poles and zeros in the strip $|\operatorname{Im} z| < (a+b)/2$, one readily deduces from the defining AΔE (2.38) that G_{hyp} extends to a meromorphic function with poles and zeros given by

$$z = -i(k+1/2)a - i(l+1/2)b, \quad k, l \in \mathbb{N}, \quad \text{(poles)}, \tag{2.41}$$

$$z = i(k+1/2)a + i(l+1/2)b, \quad k, l \in \mathbb{N}, \quad \text{(zeros)}. \tag{2.42}$$

These doubly infinite sequences are depicted in Fig. 4, which should be compared to Figs. 1–3.

In the following section we need various other properties of $G_{\mathrm{hyp}}(a, b; z)$. Some automorphy properties are immediate from (2.40): One has

$$G_{\mathrm{hyp}}(a, b; -z) = 1/G_{\mathrm{hyp}}(a, b; z), \tag{2.43}$$

$$G_{\mathrm{hyp}}(a, b; z) = G_{\mathrm{hyp}}(b, a; z), \tag{2.44}$$

$$G_{\mathrm{hyp}}(\lambda a, \lambda b; \lambda z) = G_{\mathrm{hyp}}(a, b; z), \quad \lambda > 0. \tag{2.45}$$

It is also easy to establish from the defining AΔE that the multiplicity of a pole or zero z_0 equals the number of distinct pairs $(k, l) \in \mathbb{N}^2$ giving rise to z_0, cf. (2.41), (2.42). In particular, the pole at $-i(a+b)/2$ and zero at $i(a+b)/2$ are simple, and for a/b irrational *all* poles and zeros are simple.

The remaining properties we have occasion to use are not clear by inspection, and we refer to Subsection IIIA in Ref. [**33**] for a detailed account. First, we need to know the residue at the simple pole $-i(a+b)/2$. It is given by

$$\mathrm{Res}(-i(a+b)/2) = \frac{i}{2\pi}(ab)^{1/2}. \tag{2.46}$$

(Here and below, we take positive square roots of positive quantities.)

Second, we need two distinct zero step size limits of the hyperbolic gamma function. The first one reads

$$\lim_{b\downarrow 0} \frac{G_{\mathrm{hyp}}(\pi, b; z+i\lambda b)}{G_{\mathrm{hyp}}(\pi, b; z+i\mu b)} = \exp\big((\lambda-\mu)\ln(2\,\mathrm{ch}\, z)\big), \tag{2.47}$$

where the limit is uniform on compact subsets of the cut plane

$$\mathbb{C}\setminus\{\pm iz \in [\pi/2, \infty)\}. \tag{2.48}$$

When $\lambda-\mu$ is an integer, this limit is immediate from the $(a\leftrightarrow b)$-invariance of G_{hyp} and the AΔE (2.38). For $\lambda-\mu\notin\mathbb{Z}$, the emergence of the logarithmic branch cuts on the imaginary axis may be viewed as a consequence of the coalescence of an infinite number of zeros and poles on the cuts.

The second zero step size limit yields the connection to the Γ-function. Consider the function

$$H(\rho; z) \equiv G_{\mathrm{hyp}}(1, \rho; \rho z + i/2)\exp\big(iz\ln(2\pi\rho)\big)/(2\pi)^{1/2}. \tag{2.49}$$

From the AΔE (2.38) and its $(a\leftrightarrow b)$-counterpart one sees that H satisfies the AΔE

$$\frac{H(\rho; z+i/2)}{H(\rho; z-i/2)} = \frac{i\,\mathrm{sh}(\pi\rho z)}{\pi\rho}, \tag{2.50}$$

and reflection equation

$$H(\rho; z)H(\rho; -z) = \pi^{-1}\,\mathrm{ch}\,\pi z. \tag{2.51}$$

(Recall also (2.43) to check (2.51).) Therefore, it should not come as a surprise that one has

$$\lim_{\rho\downarrow 0} H(\rho; z) = 1/\Gamma(iz+1/2), \tag{2.52}$$

uniformly for z in $\mathbb{C}$-compacts.

We have made a (convenient) choice for the first parameter a of the hyperbolic gamma function, but it should be pointed out that the scale invariance (2.45) can be used to handle arbitrary a. In particular, this yields the limit

$$\lim_{b\uparrow\infty} G_{\mathrm{hyp}}(a, b; z-ib/2)\exp\big(-iz\ln(2\pi a/b)\big)(2\pi)^{1/2} = G_{\mathrm{rat}}(a; z), \tag{2.53}$$

cf. (2.8).

Finally, we need the $|\,\mathrm{Re}\, z| \to \infty$ asymptotics of G_{hyp}. To detail this, we set

$$c \equiv \max(a, b), \quad \sigma = 1-\epsilon, \quad \epsilon > 0. \tag{2.54}$$

Then we have

$$\mp i\ln G_{\mathrm{hyp}}(a, b; z) = -\frac{\pi z^2}{2ab} - \frac{\pi}{24}\left(\frac{a}{b}+\frac{b}{a}\right) + O(\exp(\mp 2\pi\sigma\,\mathrm{Re}\, z/c), \quad \mathrm{Re}\, z \to \pm\infty. \tag{2.55}$$

Here, the bound is uniform for $\mathrm{Im}\, z$ in $\mathbb{R}$-compacts, but it is not uniform as $\epsilon \downarrow 0$, since it is false for $\sigma = 1$ and $a = b$.

To conclude this subsection we add some further information on the hyperbolic gamma function. First of all, this function was introduced in another guise and from a quite different perspective in previous literature—a fact we were not aware of at the time we wrote Ref. [**33**]. Indeed, our hyperbolic gamma function is related to a function that is nowadays referred to as Kurokawa's double sine function [**25**]. The latter is usually denoted $S_2(x|\omega_1, \omega_2)$, and the relation reads

$$G_{\mathrm{hyp}}(a, b; z) = S_2(iz + (a+b)/2|a, b). \tag{2.56}$$

The first occurrence of the double sine function is however in a series of papers by Barnes, published a century ago. He generalized the gamma function from another viewpoint to his so-called multiple gamma functions; the double sine function is then a quotient of two double gamma functions. We have reobtained and extended a number of his results via our minimal solution approach (sketched in Subsection 2.1) in a recent paper [**41**], where further bibliographical information can be found.

Multiple gamma functions show up in particular in number theory. The double gamma and sine functions were studied from this viewpoint in a paper by Shintani [**42**]. He derived a product formula that can be used to tie in the hyperbolic and trigonometric gamma functions in a quite explicit way, and we continue by presenting the pertinent formulas.

First, we should mention that $G_{\mathrm{hyp}}(a, b; z)$ admits a representation as an infinite product of Γ-functions, from which meromorphy properties in a, b and z follow by inspection (cf. Prop. III.5 in Ref. [**33**]). In particular, $G_{\mathrm{hyp}}(a, b; z)$ has an analytic continuation to $b \in i(0, \infty)$. Now for $b = i\pi/r, r > 0$, the rhs of (2.38) can be rewritten as

$$\exp(-irz)\Big(1 - \exp\big(2ir(z + \pi/2r)\big)\Big). \tag{2.57}$$

Comparing to the rhs of (2.17), we deduce that the quotient function

$$\exp(-rz^2/2a)G_{\mathrm{trig}}(r, a; z + \pi/2r)/G_{\mathrm{hyp}}(a, i\pi/r; z) \tag{2.58}$$

is ia-periodic.

From Shintani's formula one can in particular determine this ia-periodic quotient explicitly. For the case at hand his product formula amounts to

$$\begin{aligned} &G_{\mathrm{hyp}}(a, i\pi/r; z) \\ &= \exp\left(-\frac{rz^2}{2a} - \frac{1}{24}\left(ra - \frac{\pi^2}{ra}\right)\right)\prod_{k=1}^{\infty}\frac{1 + \exp(2i\pi a^{-1}[iz + i(k-1/2)\pi r^{-1}])}{1 + \exp(2ir[z + i(k-1/2)a])}. \end{aligned} \tag{2.59}$$

From the definition (2.19) of the trigonometric gamma function we then get the remarkable relation

$$G_{\mathrm{hyp}}(a, i\pi/r; z) = \exp\left(-\frac{rz^2}{2a} - \frac{1}{24}\left(ra - \frac{\pi^2}{ra}\right)\right)\frac{G_{\mathrm{trig}}(r, a; z + \pi/2r)}{G_{\mathrm{trig}}(\pi/a, \pi/r; iz + a/2)}, \tag{2.60}$$

from which the ia-periodic function (2.58) can be read off. (The reader who is familiar with modular transformation properties may find more information on this angle below Eq. (A.27) in Ref. [**38**], where we rederived Shintani's product formula.)

3. A Generalized Hypergeometric Function

The subject of this section is a function $R(a_+, a_-, \mathbf{c}; v, \hat{v})$, depending on parameters $a_+, a_- \in (0, \infty)$, couplings $\mathbf{c} = (c_0, c_1, c_2, c_3) \in \mathbb{R}^4$ and variables $v, \hat{v} \in \mathbb{C}$. It generalizes both the hypergeometric function ${}_2F_1(a, b, c; w)$ and the Askey-Wilson polynomials $p_n(q, \alpha, \beta, \gamma, \delta; \cos v)$. (For a complete account of the features of ${}_2F_1$ we use in these lectures we refer to Ref. [**45**]. For information on the Askey-Wilson polynomials, see Refs. [**4, 17**].) The R-function was introduced in Ref. [**37**]. Detailed proofs of several assertions made below can be found in our forthcoming paper Ref. [**38**].

In Subsection 3.1 we consider various features of the ${}_2F_1$-function that admit a generalization to the R-function. The latter is defined in Subsection 3.2, where we also obtain some automorphy properties. In Subsection 3.3 we introduce the four independent hyperbolic AΔOs of which the R-function is an eigenfunction. In Subsection 3.4 we derive the specialization to the Askey-Wilson polynomials. Subsection 3.5 concerns the 'nonrelativistic limit' $R \to {}_2F_1$.

3.1. Some Reminders on ${}_2F_1$. The hypergeometric function was already known to Euler in terms of an integral representation. Our generalized hypergeometric function is defined in terms of an integral as well, but this integral representation does not generalize Euler's integral representation for ${}_2F_1$, but rather the much later one due to Barnes.

The latter representation can be readily understood from Gauss' series representation,

$$(3.1) \qquad {}_2F_1(a, b, c; w) = \sum_{n=0}^{\infty} \frac{\Gamma(a+n)}{\Gamma(a)} \frac{\Gamma(b+n)}{\Gamma(b)} \frac{\Gamma(c)}{\Gamma(c+n)} \frac{w^n}{n!}.$$

Using for instance the ratio test, one sees that this power series converges for $|w| < 1$. The Barnes representation makes it possible to analytically continue ${}_2F_1$ to the cut plane $\mathrm{Arg}(-w) < \pi$. It reads

$$(3.2) \qquad \int_{\mathcal{C}} dz \, \exp\bigl(-iz \ln(-w)\bigr) \cdot \frac{\Gamma(iz)\Gamma(c)}{2\pi\Gamma(c - iz)} \cdot \frac{\Gamma(a - iz)\Gamma(b - iz)}{\Gamma(a)\Gamma(b)}.$$

Here, the logarithm branch is fixed by choosing $\ln(-w) \in \mathbb{R}$ for $w \in (-\infty, 0)$. Taking first $\mathrm{Re}\, a, \mathrm{Re}\, b > 0$, the contour $\mathcal{C}$ runs along the real axis from $-\infty$ to ∞, with a downward indentation at the origin to avoid the pole due to $\Gamma(iz)$. Thus it separates the downward pole sequences starting at $-ia$ and $-ib$ from the upward sequence starting at 0 cf. Fig. 5.

Invoking the asymptotics of the Γ-function (Stirling's formula), one sees that the integrand has exponential decay for $\mathrm{Re}\, z \to \pm\infty$, provided $\mathrm{Arg}(-w) < \pi$. Thus the integral yields an analytic function of w in the cut plane. After multiplication by $2\pi i$, the residues at the simple poles $z = in$ of the integrand are equal to the terms in the Gauss series (3.1). (Indeed, $\Gamma(iz)$ has residue $-i(-)^n/n!$ for $z = in$.) A second somewhat subtle application of the Γ-function asymptotics now shows that when one moves the contour $\mathcal{C}$ up across the poles $0, i, \ldots, in$, picking up $2\pi i$ times the residues in the process, then the integral over the shifted contour converges to 0 for $n \to \infty$, provided $|w| < 1$. Thus the integral (3.2) yields an analytic continuation to the cut plane $\mathrm{Arg}(-w) < \pi$, as advertised.

The analyticity region cannot be much improved, given that the function ${}_2F_1(a, b, c; w)$ has a logarithmic branch point at $w = 1$ for generic $a, b, c \in \mathbb{C}$.

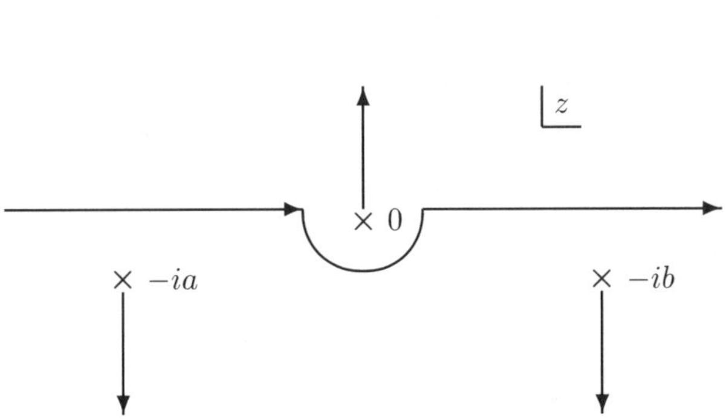

FIGURE 5. The pole sequences and contour for ${}_2F_1(a,b,c;w)$.

In this connection we should add that the representation (3.2) can be modified to handle arbitrary $a,b\in\mathbb{C}$: For $\mathrm{Re}\,a\le 0$ and/or $\mathrm{Re}\,b\le 0$ one need only shift the contour $\mathcal{C}$ up, so that the downward pole sequences starting at $-ia$ and $-ib$ stay below it, cf. Fig. 5. In this way one can demonstrate that for fixed w in the cut plane one obtains a meromorphic function of a,b and c.

Next, we detail the role of ${}_2F_1$ in the problem of finding eigenfunctions for the hyperbolic specialization of H_0 (1.20). In fact, the hypergeometric function can be used to diagonalize the two-coupling family of Schrödinger operators

$$H(g,\tilde{g})\equiv-\frac{d^2}{dx^2}+\frac{g(g-1)\nu^2}{\mathrm{sh}^2\,\nu x}-\frac{\tilde{g}(\tilde{g}-1)\nu^2}{\mathrm{ch}^2\,\nu x}. \tag{3.3}$$

Specifically, one first performs the similarity transformation

$$\tilde{H}(g,\tilde{g})\equiv w(\nu x)^{-1/2}H(g,\tilde{g})w(\nu x)^{1/2}, \tag{3.4}$$

where $w(y)$ is the 'weight function'

$$w(y)\equiv\mathrm{sh}\,y^{2g}\,\mathrm{ch}\,y^{2\tilde{g}}. \tag{3.5}$$

A straightforward calculation yields

$$\tilde{H}(g,\tilde{g})=-\frac{d^2}{dx^2}-2\nu[g\coth(\nu x)+\tilde{g}\tanh(\nu x)]\frac{d}{dx}-\nu^2(g+\tilde{g})^2. \tag{3.6}$$

Then one has the eigenvalue equation

$$\tilde{H}(g,\tilde{g})\Psi_{\mathrm{nr}}=p^2\Psi_{\mathrm{nr}}, \tag{3.7}$$

where Ψ_{nr} is the nonrelativistic wave function

$$\Psi_{\mathrm{nr}}(\nu,g,\tilde{g};x,p)\equiv{}_2F_1\left(\frac{1}{2}\left(g+\tilde{g}-\frac{ip}{\nu}\right),\frac{1}{2}\left(g+\tilde{g}+\frac{ip}{\nu}\right),g+\frac{1}{2};-\mathrm{sh}^2\,\nu x\right). \tag{3.8}$$

Indeed, (3.7) is simply the rational ODE satisfied by ${}_2F_1(a,b,c;w)$, transformed to hyperbolic form via the substitution $w=-\mathrm{sh}^2\,\nu x$.

The wave function Ψ_{nr} (3.8) is also an eigenfunction of two AΔOs, one of which acts on x, while the second one acts on the spectral variable p. We will obtain this fact (which cannot be found in the textbook literature) as a corollary of the nonrelativistic limit in Subsection 3.5.

3.2. The 'Relativistic' Hypergeometric Function. The function we are about to introduce, $R(a_+, a_-, \mathbf{c}; v, \hat{v})$, can be used in particular to diagonalize the $\mathrm{II}_{\mathrm{rel}}$-version of the A$\Delta$O H_β (1.21). But just as the nonrelativistic wave function Ψ_{nr} (3.8) serves as an eigenfunction for a 2-coupling generalization of the $\mathrm{II}_{\mathrm{nr}}$-version of the Hamiltonian H_0 (1.20), we will find that the relativistic wave function Ψ_{rel} (3.51) we associate below with the R-function is in fact an eigenfunction of a 4-coupling generalization of the $\mathrm{II}_{\mathrm{rel}}$-Hamiltonian. Moreover, it is an eigenfunction of three more independent AΔOs with a similar structure. (In the nonrelativistic limit two of these give rise to the AΔOs mentioned at the end of the previous subsection.)

We have split the integrand in (3.2) in three factors to anticipate a corresponding factorization of the integrand for the R-function. Setting

$$\hat{c}_0 \equiv (c_0 + c_1 + c_2 + c_3)/2, \tag{3.9}$$

the latter reads

$$I(a_+, a_-, \mathbf{c}; v, \hat{v}, z) \equiv F(c_0; v, z) K(a_+, a_-, \mathbf{c}; z) F(\hat{c}_0; \hat{v}, z). \tag{3.10}$$

Here, the functions F and K involve the hyperbolic gamma function $G(z) \equiv G_{\mathrm{hyp}}(a_+, a_-; z)$ from Subsection 2.5, cf. (2.40).

Specifically, F and K are defined by

$$F(d; y, z) \equiv \left(\frac{G(z + y + id - ia)}{G(y + id - ia)} \right) (y \to -y), \tag{3.11}$$

$$K(a_+, a_-, \mathbf{c}; z) \equiv \frac{1}{G(z + ia)} \prod_{j=1}^{3} \frac{G(is_j)}{G(z + is_j)}, \tag{3.12}$$

where we use the notation

$$s_1 \equiv c_0 + c_1 - a_-/2, \quad s_2 \equiv c_0 + c_2 - a_+/2, \quad s_3 \equiv c_0 + c_3, \quad a \equiv (a_+ + a_-)/2. \tag{3.13}$$

We have suppressed the dependence on a_+ and a_- in G and in F, since these functions are invariant under the interchange of a_+ and a_-. (Note K is not invariant, since s_1 and s_2 are not.)

Just as we first have chosen $\mathrm{Re}\, a, \mathrm{Re}\, b > 0$ so as to define the integration contour $\mathcal{C}$ in the Barnes representation (3.2), we begin by choosing

$$s_j \in (0, a), \quad j = 1, 2, 3, \quad c_0, \hat{c}_0, v, \hat{v} \in (0, \infty). \tag{3.14}$$

Then we choose once more the contour $\mathcal{C}$ going from $-\infty$ to ∞ in the z-plane, with a downward indentation at the origin. The choices just detailed ensure that $\mathcal{C}$ separates the four upward pole sequences coming from the four z-dependent G-functions in K (3.12) and the four downward pole sequences coming from the z-dependent G-functions in the two F-factors of the integrand (3.10), cf. Fig. 6. (At this point the reader should recall the pole-zero properties of the hyperbolic G-function, cf. (2.41), (2.42) and Fig. 4.)

Our R-function is now defined by the integral

$$R(a_+, a_-, \mathbf{c}; v, \hat{v}) \equiv \frac{1}{(a_+ a_-)^{1/2}} \int_{\mathcal{C}} dz\, I(a_+, a_-, \mathbf{c}; v, \hat{v}, z). \tag{3.15}$$

The asymptotics (2.55) of the G-function plays the same role as the Stirling formula for the Γ-function in showing that the integral converges. Indeed, using (2.55) one

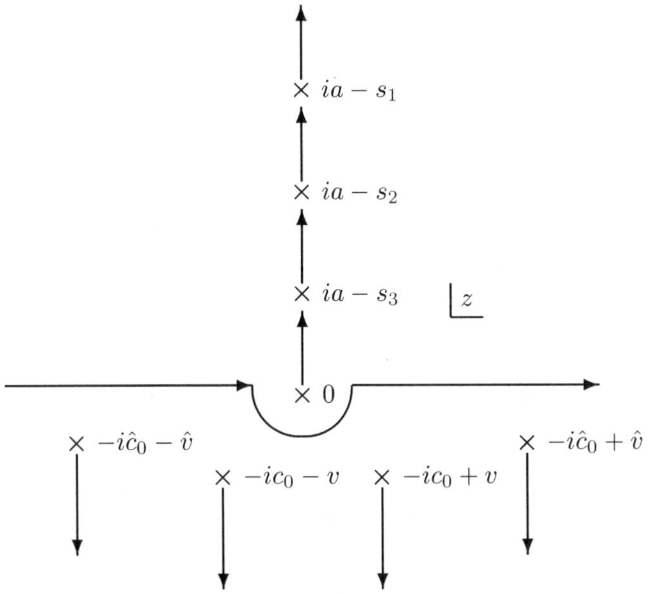

FIGURE 6. The pole sequences and contour for $R(a_+, a_-, \mathbf{c}; v, \hat{v})$.

readily obtains

$$I(a_+, a_-, \mathbf{c}; v, \hat{v}, z) = O\left(\exp\left(\mp 2\pi\left(\frac{1}{a_+} + \frac{1}{a_-}\right)\operatorname{Re} z\right)\right), \quad \operatorname{Re} z \to \pm\infty. \tag{3.16}$$

Therefore, R is well defined and analytic in v and $\hat{v}$ for $\operatorname{Re} v, \operatorname{Re} \hat{v} \neq 0$.

The R-function has in fact much stronger analyticity properties, but to demonstrate these in detail is well beyond our present scope. Thus we only summarize some results, referring to Ref. [**38**] for proofs. Briefly, the R-function extends to a function that is meromorphic in all of its eight arguments, as long as a_+ and a_- stay in the (open) right half plane. Moreover, the pole varieties and their associated orders are explicitly known. For the case of fixed positive a_+, a_- and (generic) real c_0, c_1, c_2, c_3, the R-function is meromorphic in v and $\hat{v}$, with poles that can (but need not) occur solely for certain points on the imaginary axis. These points correspond to collisions of v- and $\hat{v}$-dependent z-poles in the integrand with z-poles in the three upward s_j-pole sequences (cf. Fig. 6) and points that are given by the poles of the factors $1/G(\pm v + ic_0 - ia)$ and $1/G(\pm\hat{v} + i\hat{c}_0 - ia)$ in the integrand.

We conclude this subsection by listing some automorphy properties of the R-function. To this end we introduce the 'dual couplings'

$$\hat{\mathbf{c}} \equiv J\mathbf{c}, \quad J \equiv \frac{1}{2}\begin{pmatrix} 1 & 1 & 1 & 1 \\ 1 & 1 & -1 & -1 \\ 1 & -1 & 1 & -1 \\ 1 & -1 & -1 & 1 \end{pmatrix}, \tag{3.17}$$

whence one has (cf. (3.13))

$$c_0 + c_j = \hat{c}_0 + \hat{c}_j, \quad s_j = \hat{s}_j, \quad j = 1, 2, 3. \tag{3.18}$$

We also define the transposition

$$I\mathbf{c} \equiv (c_0, c_2, c_1, c_3). \tag{3.19}$$

Then one has the symmetries

$$R(a_+, a_-, \mathbf{c}; v, \hat{v}) = R(a_+, a_-, \hat{\mathbf{c}}; \hat{v}, v), \qquad \text{(self-duality)}, \tag{3.20}$$

$$R(a_+, a_-, \mathbf{c}; v, \hat{v}) = R(a_-, a_+, I\mathbf{c}; v, \hat{v}), \tag{3.21}$$

$$R(\lambda a_+, \lambda a_-, \lambda \mathbf{c}; \lambda v, \lambda \hat{v}) = R(a_+, a_-, \mathbf{c}; v, \hat{v}), \quad \lambda > 0, \quad \text{(scale invariance)}. \tag{3.22}$$

These features can be quite easily checked directly from the definition (3.15). (Use the G-function properties (2.44) and (2.45) to check (3.21) and (3.22), resp.)

3.3. Eigenfunction Properties. In order to detail the four AΔOs for which our R-function is a joint eigenfunction, we introduce the quantities

$$s_\delta(y) \equiv \mathrm{sh}(\pi y/a_\delta), \quad c_\delta(y) \equiv \mathrm{ch}(\pi y/a_\delta), \tag{3.23}$$

$$C_\delta(\mathbf{c}; z) \equiv \frac{s_\delta(z - ic_0)}{s_\delta(z)} \frac{c_\delta(z - ic_1)}{c_\delta(z)} \frac{s_\delta(z - ic_2 - ia_{-\delta}/2)}{s_\delta(z - ia_{-\delta}/2)} \times \frac{c_\delta(z - ic_3 - ia_{-\delta}/2)}{c_\delta(z - ia_{-\delta}/2)}, \tag{3.24}$$

$$A_\delta(\mathbf{c}; y) \equiv C_\delta(\mathbf{c}; y)(T^y_{ia_{-\delta}} - 1) + C_\delta(\mathbf{c}; -y)(T^y_{-ia_{-\delta}} - 1) + 2c_\delta(i(c_0 + c_1 + c_2 + c_3)), \tag{3.25}$$

where $\delta = +, -$, and where the superscript y on the shifts indicates the variable they act on. The eigenfunction properties of the R-function are now specified in the following proposition, whose proof we sketch.

PROPOSITION 3.1. *The function $R(a_+, a_-, \mathbf{c}; v, \hat{v})$ is a joint eigenfunction of the AΔOs*

$$A_+(\mathbf{c}; v), \quad A_-(I\mathbf{c}; v), \quad A_+(\hat{\mathbf{c}}; \hat{v}), \quad A_-(I\hat{\mathbf{c}}; \hat{v}), \tag{3.26}$$

with eigenvalues

$$2c_+(2\hat{v}), \quad 2c_-(2\hat{v}), \quad 2c_+(2v), \quad 2c_-(2v). \tag{3.27}$$

SKETCH OF PROOF. In view of the symmetries (3.20) and (3.21) we need only prove the AΔE

$$A_+(\mathbf{c}; v)R(\mathbf{c}; v, \hat{v}) = 2c_+(2\hat{v})R(\mathbf{c}; v, \hat{v}). \tag{3.28}$$

Also, due to the analyticity properties of the R-function already detailed, we may restrict the parameters and imaginary parts of v and $\hat{v}$ in such a way that R is given by

$$R(\mathbf{c}; v, \hat{v}) = \frac{1}{(a_+a_-)^{1/2}} \int_C F(c_0; v, z)K(\mathbf{c}; z)F(\hat{c}_0; \hat{v}, z)\, dz, \tag{3.29}$$

and that we may let the AΔO $A_+(\mathbf{c}; v)$ act on the integrand.

A main tool in proving the second-order AΔE (3.28) is now to exploit the first-order AΔE

$$\frac{G(z + ia_-/2)}{G(z - ia_-/2)} = 2c_+(z), \tag{3.30}$$

satisfied by the G-function. Indeed, using (3.30), one readily checks that the function F (3.11) solves the two AΔEs

$$\frac{F(d;y+ia_-/2,z)}{F(d;y-ia_-/2,z)}=\frac{s_+(y+z+id-ia_-/2)}{s_+(y-z-id+ia_-/2)}\frac{s_+(y-id+ia_-/2)}{s_+(y+id-ia_-/2)},\tag{3.31}$$

$$\frac{F(d;y,z-ia_-)}{F(d;y,z)}=\frac{1}{4s_+(y+z+id-ia_-)s_+(y-z-id+ia_-)}.\tag{3.32}$$

Using (3.31) with $d=c_0$ and $y=v$, one can calculate the quotient

$$Q(\mathbf{c};v,z)\equiv(A_+(\mathbf{c};v)F)(c_0;v,z)/F(c_0;v,z).\tag{3.33}$$

A key point is now that Q can be rewritten as

$$2c_+(2z+2i\hat{c}_0)+\frac{4\prod_{j=1}^4 c_+(z-ia_-/2+is_j)}{s_+(v+z+ic_0-ia_-)s_+(v-z-ic_0+ia_-)},\quad s_4\equiv a.\tag{3.34}$$

This fact amounts to a functional equation that can be proved by comparing residues at simple poles and $|\operatorname{Re} v|\to\infty$ asymptotics. We now observe that the denominator in (3.34) appears in (3.32) with $d=c_0, y=v$. Thus we get

$$\begin{aligned}A_+(\mathbf{c};v)F(c_0;v,z)\\ =2c_+(2z+2i\hat{c}_0)F(c_0;v,z)+F(c_0;v,z-ia_-)\Pi(\mathbf{c};z-ia_-/2),\end{aligned}\tag{3.35}$$

where we have introduced the product

$$\Pi(\mathbf{c};z)\equiv 16\prod_{j=1}^4 c_+(z+is_j).\tag{3.36}$$

The upshot of these calculations is the identity

$$\begin{aligned}A_+(\mathbf{c};v)R(\mathbf{c};v,\hat{v})=\frac{1}{(a_+a_-)^{1/2}}\int_{\mathcal{C}}dz\,[2c_+(2z+2i\hat{c}_0)I(\mathbf{c};v,\hat{v},z)\\ +F(c_0;v,z-ia_-)\Pi(\mathbf{c};z-ia_-/2)K(\mathbf{c};z)F(\hat{c}_0;\hat{v},z)].\end{aligned}\tag{3.37}$$

To proceed, we now shift $\mathcal{C}$ down by ia_- in the second term and then take $z\to z+ia_-$. Then we are in the position to exploit a critical property of the function $K(\mathbf{c};z)$: Due to (3.30) it obeys the AΔE

$$K(\mathbf{c};z+ia_-/2)=K(\mathbf{c};z-ia_-/2)/\Pi(\mathbf{c};z).\tag{3.38}$$

Therefore we obtain

$$\begin{aligned}A_+(\mathbf{c};v)R(\mathbf{c};v,\hat{v})=\frac{1}{(a_+a_-)^{1/2}}\int_{\mathcal{C}}dz\,[2c_+(2z+2i\hat{c}_0)I(\mathbf{c};v,\hat{v},z)\\ +F(c_0;v,z)K(\mathbf{c};z)F(\hat{c}_0;\hat{v},z+ia_-)].\end{aligned}\tag{3.39}$$

Finally, we use (3.32) with $d=\hat{c}_0, y=\hat{v}$ to get

$$\begin{aligned}F(\hat{c}_0;\hat{v},z+ia_-)&=4s_+(\hat{v}+z+i\hat{c}_0)s_+(\hat{v}-z-i\hat{c}_0)F(\hat{c}_0;\hat{v},z)\\ &=2[c_+(2\hat{v})-c_+(2z+2i\hat{c}_0)]F(\hat{c}_0;\hat{v},z).\end{aligned}\tag{3.40}$$

Then substitution in (3.39) yields (3.28). □

Let us now explain how the R-function gives rise to an eigenfunction for the $\mathrm{II}_{\mathrm{rel}}$ Hamiltonian. To this end we specialize to

$$a_+=\pi,\quad a_-=\beta\nu,\quad c_0=\beta\nu g,\quad c_1,c_2,c_3=0,\quad v=\nu x,\quad \hat{v}=\beta p/2.\tag{3.41}$$

Then the AΔO $A_+(\mathbf{c}; v)$ reduces to

$$A_\beta = \frac{\operatorname{sh} \nu(x - i\beta g)}{\operatorname{sh} \nu x} T^x_{i\beta} + (i \to -i), \tag{3.42}$$

and the eigenvalue $2c_+(2\hat{v})$ becomes $2\operatorname{ch}(\beta p)$. The point is now that we have

$$A_\beta = w(c_0; v)^{-1/2} H_\beta w(c_0; v)^{1/2}, \quad c_0 = \beta\nu g, \quad v = \nu x. \tag{3.43}$$

Here, H_β is the hyperbolic specialization of the AΔO (1.21) with $a = \pi/\nu$, and $w(c_0; v)$ is the 'hyperbolic weight function', defined by

$$w(b; z) = \frac{G(z + ib - i(a_+ + a_-)/2)G(z + i(a_+ + a_-)/2)}{G(z - ib + i(a_+ + a_-)/2)G(z - i(a_+ + a_-)/2)}. \tag{3.44}$$

(As before, $G(z)$ denotes $G_{\mathrm{hyp}}(a_+, a_-; z)$.) Consequently, the function

$$w(\beta\nu g; \nu x)^{1/2} R(\pi, \beta\nu, (\beta\nu g, 0, 0, 0); \nu x, \beta p/2) \tag{3.45}$$

is an H_β-eigenfunction with eigenvalue $2\operatorname{ch}\beta p$.

3.4. The Askey-Wilson Specialization. We continue by sketching how the Askey-Wilson polynomials arise as a specialization of the R-function. With the restriction (3.14) on the arguments in effect, we may and will use the representation (3.15). We are going to exploit the analyticity properties of the R-function in the variable $\hat{v}$ and the eigenvalue equation

$$A_+(\hat{\mathbf{c}}; \hat{v}) R(\mathbf{c}; v, \hat{v}) = 2c_+(2v) R(\mathbf{c}; v, \hat{v}). \tag{3.46}$$

To prevent nongeneric singularities, we choose $\hat{c}_0$ rationally independent of a_+, a_-, $\hat{c}_1$, $\hat{c}_2$ and $\hat{c}_3$. Then R has no pole at the points $\hat{v} = i\hat{c}_0 + ina, n \in \mathbb{Z}$, so we may define

$$R_n(v) \equiv R(\mathbf{c}; v, i\hat{c}_0 + ina_-), \quad n \in \mathbb{Z}. \tag{3.47}$$

PROPOSITION 3.2. *One has* $R_n(v) = P_n\big(c_+(2v)\big)$, $n \in \mathbb{N}$, *with* $P_n(u)$ *a polynomial of degree* n *in* u.

PROOF. The pole of $I(z)$ at $z = 0$ is simple and has residue $-i(a_+a_-)^{1/2}/2\pi$. (This follows from (2.46) and (2.43).) Thus we have

$$R(\mathbf{c}; v, \hat{v}) = 1 + \frac{1}{(a_+a_-)^{1/2}} \int_{\mathcal{C}^+} dz\, I(\mathbf{c}; v, \hat{v}, z), \tag{3.48}$$

where $\mathcal{C}^+$ denotes the contour $\mathcal{C}$ with an upward indentation at $z = 0$ (instead of downward).

We can now let $\hat{v}$ converge to $i\hat{c}_0$ without $\hat{v}$-dependent poles crossing $\mathcal{C}^+$, cf. Fig. 6. The factor $1/G(-\hat{v} + i\hat{c}_0 - ia)$ in $I(z)$ has a zero for $\hat{v} = i\hat{c}_0$, whereas the factor $1/G(\hat{v} + i\hat{c}_0 - ia)$ has no pole for $\hat{v} = i\hat{c}_0$ (due to the rational independence requirement). Hence we deduce

$$R_0(v) = 1. \tag{3.49}$$

Next, we write out the eigenvalue equation (3.46) for $\hat{v} = i\hat{c}_0 + ina_-$:

$$\begin{aligned} &C_+(\hat{\mathbf{c}}; i\hat{c}_0 + ina_-)[R_{n-1}(v) - R_n(v)] \\ &\quad + C_+(\hat{\mathbf{c}}; -i\hat{c}_0 - ina_-)[R_{n+1}(v) - R_n(v)] + 2c_+(2ic_0)R_n(v) = 2c_+(2v)R_n(v). \end{aligned} \tag{3.50}$$

The rational independence assumption entails that the coefficients are well defined, and that $C_+(\hat{\mathbf{c}}; -i\hat{c}_0 - ina_-)$ does not vanish for $n \in \mathbb{N}$. Since we have $C_+(\hat{\mathbf{c}}; i\hat{c}_0) =$

0 (cf. (3.24)), we may now use (3.49) as a starting point to prove the assertion recursively. □

Note that when one restricts attention to a *finite* number of the above functions $R_n(v)$, one may let $\hat{c}_0$ vary over suitable intervals without encountering singularities or zeros of the recurrence coefficients (save for $C_+(\hat{\mathbf{c}}; i\hat{c}_0)$, of course). We can now continue a_+ analytically to $-i\pi/r$, $r > 0$, to obtain polynomials $P_n\big(\cos(2rv)\big)$ with recurrence coefficients that can be read off from (3.50). The a_+-continuation turns the hyperbolic AΔO $A_+(\mathbf{c}; v)$ into a trigonometric AΔO with eigenvalue $2\,\mathrm{ch}\,2r(\hat{c}_0 + na_-)$ on $P_n\big(\cos(2rv)\big)$. In essence, the latter AΔO is the Askey-Wilson AΔO and the recurrence is the 3-term recurrence of the Askey-Wilson polynomials. More precisely, taking $r = 1/2$, the polynomials $P_n(\cos v)$ turn into the Askey-Wilson polynomials $p_n(q, \alpha, \beta, \gamma, \delta; \cos v)$ under a suitable parameter substitution and n-dependent renormalization, cf. Ref. [**38**].

3.5. The Nonrelativistic Limit $R \to {}_2F_1$. We continue by clarifying the relation between the R- and ${}_2F_1$-functions. To this end we introduce the relativistic wave function

(3.51) $$\Psi_{\mathrm{rel}}(\beta, \nu, (g_0, g_1, g_2, g_3); x, p) \equiv R(\pi, \beta\nu, \beta\nu(g_0, g_1, g_2, g_3); \nu x, \beta p/2).$$

Now we change variables $z \to \beta\nu z$ in the integral representation (3.15), and rewrite the result as

(3.52) $$\Psi_{\mathrm{rel}} = \int_C dz\, S_l M S_r,$$

where

(3.53) $$S_l \equiv \exp(2iz \ln 2) F(\beta\nu g_0; \nu x, \beta\nu z),$$

(3.54) $$S_r \equiv \exp\big(2iz \ln(2\beta\nu)\big) F(\beta\nu \hat{g}_0; \beta p/2, \beta\nu z),$$

(3.55) $$M \equiv \left(\frac{\beta\nu}{\pi}\right)^{1/2} \exp\big(-2iz \ln(4\beta\nu)\big) K(\pi, \beta\nu, \beta\nu(g_0, g_1, g_2, g_3); \beta\nu z).$$

The factorization performed here ensures that the $\beta \downarrow 0$ limit of the three factors exists. Indeed, using the two zero step size limits (2.47) and (2.52), we obtain

(3.56) $$\lim_{\beta\downarrow 0} S_l = \exp\big(-iz \ln(\mathrm{sh}^2\, \nu x)\big),$$

(3.57)
$$\lim_{\beta\downarrow 0} S_r = \left(\frac{\Gamma(-ip/(2\nu) + 1/2(g+\tilde{g}) - iz)}{\Gamma(-ip/(2\nu) + 1/2(g+\tilde{g}))}\right)(p \to -p), \quad g \equiv g_0 + g_2, \quad \tilde{g} \equiv g_1 + g_3,$$

(3.58) $$\lim_{\beta\downarrow 0} M = \frac{\Gamma(iz)\Gamma(g+1/2)}{2\pi\Gamma(g+1/2-iz)},$$

where the limits are uniform on sufficiently small discs around any point on the contour.

When we now interchange these $\beta \downarrow 0$ limits with the contour integration, we obviously get

(3.59) $$\lim_{\beta\downarrow 0} \Psi_{\mathrm{rel}}(\beta, \nu, (g_0, g_1, g_2, g_3); x, p) = \Psi_{\mathrm{nr}}(\nu, g, \tilde{g}; x, p),$$
$$g \equiv g_0 + g_2, \quad \tilde{g} \equiv g_1 + g_3,$$

cf. (3.8), (3.2). To date, we have no justification for this interchange. A uniform L^1 tail bound as $\beta \downarrow 0$ would suffice (by dominated convergence), but it remains to supply such a bound. In any case, we conjecture that the limit (3.59) holds true uniformly on x-compacts in $\{\mathrm{Re}\, x > 0, |\,\mathrm{Im}\, x| < \pi/2\nu\}$ and p-compacts in $\mathbb{C}$.

Let us now consider the $\beta \downarrow 0$ limits of the above four AΔOs with parameters and variables

$$a_+ = \pi, \quad a_- = \beta\nu, \quad \mathbf{c} = \beta\nu(g_0, g_1, g_2, g_3), \quad v = \nu x, \quad \hat{v} = \beta p/2. \tag{3.60}$$

Clearly, $A_-(\hat{\mathbf{c}}; \hat{v})$ and its eigenvalue $\hat{E}_- = 2\,\mathrm{ch}(2\pi x/\beta)$ diverge for $\beta \downarrow 0$. For the remaining AΔOs and their eigenvalues one readily verifies the following limiting behavior:

$$A_+(\mathbf{c}; v) = 2 + \beta^2 \tilde{H}(g, \tilde{g}) + O(\beta^4), \quad \beta \downarrow 0, \tag{3.61}$$

$$E_+ = 2\,\mathrm{ch}(\beta p) = 2 + \beta^2 p^2 + O(\beta^4), \quad \beta \downarrow 0, \tag{3.62}$$

$$\lim_{\beta\downarrow 0} A_-(\mathbf{c}; v) = \exp\bigl(-i\pi(g + \tilde{g})\bigr) T^x_{i\pi/\nu} + (i \to -i), \quad \mathrm{Re}\, x > 0, \tag{3.63}$$

$$E_- = 2\,\mathrm{ch}(\pi p/\nu), \tag{3.64}$$

$$\lim_{\beta\downarrow 0} A_+(\hat{\mathbf{c}}; \hat{v}) = \frac{[p - i\nu(g + \tilde{g})]}{p} \cdot \frac{[p - i\nu - i\nu(g - \tilde{g})]}{p - i\nu}(T^p_{2i\nu} - 1) + (i \to -i) + 2, \tag{3.65}$$

$$\hat{E}_+ = 2\,\mathrm{ch}(2\nu x). \tag{3.66}$$

It can be shown directly that the limiting operators do have the pertinent eigenvalues on Ψ_{nr} (3.8). Indeed, for the AΔO $A_+(\mathbf{c}; v)$ this amounts to (3.7). The limit AΔO on the rhs of (3.63) does have eigenvalue $2\,\mathrm{ch}(\pi p/\nu)$ by virtue of the known analytic continuation of ${}_2F_1(a, b, c; w)$ across the logarithmic branch cut $w \in [1, \infty)$. Finally, the eigenvalue $2\,\mathrm{ch}(2\nu x)$ for the AΔO on the rhs of (3.65) can be verified by using the contiguous relations of the hypergeometric function, cf. Ref. [**45**].

4. Generalized Lamé Functions

This section is concerned with eigenfunctions of the Hamiltonians H_0 (1.20) and H_β (1.21) for special choices of parameters, emphasizing however the most general (elliptic) regime. As such it is mostly based on our paper Ref. [**34**]. We present far less detail on the hyperbolic and trigonometric specializations studied in Ref. [**35**], but we do clarify the relation to the arbitrary-parameter hyperbolic eigenfunction (3.45). Our account in this section inevitably overlaps with our conference contribution Ref. [**39**], which contains more information on some items (in particular on related work [**24, 12, 15, 43**]).

In Subsection 4.1 we recall what is known concerning eigenfunctions for the $\mathrm{IV}_{\mathrm{nr}}$ Hamiltonian (1.20). Since we concentrate on those features that turn out to admit $\mathrm{IV}_{\mathrm{rel}}$-analogs, we specialize to integer couplings $g = N + 1$ with $N \in \mathbb{N}^*$. (The cases $g = 0, 1$ are of course elementary.) Subsection 4.2 deals with relativistic eigenfunctions for integer g, which generalize their nonrelativistic counterparts first introduced by Hermite more than a century ago.

A new feature of our relativistic eigenfunctions is however that they are explicitly known for a dense set in the natural parameter space $r, a, \beta > 0$, $g \in \mathbb{R}$. These more general eigenfunctions are presented in Subsection 4.3. In Subsection 4.4 we discuss the question whether a continuous interpolation for arbitrary parameters exists.

We should perhaps add that we use the expression 'explicitly known' for lack of a more telling mathematical term. Indeed, our 'explicit' representation leaves several obvious questions unanswered, in particular the interpolation question. Thus, one would like to have other representations available from which the answers can be readily established. In the hyperbolic case such representations are now known, but for the elliptic case we have nothing more to offer.

As we will explain in some detail below, the eigenfunctions that are relevant for the Hilbert space theory associated to the AΔO H_β (1.21) do not have the orthogonality properties one might have expected from experience with ordinary differential and discrete difference operators. Specifically, one must impose a constraint on the parameters to prevent a breakdown of orthogonality (and hence of self-adjointness of the Hilbert space operator associated to H_β). Another remarkable feature of the elliptic eigenfunctions is that they are invariant under interchanging a and β. (This feature generalizes the hyperbolic symmetry (3.21) for the relevant parameter domain.)

We conclude these introductory remarks by commenting on 'finite-gap integration' aspects and on the *discrete* difference operators that can be associated with the commuting operators (1.13). In these contexts one is dealing with regimes that are both physically and mathematically quite different from the regime we concentrate on, though there is a connection via analytic continuation. To begin with H_0, one needs to choose not only $g = N + 1 \in \mathbb{N}^*$, but also $x \to x + ia/2$ so as to get a smooth real-valued π/r-periodic potential for which the corresponding self-adjoint Schrödinger operator $\tilde{H}_0$ on $L^2(\mathbb{R}, dx)$ has exactly N gaps in its spectrum. This choice of x has no integrable many-particle analog, in the sense that the pair potential $\wp(x_j - x_k + ia/2)$ does not yield an integrable N-particle system for $N > 2$.

Of course, this x-choice is of considerable interest in itself, both for H_0 (1.20) and for H_β (1.21). Even so, we will not discuss it in the following subsections, since we concentrate on properties that have a chance of generalizing to the $N > 2$ case. In order to clear up some discrepancies in the literature, however, we add the following.

Taking $x \to x + ia/2$ (and as always $r, a > 0$) in the operator H_β and denoting the result by $\tilde{H}_\beta$, there are various essentially different ways to associate formally self-adjoint operators to $\tilde{H}_\beta$. Our viewpoint is inspired by 'relativistic invariance', so that we choose $\beta > 0$ and wind up with a formally self-adjoint AΔO on $L^2(\mathbb{R}, dx)$. But one can also choose $\beta = i\alpha$ with $\alpha \in (0, \infty)$, and associate to $\tilde{H}_{i\alpha}$ a family of *discrete* difference operators on $l^2(\alpha\mathbb{Z} + \delta)$, where $\delta \in [0, \alpha)$. The latter are once more formally self-adjoint. Since they are manifestly bounded as well, they are in fact well-defined and unambiguous self-adjoint operators acting on all of $l^2(\alpha\mathbb{Z}+\delta)$.

Somewhat implicitly, these discrete difference operators arose in a paper by Krichever and Zabrodin [**24**], and quite explicitly in later work by Zabrodin, cf. Ref. [**46**]. For these operators the number of gaps in their spectrum is twice as large as for their $\alpha \to 0$ limits, as is shown in the papers just cited.

Choosing $\beta > 0$, however, one is dealing with (unbounded) AΔOs, and the functional-analytic aspects are far less accessible. We have studied the $g = 2$ case in Ref. [**36**], the conclusion being that the number of gaps remains the same (namely one). More precisely, provided $\beta \in (0, a/2)$ and a certain completeness property is assumed, the AΔO $\tilde{H}_\beta$ gives rise to a self-adjoint operator on $L^2(\mathbb{R}, dx)$ with purely absolutely continuous spectrum $[E_1, E_2] \cup [E_3, \infty), E_1 < E_2 < E_3$, of multiplicity two, just as the operator $\tilde{H}_0$. It is quite likely (but unproven) that a similar picture holds true for arbitrary integer $g \geq 2$.

Of course, one can also choose $\beta = i\alpha$, $\alpha \in (0, \infty)$, in the many-particle AΔOs (1.13) and view the resulting operators as acting on a lattice with lattice constant α. (This is done in Ref. [**22**], for instance.) But then one looses not only relativistic invariance, but in general self-adjointness as well. Indeed, one must discretize in a quite special way (entailing a bound on the particle number) for the resulting discrete difference operators to be self-adjoint. (See Ref. [**9**] for the trigonometric specialization of the pertinent discretization.)

4.1. The Nonrelativistic Case. To obtain the eigenfunctions of H_0 (1.20), we should solve the ODE

$$-F''(x) + g(g-1)\wp(x)F(x) = EF(x). \tag{4.1}$$

Since this is a second order ODE, it can be rewritten as a first order two-by-two matrix ODE. In turn, the latter can be replaced by a Volterra type integral equation (requiring certain initial values $F(x_0), F'(x_0)$). The iteration of this equation converges, yielding a two-dimensional solution space. Of course, this is (nowadays) standard fare. The procedure applies to a wide class of potentials, but the generality is paid for by the Volterra series solution yielding little information.

For the potential $g(g-1)\wp(x)$ at hand, there is a second less general method to construct solutions, which yields more information. Indeed, one can solve the ODE near $x = 0$ with a convergent fractional power series Ansatz, which gives rise to two solutions behaving as x^g and x^{1-g} for $x \to 0$ (for generic g). The insights coming from this method ('Fuchs theory') are quite useful for studying the Hilbert space aspects associated with the singular Sturm-Liouville equation (4.1).

Unfortunately, there appear to be no generalizations of these standard ODE techniques to the AΔE setting. (In quite special cases iteration solutions do converge, as we have seen for instance for the trigonometric gamma function, cf. (2.17)–(2.19).) But for g integer there is yet a third way to obtain a basis for the solution space of (4.1) (for generic E). The pertinent formulas were first derived by Hermite, but are more readily accessible in the (last pages of the) monograph by Whittaker and Watson [**45**]. (As we will see in the next section, these formulas can be generalized to the relativistic case.)

Specifically, the basis functions for $g = N + 1 \in \mathbb{N}^*$ can be chosen to be $\mathcal{F}(\pm x, y)$, where $\mathcal{F}(x, y)$ is of the form

$$\mathcal{F}(x, y) = \prod_{j=1}^{N} \frac{s(x + z_j)}{s(x)} \cdot \exp[irx(N+1) + ixy]. \tag{4.2}$$

The 'zero parameters' $z_1, \ldots, z_N$ and the 'spectral variable' y are related by the constraint system

$$y = -(N+1)r + i \sum_{j=1}^{N} \frac{s'(z_j)}{s(z_j)}, \tag{4.3}$$

$$N\frac{s'(z_k)}{s(z_k)}+\sum_{\substack{j=1\\ j\neq k}}^{N}\frac{s'(z_j-z_k)}{s(z_j-z_k)}-\sum_{j=1}^{N}\frac{s'(z_j)}{s(z_j)}=0,\quad k=1,\ldots,N, \tag{4.4}$$

and the energy eigenvalue is then given by

$$E=-(2N-1)\sum_{j=1}^{N}\wp(z_j). \tag{4.5}$$

As it turns out, the system (4.3)–(4.4) admits a solution curve with curve parameter y; for $y\in(K,\infty)$ with K sufficiently large, one has $z_j=i\epsilon_j(y)\in i(0,\infty)$, with $\epsilon_j\downarrow 0$ for $y\uparrow\infty$.

In agreement with Fuchs theory, one reads off from (4.2) that one has

$$\mathcal{F}(\pm x,y)\sim x^{-N},\quad x\to 0. \tag{4.6}$$

Since the leading singularity is taken out in

$$\Phi(x,y)\equiv\mathcal{F}(x,y)-(-)^N\mathcal{F}(-x,y), \tag{4.7}$$

it then *follows* from Fuchs theory that

$$\Phi(x,y)\sim x^{N+1},\quad x\to 0. \tag{4.8}$$

(Notice that this behavior is very non-obvious from the formula (4.2).)

From (4.2) one also sees by inspection that

$$\mathcal{F}(x+\pi/r,y)=(-)^{N+1}\exp(i\pi y/r)\mathcal{F}(x,y). \tag{4.9}$$

Therefore, $\mathcal{F}(x,nr),n\in\mathbb{N}$, is π/r-periodic or -antiperiodic. As a consequence, the functions

$$\Phi_n(x)\equiv\Phi(x,nr),\quad n\in\mathbb{N}, \tag{4.10}$$

vanish not only at $x=0$, but also at $x=\pi/r$. Thus they belong to the Hilbert space

$$\mathcal{H}\equiv L^2\big((0,\pi/r),dx\big) \tag{4.11}$$

of square-integrable functions on the interval $(0,\pi/r)$. For $n\to\infty$ one gets $E_n=E(nr)\uparrow\infty$, so the eigenvalues of H_0 on Φ_n are distinct for n large. It now follows from a well-known argument that the functions Φ_n are pairwise orthogonal for n large: One need only use

$$0=(H_0\Phi_n,\Phi_m)-(\Phi_n,H_0\Phi_m)=\big(E(nr)-E(mr)\big)(\Phi_n,\Phi_m). \tag{4.12}$$

More generally, the Weyl-Titchmarsh-Kodaira theory yields essential self-adjointness of H_0 on $C_0^\infty\big((0,\pi/r)\big)$ for $g\geq 3/2$ and the existence of an orthonormal base of eigenfunctions. Undoubtedly, the latter is given by the functions $\{\Phi_n\}_{n=0}^{\infty}$ (up to normalization), but we are not aware of a complete proof, even for $g=2$.

In this connection it should be noted that the $g=2$ case is especially accessible, since the constraint system (4.4) is trivial for $N=1$. From (4.3) one then sees that the sequence of values $y=0,r,2r,\ldots$ yields a sequence of distinct z_1-values $z_1(nr)$ in the interval $i(0,a/2)$. The corresponding energies E_n are obviously distinct, too (cf. (4.5)), so the functions $\Phi_0,\Phi_1,\ldots$ are well-defined, non-zero, and pairwise orthogonal. (But it is not clear that they are *complete*, cf. also Section 6 in [**36**].)

By contrast, the system (4.4) for $N>1$ is quite inaccessible. Though results for generic $y\in\mathbb{C}$ can be gleaned from [**45**] and [**11**], the Hilbert space aspects involve the *non-generic* values $y=nr$. (It is not clear e.g. whether the functions

Φ_n are non-zero for n small and whether the eigenvalues E_n are distinct.) Thus, even in the nonrelativistic integer g context the Hilbert space questions have not been completely elucidated.

For the trigonometric specialization the above functions $\Phi_n(x)$ and eigenvalues E_n can be seen to be of the form

$$\Phi_n(x) = w(x)^{1/2} P_n(\cos rx), \quad n \in \mathbb{N}, \tag{4.13}$$

$$E_n = (n+N+1)^2 r^2 - (N+1)Nr^2/3. \tag{4.14}$$

Here, the functions $P_n(u)$ are Gegenbauer polynomials, of degree n and parity $(-)^n$, and

$$w(x) = (\sin rx)^{2N+2} \tag{4.15}$$

amounts to the weight function w.r.t. which they are orthogonal. Therefore, completeness follows from the Weierstrass density theorem, and this is one important reason to conjecture that for $a < \infty$ the functions $\Phi_0, \Phi_1, \ldots$ are still complete.

4.2. The Relativistic Integer g Case. The time-independent Schrödinger equation associated with H_β (1.21) reads

$$\left(\frac{s(x - i\beta g)}{s(x)} \cdot \frac{s(x - i\beta + i\beta g)}{s(x - i\beta)} \right)^{1/2} F(x - i\beta) + (i \to -i) = EF(x). \tag{4.16}$$

The problem is now to invent/discover solutions to this second order AΔE that can be used to define the AΔO H_β (1.21) as a genuine self-adjoint operator on $\mathcal{H}$ (4.11). The point is that there is no obvious way to define H_β first as a symmetric operator on a dense subspace, by contrast to H_0, where for instance $C_0^\infty\big((0, \pi/r)\big)$ serves this purpose. Since no general Hilbert space theory for AΔOs exists at the present time, one may instead try to find sufficiently explicit pairwise orthogonal eigenfunctions $\Phi_n \in \mathcal{H}$ with real eigenvalues E_n. Setting then $H\Phi_n \equiv E_n \Phi_n$, extending linearly, and taking the Hilbert space closure $\overline{H}$ of the symmetric operator H thus defined, one obtains a self-adjoint operator $\overline{H}$ on (a dense subspace of) the closed subspace spanned by the pertinent eigenfunctions.

As it turns out, this scenario can be realized to a large extent. We continue by describing the eigenfunctions that generalize the above eigenfunctions $\mathcal{F}(x, y)$ and that play the desired role in rigorously redefining H_β as a self-adjoint quantum dynamics. Choosing as before $g = N + 1$ with $N \in \mathbb{N}^*$, and requiring first

$$2N\beta \in (0, a), \tag{4.17}$$

they are of the form

$$\mathcal{F}(x, y) = \prod_{j=1}^{N} \frac{s(x + z_j)}{[s(x + ij\beta) s(x - ij\beta)]^{1/2}} \cdot \exp[irx(N+1) + ixy]. \tag{4.18}$$

Here, the spectral variable and zero functions are related by the constraint system

$$y = -(N+1)r - \frac{1}{2\beta} \ln \left(\prod_{j=1}^{N} \frac{s(z_j - i\beta)}{s(z_j + i\beta)} \right), \tag{4.19}$$

$$s(z_k - iN\beta) \prod_{j \neq k} s(z_j - z_k - i\beta) \prod_{j} s(z_j + i\beta) - (\beta \to -\beta) = 0, \quad k = 1, \ldots, N. \tag{4.20}$$

It is clear that for $\beta \downarrow 0$ these equations yield the nonrelativistic counterparts (4.2)–(4.4). But in contrast to (4.4) (where the sum of the N equations obviously vanishes), it is by no means clear that one of the N equations for the N unknowns $z_1, \ldots, z_N$ is a consequence of the remaining $N-1$ equations. This is however true, and it is important to understand the reason. Viewing (4.18) as an Ansatz for solving (4.16) with $g = N+1$, one obtains the function

$$(4.21)\quad E \equiv \frac{1}{s(x)}\left(s(x+iN\beta)\exp[(N+1)\beta r+\beta y]\prod_{j=1}^{N}\frac{s(x-i\beta+z_j)}{s(x+z_j)}+(\beta\to-\beta)\right).$$

Of course, this function depends on x when one gives $z_1, \ldots, z_N$ arbitrary values. But since the function is elliptic in x with periods π/r and ia, one need only require that the residues at N of its $N+1$ (generically) simple poles in a period cell vanish to ensure that it is constant. Now the requirement that the residue at $x=0$ vanish yields (4.19), whereas the residues at $x=-z_k$ give rise to (4.20). Thus we need only prove that the system (4.20) with $k = 2, \ldots, N$ (for instance) admits a solution curve to infer that all of the equations are solved.

Now it is obvious that all of the N equations are solved by choosing

$$(4.22)\qquad z_j = ij\beta, \quad j = 1, \ldots, N.$$

An application of the implicit function theorem then shows that the equations with $k = 2, \ldots, N$ have a unique holomorphic solution $z_k(z_1), k = 2, \ldots, N$, near (4.22). Moreover, taking $z_1(t) = i\beta + it$ with $t \in [0, \epsilon)$, the functions $z_j\big(z_1(t)\big)$ are real-analytic functions from $[0,\epsilon)$ to $i(0,\infty)$ for ϵ small enough. From (4.19) it is then clear that (eventually decreasing ϵ) $y = y(t)$ is real-analytic and real-valued on $(0,\epsilon)$, and that one has $y \uparrow \infty$ for $t \downarrow 0$.

As a consequence, one can trade t for y in a neighborhood (K,∞) of ∞. Since we know very little about the minimal K satisfying various requirements, we may increase K as the need arises. In particular, we can choose it sufficiently large so that the functions

$$(4.23)\qquad \Phi_n(x) \equiv \mathcal{F}(x, nr) - \mathcal{F}(-x, nr), \quad nr > K, \quad n \in \mathbb{N},$$

are well defined and non-zero. Indeed, the above functions $z_j\left(z_1\big(t(y)\big)\right)$ (denoted simply $z_j(y)$ from now on) satisfy

$$(4.24)\qquad y \uparrow \infty \Rightarrow z_j(y) \to ij\beta, \quad j = 1, \ldots, N,$$

so that the summands on the rhs of (4.23) have distinct zeros for y large enough. (Recall our standing assumption (4.17).) Moreover, taking $x = iN\beta$ in (4.21) (which we may do, since E is x-independent), one deduces that an eventual increase of K ensures $E(y)$ is increasing on (K,∞). Then H_β has distinct eigenvalues on (K,∞).

The crux is now that all of the functions $\Phi_n(x)$ just defined belong to a dense subspace $\mathcal{A} \subset \mathcal{H}$ such that $H_\beta\mathcal{A} \subset \mathcal{H}$ and such that H_β is symmetric on $\mathcal{A}$. Specifically, $\mathcal{A}$ is defined by

$$(4.25)\quad \mathcal{A} \equiv \left\{\prod_{j=1}^{N}[s(x+ij\beta)s(x-ij\beta)]^{-1/2}\cdot f(x)\;\middle|\; \begin{array}{l} f(x)\text{entire, odd, } 2\pi/r\text{-periodic} \\ f(x+ij\beta) = f(x+ij\beta+\pi/r) = 0,\ |j| \le N \end{array}\right\}.$$

It is important to point out that this definition of $\mathcal{A}$ is not directly motivated by H_β, but rather by properties of the above H_β-eigenfunctions. (The 'extra' zeros of the eigenfunctions $\Phi_n(x)$ featuring in (4.25) can be derived from the AΔE (4.16).) Since the eigenvalues E_n and E_m are distinct for $n \neq m$, it now follows from symmetry that Φ_n and Φ_m are orthogonal, cf. (4.12) with $H_0 \to H_\beta$. Thus we obtain a self-adjoint operator (denoted again H_β) on the closed subspace $\mathcal{H}_K \subset \mathcal{H}$ spanned by the functions (4.23).

We expect that $\mathcal{H}_K$ equals $\mathcal{H}$ whenever K can be chosen negative. Put differently, we conjecture that for $K < 0$ the functions $\Phi_0, \Phi_1, \ldots$ are an orthogonal base for $\mathcal{H}$. In the special case $N = 1$ one can choose $K = -r$ (cf. [**36**], Eq. (2.35)), but completeness is still open, even in this simple case. More generally, we expect that the orthocomplement of $\mathcal{H}_K$ is spanned by functions $\Phi_0, \ldots, \Phi_{[K/r]}$ that are eigenfunctions of H_β with real eigenvalues.

Next, recall that we have restricted β by (4.17) in the above account. But a substantial part of our results continues to be valid under the restriction

$$k\beta \notin \mathbb{N}a, \quad k = 1, \ldots, 2N. \tag{4.26}$$

In particular, this suffices to infer the existence of eigenfunctions of the form (4.18)–(4.20). (Note that this more general restriction still guarantees that for y large the zeros $z_1(y), \ldots, z_N(y), -z_1(y), \ldots, -z_N(y)$ are distinct modulo the period ia, cf. (4.24).)

A key difference is however that for $N\beta > a$ the functions Φ_n (4.23) are most likely no longer pairwise orthogonal. More precisely, our symmetry proof breaks down for $N\beta > a$, and orthogonality is indeed violated in all cases where this could be tested. For $N = 1$ and $\beta > a$ a breakdown of orthogonality occurs in the strongest possible form: One has $(\Phi_n, \Phi_m) \neq 0$ for all $n \neq m$ with $n - m$ even. (Here, we still assume (4.26); note that H_β becomes 'free' for $N = 1$ and $\beta = la/2, l \in \mathbb{N}^*$.)

We prove the latter assertion in [**36**], which is concerned with the $g = 2$ case. We continue by describing another remarkable result from this paper. Taking $\beta \uparrow a$ (the edge of the unitarity region) and simultaneously $a \downarrow 0$ in a certain way, the above eigenfunctions $\Phi_n(x)$ converge to the Lieb-Liniger eigenfunctions [**27**] for the ($M =$ 2, center-of-mass) repulsive delta-function Bose gas. The role of the finite volume in [**27**] is played by the elliptic period $\pi/r < \infty$. This limiting transition generalizes the connection between the $g = 2$ *hyperbolic* relativistic $M = 2$ eigenfunctions and the *infinite-volume* delta-function eigenfunctions, which we already pointed out at the end of [**32**].

Just as in the nonrelativistic case, the orthogonality and completeness problems are trivial for the trigonometric specialization, since one winds up once again with orthogonal polynomials. Specifically, for $a = \infty$ the functions Φ_n are still of the form (4.13), with the weight function (4.15) now given by

$$w(x) = \sin^2(rx) \prod_{j=1}^{N} \sin r(x - ij\beta) \sin r(x + ij\beta). \tag{4.27}$$

The associated orthogonal polynomials are then q-Gegenbauer polynomials (cf. e.g. [**3**]), with $q = \exp(-2\beta r)$.

For the hyperbolic specialization the orthogonality and completeness analysis of the elliptic case can be greatly improved as well. This hinges on a second, far more elementary representation of the integer-g eigenfunctions (first presented in Ref. [**32**]), which we have studied from an algebraic viewpoint in Ref. [**35**]. With these algebraic results at our disposal, we recently presented a quite detailed clarification of the functional-analytic aspects in Ref. [**40**]. The upshot is that the suitably normalized eigenfunction transform is orthogonal and complete on the odd subspace of $L^2(\mathbb{R}, dx)$, provided $\beta \in (0, a/N)$. On the even subspace and for (generic) $\beta > a/N$, orthogonality and completeness break down in a way that can be made completely explicit [**40**].

4.3. Eigenfunctions for a Dense Parameter Set. Thus far, we have viewed the functions $\mathcal{F}(x,y)$ (4.18) as eigenfunctions of H_β (1.21). However, they also satisfy the quasi-periodicity relations (cf. (1.9))

$$\mathcal{F}(x+ia,y) = \exp\left[-2ir\sum_{j=1}^{N} z_j(y) - (N+1)ar - ay\right]\mathcal{F}(x,y), \tag{4.28}$$

$$\mathcal{F}(x+\pi/r,y) = -\exp(i\pi y/r)\mathcal{F}(x,y). \tag{4.29}$$

Thus, they can also be regarded as eigenfunctions of the AΔOs $T_{\pm ia}$ and $T_{\pm\pi/r}$, where we use the notation (1.22). Now this is true for $\mathcal{F}(-x,y)$, too, but then we obtain different eigenvalues. On the other hand, introducing the 'extra' AΔO

$$H_e \equiv T_{ia} + T_{-ia}, \tag{4.30}$$

and the 'quasi-periodicity' AΔO

$$Q = T_{\pi/r} + T_{-\pi/r}, \tag{4.31}$$

we obtain the *same* eigenvalues for $\mathcal{F}(x,y)$ and $\mathcal{F}(-x,y)$. Hence the functions $\mathcal{F}(\pm x, y)$ are joint eigenfunctions of the triple of independent AΔOs (H_β, H_e, Q).

As will now be detailed, we have found eigenfunctions $\mathcal{F}(\pm x,y)$ of H_β for a set that is dense in the parameter space $r, \beta, a > 0, g \in \mathbb{R}$. These H_β-eigenfunctions are once more eigenfunctions of Q with eigenvalue $-2\cos(\pi y/r)$ and of an extra AΔO H_e involving the shifts $T_{\pm ia}$. But the latter operator is no longer 'free': It involves the functions (1.14) with an interchange of a and β. Within this more general setting, H_β and H_e are on the same footing from a mathematical viewpoint, and we take this into account by switching to notation that makes this symmetry manifest.

Specifically, we work from now on with parameters a_+, a_- and b defined by

$$a_+ = \beta, \quad a_- = a, \quad b = \beta g. \tag{4.32}$$

The pertinent parameter domain at the elliptic level is then

$$\mathcal{E} \equiv \{(r, a_+, a_-, b) \in (0,\infty)^3 \times \mathbb{R}\}. \tag{4.33}$$

Introducing the notation

$$s_\delta(x) \equiv s(r, a_\delta; x), \quad \delta = +,-, \tag{4.34}$$

we now define the AΔOs

$$H_\delta \equiv e^{-br}\left(\left(\frac{s_\delta(x-ib)}{s_\delta(x)}\right)^{1/2} T_{ia_{-\delta}} \left(\frac{s_\delta(x+ib)}{s_\delta(x)}\right)^{1/2} + (i \to -i)\right), \quad \delta = +,-. \tag{4.35}$$

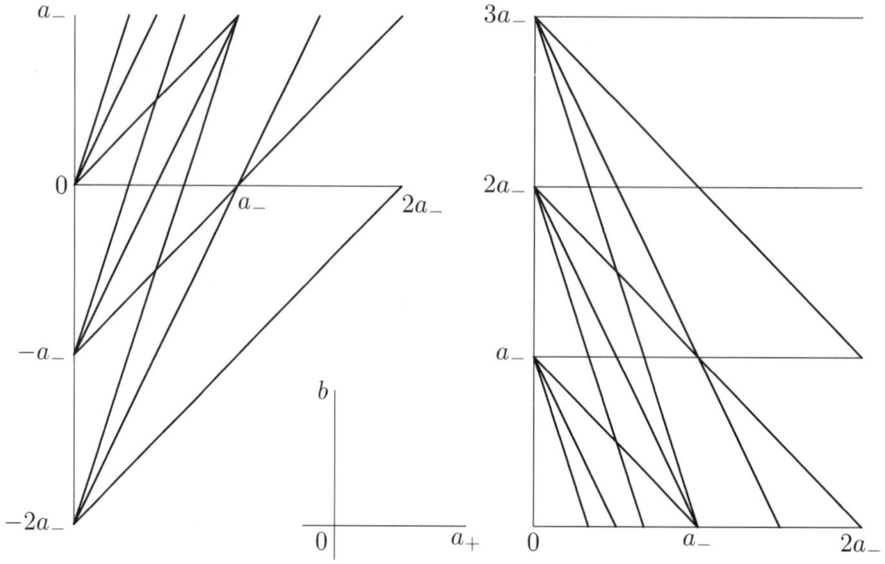

FIGURE 7. Some lines belonging to $\mathcal{D}_+$ (left) and $\mathcal{D}_-$ (right) with a_- fixed.

Comparing to H_β (1.21), one sees that H_- arises from H_β via the substitutions (4.32), but for the prefactor $\exp(-br)$. Likewise, taking $g = N + 1$, the AΔO H_+ reduces to a positive multiple of H_e (4.30). The choice of prefactor in (4.35) ensures that the operators thus defined satisfy the invariance property

$$H_\delta(a_+ + a_- - b) = H_\delta(b), \quad \delta = +, -. \tag{4.36}$$

The assertions just made can be easily verified by using the AΔE (1.9). Similarly, this AΔE can be used to check that H_+ and H_- *commute.* Now this is in accord with the existence of joint eigenfunctions, but there are no general results to the effect that commutativity of two AΔOs *implies* the existence of joint eigenfunctions. (In this connection it is important to observe that when the two summands of H_+ are multiplied by meromorphic functions with period ia_+, then the resulting AΔO still commutes with H_-.)

Even so, we have found joint eigenfunctions $\mathcal{F}(\pm x, y)$ of the three independent commuting AΔOs H_+, H_- and Q for a dense set $\mathcal{D}$ in $\mathcal{E}$ (4.33). For expository simplicity, we will specify these functions for a subset $\mathcal{D}_+ \cup \mathcal{D}_-$ of $\mathcal{D}$ that is already dense. The two sets $\mathcal{D}_\alpha, \alpha \in \{+, -\}$, are defined by

$$\mathcal{D}_\alpha \equiv \{(r, a_+, a_-, b) \in \mathcal{E} \mid b = (N_\alpha + 1)a_\alpha - N_{-\alpha}a_{-\alpha}, N_+, N_- \in \mathbb{N} \\ , a_+/a_- \notin \mathbb{Q}\}. \tag{4.37}$$

Since the quotient a_+/a_- is allowed to be an arbitrary positive irrational number, the b-values occurring here are dense in $\mathbb{R}$. Hence each of the two (disjoint) sets $\mathcal{D}_+$ and $\mathcal{D}_-$ is dense in $\mathcal{E}$. To vizualize the situation, it may be helpful to inspect Fig. 7, where we have fixed a_- and drawn some of the pertinent lines in the (a_+, b)-plane.

The two sets $\mathcal{D}_+, \mathcal{D}_-$ are interchanged under the transformation $b \to a_+ + a_- - b$. Fixing $r > 0, a_+/a_- \notin \mathbb{Q}$ and $N_+, N_- \in \mathbb{N}$, we get a point in $\mathcal{D}_+$ by taking

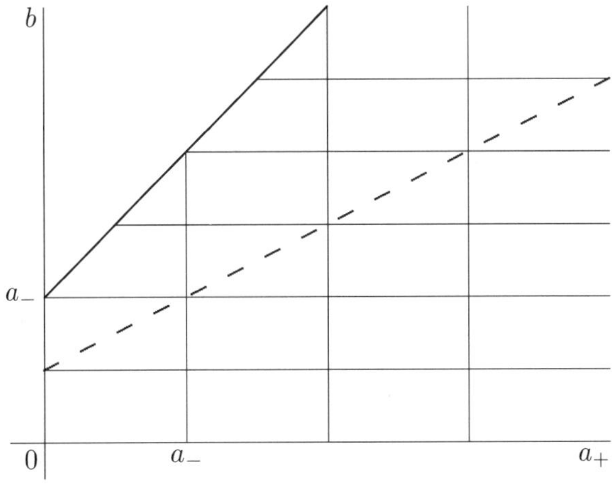

FIGURE 8. The unitarity region and the symmetry line $b = (a_+ + a_-)/2$ (dashed).

$b = (N_+ + 1)a_+ - N_- a_-$ and a point in $\mathcal{D}_-$ by taking $b = (N_- + 1)a_- - N_+ a_+$. In agreement with the invariance property (4.36), the joint eigenfunctions $\mathcal{F}(\pm x, y)$ are the same in both points. They read

$$\mathcal{F}(x,y) = \prod_{\delta=+,-} \prod_{j=1}^{N_\delta} \frac{s_{-\delta}\big(x + z_j^\delta(y)\big)}{[s_{-\delta}(x + ija_\delta)s_{-\delta}(x - ija_\delta)]^{1/2}} \times \exp[irx(2N_+N_- + N_+ + N_- + 1) + ixy]. \tag{4.38}$$

The zero functions $z_j^\delta, j = 1, \ldots, N_\delta$, satisfy the constraint system (4.20) with $N \to N_\delta, \delta = +, -$, and the two systems are coupled via the spectral variable y in a somewhat involved way that we will not detail here.

We do specify the asymptotics of the zero functions and eigenvalues:

$$y \uparrow \infty \Rightarrow z_j^\delta(y) \to ija_\delta, \quad j = 1, \ldots, N_\delta, \quad \delta = +, -, \tag{4.39}$$

$$y \uparrow \infty \Rightarrow E_\delta(y) \sim \exp(a_{-\delta}y), \quad dE_\delta(y)/dy \sim a_{-\delta}\exp(a_{-\delta}y), \quad \delta = +, -. \tag{4.40}$$

In view of the eigenvalue asymptotics, we can choose K such that on (K, ∞) the two eigenvalues separate points:

$$K < y_1 < y_2 \Rightarrow \big(E_+(y_1), E_-(y_1)\big) \neq \big(E_+(y_2), E_-(y_2)\big). \tag{4.41}$$

Another important feature is that the joint eigenspace is two-dimensional. (Again, we can only prove this for $y \in (L, \infty)$ and sufficiently large $L \geq K$, cf. Appendix B in Ref. [**34**].) Notice that this is false when a_+/a_- is rational: In that case joint eigenspaces are infinite-dimensional whenever they contain one non-trivial function.

We now turn to Hilbert space aspects. To this end we study the functions (4.23), with $\mathcal{F}$ given by (4.38). These functions are clearly π/r-periodic (-antiperiodic) for n odd (even). It is far from clear, but true that they are also pairwise orthogonal in $\mathcal{H}$ (4.11), provided b belongs to the unitarity interval $(0, a_+ + a_-)$. Fixing once more a_-, we have depicted the resulting unitarity region in the (a_+, b)-plane in Fig. 8.

As a consequence, H_+ and H_- can be redefined as commuting self-adjoint operators on the closed subspace $\mathcal{H}_K$ spanned by the functions (4.23): We need only set

$$H_\delta \Phi_n \equiv E_\delta(nr)\Phi_n, \quad nr > K, \quad n \in \mathbb{N}, \tag{4.42}$$

extend linearly, and take the closure.

Just as in the special case $N_- = 0$ (cf. Subsection 4.2), orthogonality is most likely violated for $b < 0$ and $b > a_+ + a_-$. We also expect that for $b \in (0, a_+ + a_-)$ the subspace $\mathcal{H}_K^\perp$ has dimension $[K/r] + 1$ and is spanned by joint eigenfunctions $\Phi_0, \ldots, \Phi_{[K/r]}$ with real eigenvalues.

We conclude this subsection by explaining a key feature in the symmetry/orthogonality analysis. To this end we first rewrite $\mathcal{F}(x, y)$ as

$$\mathcal{F}(x,y) = \mathcal{H}(x,y) \Big/ \prod_{\delta=+,-} \prod_{j=1}^{N_\delta} [s_{-\delta}(x + ija_\delta)s_{-\delta}(x - ija_\delta)]^{1/2}, \tag{4.43}$$

so that we have

$$\mathcal{H}(x,y) = \prod_{\delta=+,-} \prod_{j=1}^{N_\delta} s_{-\delta}(x + z_j^\delta(y)) \cdot \exp[irx(2N_+N_- + N_+ + N_- + 1) + ixy]. \tag{4.44}$$

Note that this similarity transformation yields *holomorphic* eigenfunctions. By exploiting the transformed AΔEs it can now be proved that one has the identities

$$\begin{gathered} \mathcal{H}(ik_+a_+ + ik_-a_-, y) = \mathcal{H}(-ik_+a_+ - ik_-a_-, y), \\ k_\delta \in \{-N_\delta, \ldots, 0, \ldots, N_\delta\}, \quad \delta = +, -. \end{gathered} \tag{4.45}$$

Therefore, the function $\mathcal{H}(x, y) - \mathcal{H}(-x, y)$ has $(2N_+ + 1)(2N_- + 1)$ explicitly known zeros on the imaginary axis. Since the functions

$$\psi_n(x) \equiv \mathcal{H}(x, nr) - \mathcal{H}(-x, nr) \tag{4.46}$$

are π/r-periodic (n odd) or π/r-antiperiodic (n even), the above zeros repeat under a shift by π/r:

$$\begin{gathered} \psi_n(ik_+a_+ + ik_-a_- + l\pi/r) = 0, \\ k_\delta \in \{-N_\delta, \ldots, 0, \ldots, N_\delta\}, \quad \delta = +, -, \quad l \in \mathbb{Z}. \end{gathered} \tag{4.47}$$

These explicit zeros of $\psi_n(x)$ are crucial in canceling poles arising from the squared denominator in (4.38). But only for $b \in (0, a_+ + a_-)$ one gets a pole-zero cancellation in a strip around the real x-axis that suffices to deduce pairwise orthogonality. For $b < 0$ and $b > a_+ + a_-$ one obtains instead a residue sum that has no reason to vanish.

4.4. The Interpolation Problem. As we have seen in the previous subsection, there exist joint eigenfunctions of the commuting AΔOs H_+, H_- (4.35) for a dense subset of $\mathcal{E}$ (4.33). A natural question is, therefore, whether there exist continuous interpolating joint eigenfunctions for all of $\mathcal{E}$. Clearly, such interpolations are uniquely determined up to multipliers depending solely on y and the parameters. Now a crucial feature of the above functions $\mathcal{F}(x, y)$ is that their $y \to \infty$ asymptotics is given in terms of a scattering function $u(r, a_+, a_-, b; x)$ that has a real-analytic extension to all of $\mathcal{E}$ (taking $x \in \mathbb{R}$), as will now be detailed.

First, we note that (4.39) entails

$$\mathcal{F}(x,y) \sim \zeta(N_+, N_-)[-\exp(2irx)u(x)]^{1/2} \exp(ixy), \quad y \to \infty, \tag{4.48}$$

where ζ is a suitable phase and where

$$(4.49)\quad u(x) = (-)^{N_+ + N_-} \prod_{\delta=+,-} \prod_{j=1}^{N_\delta} \frac{s_\delta(x + ija_{-\delta})}{s_\delta(x - ija_{-\delta})} \cdot \exp[2irx(2N_+N_- + N_+ + N_-)].$$

To explain why the u-function extends to all of $\mathcal{E}$, we introduce the c-function

$$(4.50)\qquad c(r, a_+, a_-, b; x) \equiv \frac{G(r, a_+, a_-; x - ib + i(a_+ + a_-)/2)}{G(r, a_+, a_-; x + i(a_+ + a_-)/2)}.$$

Here, $G(r, a_+, a_-; z)$ is the elliptic gamma function from Subsection 2.4. Now the extension to $\mathcal{E}$ of the function $u(x) = u(r, a_+, a_-, (N_\alpha + 1)a_\alpha - N_{-\alpha}a_{-\alpha}; x)$ given by (4.49) reads

$$(4.51)\qquad u(x) = -e^{-2irx} \frac{c(x)}{c(-x)}.$$

For real x this function is real-analytic on $\mathcal{E}$, as advertized. In particular, it is uniquely determined by (4.49). Thus the *asymptotics* of the joint eigenfunctions admits a unique real-analytic interpolation. (The 'elementary' appearance (4.49) of the u-function for the pertinent parameters can be readily derived from the two AΔEs satisfied by the elliptic gamma function, cf. Ref. [**33**], Subsection IVB.)

The similarity transformation (4.43) turning the two-valued function $\mathcal{F}(x, y)$ into the holomorphic function $\mathcal{H}(x, y)$ is readily seen *not* to admit a continuous interpolation. But when we introduce a generalized weight function

$$(4.52)\qquad w(x) \equiv \frac{1}{c(x)c(-x)},$$

then the similarity transformation to a new, meromorphic function

$$(4.53)\qquad \Psi(x, y) \equiv \mathcal{F}(x, y)/\zeta(N_+, N_-)w(x)^{1/2},$$

does admit an interpolation, since $w(x)$ does. The function Ψ has asymptotics

$$(4.54)\qquad \Psi(x, y) \sim c(x)e^{ixy}, \quad y \to \infty,$$

and correspondingly $c(x)$ may be viewed as a generalization of the Harish-Chandra c-function for symmetric spaces of rank 1 (cf. Refs. [**28**] and [**32**]).

Of course, this does not answer the question whether the meromorphic joint eigenfunctions $\Psi(\pm x, y)$ of the similarity-transformed AΔOs

$$(4.55)\quad A_\delta \equiv w(x)^{-1/2} H_\delta w(x)^{1/2} = e^{-br} \frac{s_\delta(x - ib)}{s_\delta(x)} T_{ia_{-\delta}} + (i \to -i), \quad \delta = +, -,$$

admit an extension to $\mathcal{E}$. This turns out to be a quite delicate issue. For the even combination

$$(4.56)\qquad \chi(x, y) \equiv \Psi(x, y) + \Psi(-x, y)$$

the identities (4.45) give rise to pole-zero cancellations on the imaginary axis, but as before one needs to choose $y = nr, n \in \mathbb{N}$, to ensure the same cancellation on the lines $\mathrm{Re}\, x = k\pi/r,\ k \in \mathbb{Z}^*$.

These cancellations are not only crucial for the orthogonality issue, but they are also relevant for the question of meromorphic interpolations. Indeed, for convergence to points in $\mathcal{E}$ for which a_+/a_- is irrational and b not equal to $ka_+ + la_-$ with k, l integers (e.g.), one needs to let N_+ and N_- go to ∞; hence cancellations are needed to prevent poles from becoming dense.

Even so, it appears hard to *exclude* the existence of a meromorphic interpolation for $\Psi(x, y)$. (This is because poles of meromorphic functions can exhibit drastic changes under convergence.) At any rate, for Hilbert space purposes it would suffice to control the convergence to arbitrary points in $\mathcal{E}$ for the functions $\chi(x, nr)$, $n \in \mathbb{N}$.

Though we have discussed these questions in more detail in Ref. [**34**], we have obtained no clear-cut answers. By contrast, at the hyperbolic level the existence of an interpolation of $\chi(x, y)$ is known. Indeed, the interpolation amounts to a special case of the R-function from Section 3.

To explain this in some detail, let us begin by pointing out that in view of (1.12) the algebraic aspects discussed in this section have an immediate specialization to the hyperbolic case, yielding simplifications at various points. (The Hilbert space aspects of the hyperbolic specialization, however, are far more difficult, since continuous spectrum is involved.) But in fact there exists a second much more elementary representation for the joint eigenfunctions $\Psi(\pm x, y)$, which does not involve the above constraint system. In particular, one may take $y \in \mathbb{C}$ in this representation, and one obtains

$$E_+(y) = 2\,\mathrm{ch}(a_- y), \quad E_-(y) = 2\,\mathrm{ch}(a_+ y), \tag{4.57}$$

for the joint eigenvalues, cf. Ref. [**35**].

The essential point is now that the AΔOs A_δ (4.55) specialize to the AΔOs $A_\delta\big((b, 0, 0, 0); x\big)$ from Subsection 3.3 (cf. (3.25)), and that for $\mathrm{Re}\, y > 0$ the functions $\Psi(\pm x, y)$ yield a base for the space of joint eigenfunctions with eigenvalues $E_\delta(y), \delta = +, -$. (This follows from the hyperbolic specialization of Appendix B in Ref. [**34**], combined with the elementary representation for $\Psi(\pm x, y)$, derived in Ref. [**35**].) Comparing the eigenvalues $E_\delta(y)$ (4.57) and $2c_\delta(2\hat{v})$, and using evenness in x, we then deduce the interpolation formula

$$R(a_+, a_-, (b, 0, 0, 0); x, a_+ a_- y/\pi) = r(a_+, a_-, b; y)\chi(a_+, a_-, b; x, y), \tag{4.58}$$

where $\mathrm{Re}\, y > 0$ and b takes the values occurring in (4.37). (The y-dependent proportionality factor r can be made explicit by using duality properties.)

The question whether the odd combination of the joint eigenfunctions $\Psi(\pm x, y)$ also admits a continuous interpolation to all of the hyperbolic parameter domain $a_+, a_- > 0$, $b \in \mathbb{R}$, is wide open. For further details on the interpolation issue we refer to our papers Refs. [**34, 35**].

Acknowledgments

We would like to thank J. Harnad and P. Winternitz for inviting us to present this material at the Séminaire de Mathématiques Supérieures (Montréal, 1999). A shorter version of these lectures was presented at the Workshop on Invariant Differential Operators, Special Functions and Representation Theory (R.I.M.S., Kyoto, 1997); it is a pleasure to thank M. Kashiwara and T. Oshima for the invitation.

References

1. A. Antonov, K. Hasegawa, and A. Zabrodin, *On trigonometric intertwining vectors and non-dynamical R-matrix for the Ruijsenaars model*, Nuclear Phys. B **503** (1997), 747–770.
2. G. E. Arutyunov, L. O. Chekhov, and S. A. Frolov, *R-matrix quantization of the elliptic Ruijsenaars-Schneider model*, Comm. Math. Phys. **192** (1998), 405–432.
3. R. Askey and M. E.-H. Ismail, *A generalization of ultraspherical polynomials*, Studies in Pure Math. (P. Erdös, ed.), Birkhäuser, Boston, 1983, pp. 55–79.

4. R. Askey and J. Wilson, *Some basic hypergeometric orthogonal polynomials that generalize Jacobi polynomials*, Mem. Amer. Math. Soc. **54**, no. 319 (1985).
5. Yu. M. Berezanskii, *Expansions in eigenfunctions of self-adjoint operators*, Transl. Math. Monogr., vol. 17, Amer. Math. Soc., Providence, 1968.
6. E. Billey, *Algebraic nested Bethe ansatz for the elliptic Ruijsenaars model*, math.QA/9806068.
7. O. A. Chalykh, *The duality of the generalized Calogero and Ruijsenaars problems*, Russian Math. Surveys **52** (1997), 1289–1291.
8. I. Cherednik, *Difference-elliptic operators and root systems*, Internat. Math. Res. Notices **1** (1995), 43–59 (1995).
9. J. F. van Diejen and L. Vinet, *The quantum dynamics of the compactified trigonometric Ruijsenaars-Schneider model*, Comm. Math. Phys. **197** (1998), 33–74.
10. N. Dunford and J. T. Schwartz, *Linear operators.* Part II: *Spectral theory*, Interscience Publ., New York, 1963.
11. G. Felder and A. Varchenko, *Integral representation of solutions of the elliptic Knizhnik-Zamolodchikov-Bernard equations*, Internat. Math. Res. Notices **5** (1995), 221–233.
12. G. Felder and A. Varchenko, *Algebraic Bethe Ansatz for the elliptic quantum group $E_{\tau,\eta}(sl_2)$*, Nuclear Phys. B **480** (1996), 485–503.
13. G. Felder and A. Varchenko, *Elliptic quantum groups and Ruijsenaars models*, J. Statist. Phys. **89** (1997), 963–980.
14. G. Felder and A. Varchenko, *Three formulas for eigenfunctions of integrable Schrödinger operators*, Compositio Math. **107** (1997), 143–175.
15. G. Felder and A. Varchenko, *Algebraic integrability of the two-body Ruijsenaars operator*, Functional Anal. Appl. **32** (1998), 81–92.
16. S. A. Fulling, *Aspects of quantum field theory in curved space-time*, London Math. Soc. Stud. Texts, vol. 17, Cambridge University Press, Cambridge, 1989 .
17. G. Gasper and M. Rahman, *Basic hypergeometric series*, Encyclopedia Math. Appl., vol. 35, Cambridge Univ. Press, Cambridge, 1990.
18. K. Hasegawa, *Ruijsenaars' commuting difference operators as commuting transfer matrices*, Comm. Math. Phys. **187** (1997), 289–325.
19. K. Hikami and Y. Komori, *Integrability, fusion, and duality in the elliptic Ruijsenaars model*, Modern Phys. Lett. A **12** (1997), 751–761.
20. K. Hikami and Y. Komori, *Diagonalization of the elliptic Ruijsenaars model—Correspondence with the Belavin model*, European Phys. J. B **5** (1998), 583–588.
21. A. Jirari, *Second-order Sturm-Liouville difference equations and orthogonal polynomials*, Mem. Amer. Math. Soc. **113**, no. 542, Amer. Math. Soc., Providence, 1995.
22. B. Jurčo and P. Schupp, *Adler-Kostant-Symes scheme for face and Calogero-Moser-Sutherland-type models*, J. Math. Phys. **39** (1998), 3577–3588.
23. Y. Komori and K. Hikami, *Quantum integrability of the generalized elliptic Ruijsenaars models*, J. Phys. A **30** (1997), 4341–4364.
24. I. Krichever and A. Zabrodin, *Spin generalization of the Ruijsenaars-Schneider model, non-abelian 2D Toda chain and representations of Sklyanin algebra*, Russian Math. Surveys **50** (1995), 1101–1150.
25. N. Kurokawa, *Multiple sine functions and Selberg zeta functions*, Proc. Japan Acad. Ser. A Math. Sci. **67** (1991), 61–64.
26. V. B. Kuznetsov and E. K. Sklyanin, *Separation of variables for the A_2 Ruijsenaars model and a new integral representation for the A_2 Macdonald polynomials*, J. Phys. A **29** (1998), 2779–2804.
27. E. H. Lieb and W. Liniger, *Exact analysis of an interacting Bose gas.* I. *The general solution and the ground state*, Phys. Rev. **130** (1963), 1605–1616.
28. M. A. Olshanetsky and A. M. Perelomov, *Quantum integrable systems related to Lie algebras*, Phys. Rep. **94** (1983), 313–404.
29. E. M. Opdam, *Harmonic analysis for certain representations of graded Hecke algebras*, Acta Math. **175** (1995), 75–121.
30. M. Reed and B. Simon, *Methods of modern mathematical physics.* II. *Fourier analysis, self-adjointness*, Academic Press, New York, 1975.
31. S. N. M. Ruijsenaars, *Complete integrability of relativistic Calogero-Moser systems and elliptic function identities*, Comm. Math. Phys. **110** (1987), 191–213.

32. S. N. M. Ruijsenaars, *Finite-dimensional soliton systems*, Integrable and Superintegrable Systems (B. Kupershmidt, ed.), World Scientific, Singapore, 1990, pp. 165–206.
33. S. N. M. Ruijsenaars, *First order analytic difference equations and integrable quantum systems*, J. Math. Phys. **38** (1997), 1069–1146.
34. S. N. M. Ruijsenaars, *Generalized Lamé functions.* I. *The elliptic case*, J. Math. Phys. **40** (1999), 1627–1663.
35. S. N. M. Ruijsenaars, *Generalized Lamé functions.* II. *Hyperbolic and trigonometric specializations*, J. Math. Phys. **40** (1999), 1595–1626.
36. S. N. M. Ruijsenaars, *Relativistic Lamé functions: The special case* $g = 2$, J. Phys. A **32** (1999), 1737–1772.
37. S. N. M. Ruijsenaars, *Systems of Calogero-Moser type*, Particles and Fields, CRM Ser. in Math. Phys. (G. Semenoff, L. Vinet, eds.), Springer, New York, 1999, pp. 251–352.
38. S. N. M. Ruijsenaars, *A generalized hypergeometric function satisfying four analytic difference equations of Askey-Wilson type*, Comm. Math. Phys. **206** (1999), 639–690.
39. S. N. M. Ruijsenaars, *On relativistic Lamé functions*, Calogero-Moser-Sutherland Models, CRM Ser. in Math. Phys. (J. F. van Diejen, L. Vinet, eds.), Springer, New York, 2000, pp. 421–440.
40. S. N. M. Ruijsenaars, *Hilbert space theory for reflectionless relativistic potentials*, to appear.
41. S. N. M. Ruijsenaars, *On Barnes' multiple zeta and gamma functions*, to appear.
42. T. Shintani, *On a Kronecker limit formula for real quadratic fields*, J. Fac. Sci. Univ. Tokyo Sect. IA Math. **24** (1977), 167–199.
43. E. K. Sklyanin, *Some algebraic structures connected with the Yang-Baxter equation. Representations of quantum algebras*, Functional Anal. Appl. **17** (1983), 273–284.
44. E. C. Titchmarsh, *Eigenfunction expansions associated with second-order differential equations.* Part I, Oxford University Press, Oxford, 1962.
45. E. T. Whittaker and G. N. Watson, *A course of modern analysis*, Cambridge Univ. Press, Cambridge, 1973.
46. A. Zabrodin, *On the spectral curve of the difference Lamé operator*, Internat. Math. Res. Notices **11** (1999), 589–614.

CENTRE FOR MATHEMATICS AND COMPUTER SCIENCE, P.O.BOX 94079, 1090 GB AMSTERDAM, THE NETHERLANDS

E-mail address: siru@wxs.nl

Centre de Recherches Mathématiques
CRM Proceedings and Lecture Notes
Volume **26**, 2000

Bäcklund Transformations and Baxter's Q-Operator

E. K. Sklyanin

Abstract. This course of five lectures given at the seminar "Integrable Systems: from Classical to Quantum" (Université de Montréal, Jul 26–Aug 6, 1999) contains a detailed comment on the recently discovered (Gaudin-Pasquier, 1992) connection between Bäcklund transformations in the theory of classical integrable systems on the one hand, and Baxter's Q-operator for quantum integrable systems, on the other hand. We restrict our attention to the systems with a finite number of degrees of freedom. Our main illustrative example is the periodic Toda lattice. We present a general construction of the Q-operator for models governed by the SL(2)-invariant R-matrix and apply it to our example. We also discuss applications of BT and Q-operators to the separation of variables and the theory of special functions.

1. Introduction

1.1. A Bit of History. The Bäcklund and Darboux transformations appeared in the 19th century in the study of the problems of differential geometry. With the advent of the Inverse Scattering Method in the 1960s, their relevance to the integrable nonlinear evolution equations was quickly recognized, and the amount of literature accumulated since then is enormous. See, for example, the monograph [**25**]. Especially important for these lectures is the Hamiltonian interpretation of Bäcklund transformations discovered by Flaschka and McLaughlin [**14**].

The Q-operator belongs to the realm of quantum integrability, and, compared to BT, is a relatively new invention. The Q-operator was introduced first by Rodney Baxter in his seminal study [**7, 8, 9**], see also [**10**], of the integrable quantum XYZ spin chain as an ingenious device which allowed to determine the spectrum of the model—a problem which was intractable by other known methods, such as the Bethe ansatz. The Q-operator is actually a one-parameter family of operators Q_λ commuting with the Hamiltonians of the integrable system. Its main characteristic property is that its eigenvalues satisfy a certain finite-difference, or, depending on the integrable model, differential equation with respect to the parameter λ, known nowadays as the *Baxter equation.* The Baxter equation, together with appropriate boundary conditions, provides a one-dimensional multiparameter spectral problem

2000 *Mathematics Subject Classification.* 37J35, 70H06, 81R50.
This is the final form of the paper.

which allows to determine the spectrum of the commuting Hamiltonians of the model in question. Thus an originally multidimensional spectral problem is reduced to a one-dimensional one—a phenonomenon similar to the separation of variables (SoV). This coincidence is not an acccident. Indeed, as shown in [**22, 24**], for classical Hamiltonian systems there exists an intimate relation between SoV and the Bäcklund transformation (BT), the latter being the classical analog of the Q-operator [**22, 28**].

For a long time the XYZ model remained the only model for which a Q-operator was known. In 1992 Pasquier and Gaudin [**28**] constructed a Q-operator for the quantum periodic Toda lattice using a somewhat different approach from Baxter's. They described the Q-operator explicitly, as an integral operator, and have found an important relation between the Q-operator and the Bäcklund transformation from the classical Toda chain. Namely, the Bäcklund transformation, as a canonical transformation, coincides with the classical limit $\hbar \to 0$ of the automorphism $\mathcal{O} \mapsto Q_\lambda \mathcal{O} Q_\lambda^{-1}$ of the associative algebra of quantum observables generated by Q_λ. The generating function $F_\lambda(y \mid x)$ of the Bäcklund transformation is obtained from the semiclassical asymptotics $Q_\lambda(y \mid x) \sim \exp\bigl(i\hbar^{-1} F_\lambda(y \mid x)\bigr)$ of the kernel $Q_\lambda(y \mid x)$ of Q_λ considered as an integral operator in the coordinate representation.

Later on, in [**11**], Bazhanov, Lukyanov and Zamolodchikov gave a boost to Baxter's original idea of constructing Q_λ as the trace of the monodromy matrix constructed of Lax operators corresponding to a special representation of the relevant quantum group in the auxiliary space. Setting the problem in the context of representation theory for quantum groups, they have taken as such a representation of $\widehat{\mathrm{sl}}_q(2)$ the so-called q-oscillator representation, and have managed to construct a pair of Q-operators for the massless sine-Gordon quantum field theory in a periodic box. It seems that the same q-oscillator representation allows one to construct a Q-operator for any integrable model governed by the quantum group $\widehat{\mathrm{sl}}_q(2)$, see [**4**].

In the paper [**23**] devoted to the construction of a Q-operator for the so-called DST (dimer self-trapping) model (a degenerate case of the XXX magnetic chain) the combination of the approaches due to Baxter (Q_λ as a trace of monodromy) and Pasquier-Gaudin (Q_λ as an integral operator) made it possible to describe the structure of the Q-operator in the greatest detail. Besides an explicit expression for the kernel of the integral operator Q_λ in several equivalent forms one can calculate explicitly the matrix elements of Q_λ in the natural monomial basis.

In the last few years considerable progress has been achieved in the understanding of the Hamiltonian properties of the BT for the classical integrable systems which parallel those of the Q-operator in the quantum case. It is worth noticing that the classical counterparts of some of the properties of the quantum Q-operator were unknown before. As an example one can mention the so-called *spectrality property* of BT discovered in [**22**] which corresponds to Baxter's finite-different equation for the quantum case. Baxter's construction of Q_λ as a trace of the monodromy matrix has led to a new construction of BT from symplectic leaves of the quadratic r-matrix Poisson bracket [**38**].

A special topic actively studied in recent years is the relation of BT and SoV, which turns out to be twofold. On the one hand, SoV can be obtaind from a composition of BT or Q, see [**22**]. On the other hand, a BT can, in turn, be obtained as the transformation intertwining a pair of SoV [**24**]. In this case, the quantum interpretation has not yet been found.

The growing interest in studying various properties of the Q-operator for a variety of quantum integrable systems is indicated by a surge of recent publications, see [**12, 29, 30, 41**].

1.2. Plan of the Lectures. In these lectures I will concentrate on the parallels between the Bäcklund transformation for the classical Hamiltonian sytems on the one hand and the Q-operator for the quantum integrable systems, on the other. As it frequently happens, when two theories merge after having been developed for a considerable time independently, the resulting cross-fertilization is quite useful for both. The most recent example is the classical r-matrix and the Lie-Poisson groups [**33, 34**] whose invention was inspired by the quantum theory.

We shall restrict our attention to systems with a finite number of degrees of freedom (pure quantum mechanics, no field theory), and put special stress on Hamiltonian mechanics which is essential for quantization. All the new notions and techniques will be introduced on the example of the periodic Toda lattice and accompanied with a short discussion of possible generalizations.

The lectures can be considered as an extended commentary to the paper by Pasquier and Gaudin [**28**], accompanied by the original results obtained by V. Kuznetsov and myself [**22, 23, 38, 39**].

2. Classical Periodic Toda Lattice

2.1. Description of the Model. The periodic Toda lattice [**42, 26**] is a system of n degrees of freedom described in terms of canonical coordinates $x \equiv (x_1, \dots, x_n)$ and momenta $X \equiv (X_1, \dots, X_n)$, having the standard Poisson brackets

$$\{X_j, X_k\} = \{x_j, x_k\} = 0, \quad \{X_j, x_k\} = \delta_{jk}. \tag{2.1}$$

In what follows we always denote canonical coordinates with small letters, e.g. x, y, s, t, φ and the corresponding canonical momenta with the respective capital letters: X, Y, S, T, Φ. Such a convention helps to deal with several sets of canonical variables.

The physical Hamiltonian H of the Toda lattice

$$H = \sum_{j=1}^{n} \left(\frac{1}{2} X_j^2 + \mathrm{e}^{x_{j+1} - x_j} \right) \tag{2.2}$$

describes the system of n one-dimensional non-relativistic particles of equal mass interacting via exponential potential between the nearest neighbors. In formulas like (2.2) we always assume the periodicity convention: $j + n \equiv j$. The Hamiltonian H is thus invariant with respect to the translation $j \mapsto j + 1$, hence the name 'periodic Toda lattice'.

The equations of motion $\dot{f} = \{H, f\}$ corresponding to the Hamiltonian (2.2) are

$$\dot{x}_j = X_j, \tag{2.3a}$$

$$\dot{X}_j = -\mathrm{e}^{x_j - x_{j-1}} + \mathrm{e}^{x_{j+1} - x_j}. \tag{2.3b}$$

2.2. Integrability. The periodic Toda lattice is an example of a *completely integrable Hamiltonian system* in the Liouville-Arnold sense [**5**]. It means that the Hamiltonian H (2.2) is an element of a ring generated by n independent Hamiltonians $H_1, \ldots, H_n$ which commute

$$\{H_{j_1}, H_{j_2}\} = 0 \tag{2.4}$$

with respect to the Poisson bracket (2.1). As a consequence, the Hamiltonian flow $\dot{f} = \{H, f\}$ leaves the Hamiltonians H_j invariant $\dot{H}_j = 0$ and, therefore, leaves invariant the level manifolds P_h obtained by fixing the values of the Hamiltonians $H_j = h_j$.

The fundamental result in the theory of Hamiltonian integrable systems is the Liouville theorem [**5**] which claims that the level manifolds P_h, if compact, are diffeomorphic to n-dimensional tori T^n. Moreover, there exist canonical *action-angle* variables Φ, φ such that the action variables Φ_j are functions of the Hamiltonians $H_1, \ldots, H_n$, and the Hamiltonian flows linearize in the angle variables: $\{H_j, \dot{\varphi}_k\} = \omega_{jk}(\Phi)$.

The easiest way to demonstrate the integrability of the periodic Toda lattice is to make use of the Inverse Scattering method (or, the Isopectral Deformation method) in its Hamiltonian version, see for example [**13**] and J. Harnad's lectures in this volume. Within the ISM framework, the commuting Hamiltonians are obtained from the spectral invariants of the *Lax matrix* $L(u; X, x)$ which is a square matrix of order N (generally speaking, different from the number of degrees of freedom n), depending on a complex parameter u called *spectral parameter* and whose matrix elements are functions on the phase space.

The spectral invariants $t_k(u)$ of $L(u)$ defined as the coefficients of the *characteristic polynomial*

$$W(v, u) \equiv \det\bigl(v - L(u)\bigr) = v^N + \sum_{k=1}^{N} (-1)^k v^{N-k} t_k(u) \tag{2.5}$$

are elementary symmetric polynomials of the eigenvalues of $L(u)$, or, in terms of matrix elements of $L(u)$, sums of principal minors of order k (determinants of submatrices of $L(u)$ of order k whose diagonal is contained in the diagonal of $L(u)$). For example, $t_1(u) = \operatorname{tr} L(u)$, $t_N(u) = (-1)^N \det L(u)$. The commuting Hamiltonians H_j are usually obtained as coefficients of $t_k(u)$, when $t_k(u)$ are polynomials in u, or coefficients of expansions of $t_k(u)$ in some other bases of functions of u (e.g. trigonometric or elliptic ones).

Proving the commutativity (2.4) of the Hamiltonians H_j is thus equivalent to proving the commutativity

$$\{t_{k_1}(u_1), t_{k_2}(u_2)\} = 0 \tag{2.6}$$

of the spectral invariants $t_k(u)$. The following theorem due to Babelon and Viallet provides a technical means for doing it. As proven in [**6**], the the commutativity (2.6) of the spectral invariants of $L(u)$ is equivalent to the existence of a so-called $\mathfrak{r}$-matrix represenation for the Poisson brackets $\{L_{a_1b_1}(u_1), L_{a_2b_2}(u_2)\}$ between the matrix elements $L_{ab}(u)$. To write down the representation in a compact form, we introduce the tensor product notation

$$\overset{1}{L} \equiv L \otimes \mathbf{1}, \quad \overset{2}{L} \equiv \mathbf{1} \otimes L, \tag{2.7}$$

where **1** is the unit matrix of order N. Respectively, $\{\overset{1}{L}(u_1), \overset{2}{L}(u_2)\}$ is the matrix of order $N^2 \times N^2$ of all Poisson brackets between the matrix elements of $L(u_1)$ an $L(u_2)$. The theorem of Babelon and Viallet claims that the commutativity (2.6) is equivalent to the existence of two matrices, $\mathfrak{r}_{12}$ and $\mathfrak{r}_{21}$, of order $N^2 \times N^2$ such that the equality

$$\{\overset{1}{L}(u_1), \overset{2}{L}(u_2)\} = [\mathfrak{r}_{12}, \overset{1}{L}(u_1)] - [\mathfrak{r}_{21}, \overset{2}{L}(u_2)] \tag{2.8}$$

holds for any u_1, u_2. Note that $\mathfrak{r}_{12}$ and $\mathfrak{r}_{21}$ depend on u_1 and u_2 and, generally speaking, contain the dynamical variables X, x. Actually, one can always choose the $\mathfrak{r}$-matrices in such a way that $\mathfrak{r}_{21}(u_1, u_2) = \mathcal{P}_{12}\mathfrak{r}_{12}(u_2, u_1)\mathcal{P}_{12}$ where $\mathcal{P}_{12}$ is the permutation matrix: $\mathcal{P}_{12} x \otimes y = y \otimes x$.

Speaking again about the periodic Toda chain, in order to construct the commutative Hamiltonians, we have to produce a Lax matrix and the corresponding $\mathfrak{r}$-matrices. There are at least two possible Lax matrices for the periodic Toda chain, one of order 2×2, another one of order $n \times n$, see [**13, 42, 26, 1**]. In this subsection we shall work with the 2×2 matrix.

The 2×2 Lax matrix (or, monodromy matrix [**13**]) $L(u; X, x)$ is defined as the product of local Lax matrices $\ell_j(u)$ depending on the variables X_j and x_j only:

$$L(u) = \ell_n(u) \ldots \ell_2(u)\ell_1(u), \tag{2.9}$$

$$\ell_j(u) \equiv \ell_j(u; X_j, x_j) = \begin{pmatrix} u + X_j & -\mathrm{e}^{x_j} \\ \mathrm{e}^{-x_j} & 0 \end{pmatrix}. \tag{2.10}$$

The characteristic polynomial of $L(u)$ is quadratic in v, having thus two spectral invariants: $t_1(u)$ and $t_2(u)$. However, $t_2(u) = \det L(u) \equiv 1$ by virtue of $\det \ell(u) = 1$ which leaves $t(u) \equiv t_1(u) = \operatorname{tr} L(u)$ as the only nontrivial spectral invariant:

$$W(u, v) \equiv \det\bigl(v - L(u)\bigr) = v^2 - t(u)v + 1. \tag{2.11}$$

The Hamiltonians H_j are then obtained from the expansion of $t(u)$:

$$t(u) = u^n + H_1 u^{n-1} + \cdots + H_n. \tag{2.12}$$

In particular, $H_1 = X_1 + \cdots + X_n$ is the total momentum. It is easy to see that the physical Hamiltonian (2.2) is given by the formula $H = \frac{1}{2}H_1^2 - H_2$ and thus belongs to the polynomial ring generated by $H_1, \ldots, H_n$.

To prove the commutativity (2.4) of the Hamiltonians H_j, or, equivalently, the commutativity

$$\{t(u_1), t(u_2)\} = 0 \tag{2.13}$$

of their generating function $t(u)$, it is sufficient to find the corresponding $\mathfrak{r}$-matrices. Actually, what we are able to prove is a much more special representation [**13**] for the left-hand-side of (2.8):

$$\{\overset{1}{L}(u), \overset{2}{L}(v)\} = [r_{12}(u - v), \overset{1}{L}(u)\overset{2}{L}(v)] \tag{2.14}$$

where

$$r_{12}(u_1 - u_2) = \frac{\mathcal{P}_{12}}{u_1 - u_2} \tag{2.15}$$

is the SL(2)-invariant solution to the classical Yang-Baxter equation

$$[r_{12}(u), r_{13}(u + v)] + [r_{12}(u), r_{23}(v)] + [r_{13}(u + v), r_{23}(v)] = 0. \tag{2.16}$$

One can easily transform the formula (2.14) to the Babelon-Viallet form by setting in (2.8)

$$\mathfrak{r}_{12} = \frac{1}{2}\big(r_{12}(u_1 - u_2)\overset{2}{L}(u_2) + \overset{2}{L}(u_2)r_{12}(u_1 - u_2)\big), \tag{2.17a}$$

$$\mathfrak{r}_{21} = -\frac{1}{2}\big(r_{12}(u_1 - u_2)\overset{1}{L}(u_1) + \overset{1}{L}(u_1)r_{12}(u_1 - u_2)\big). \tag{2.17b}$$

To prove (2.14) we shall make use of the so-called *comultiplication* property of the quadratic Poisson bracket. It is a simple exercise to verify that if two Lax matrices $L_1(u)$ and $L_2(u)$ defined, respectively, on different phase spaces P_1 and P_2 each satisfy the identity (2.14), then their matrix product $L(u) = L_1(u)L_2(u)$ defined on the direct product $P_1 \times P_2$ satisfies the same identity. The proof uses nothing but the identity (2.14) for L_1 and L_2, and the identity

$$\{\overset{1}{L}_1(u_1), \overset{2}{L}_2(u_2)\} = 0. \tag{2.18}$$

Thus it is sufficient to verify the identity (2.14) for the local Lax matrices $\ell_j(u)$ given by (2.10) which is a matter of direct calculation.

Strictly speaking, to establish the integrability, besides proving the commutativity of the Hamiltonians H_j we also need to prove their independence. For a proof, see [**32**]. It is also possible to verify that, modulo center-of-mass motion, which is easily separated, the level manifolds P_h are compact and thus, by virtue of Liouville's theorem are isomorphic to tori.

2.3. Quadratic Poisson Bracket. Before starting the discussion of Bäcklund transformations we need to learn some more facts about the r-matrix quadratic Poisson bracket (2.14) and the class of integrable models it generates.

Let us suppose that the 2×2 matrix $L(u)$ is a polynomial in u of degree n,

$$L(u; X, x) = L^{(n)}u^n + L^{(n-1)}u^{n-1} + \cdots + L^{(0)}, \tag{2.19}$$

and regard the equality (2.14) with the r-matrix given by (2.15) as introducing a Poisson bracket on the $4(n+1)$ variables $L^{(j)}_{ab}$, $j = 0, \ldots, n$, $a, b = 1, 2$. For the sake of simplicity we think of $L^{(j)}_{ab}$ as complex variables and do not consider here the question of choosing an appropriate $*$-conjugation.

It is easy to see that, despite the denominator $(u_1 - u_2)$ present in the r-matrix (2.15), the right-hand-side of (2.14) is polynomial both in u_1 and u_2 because of the identity $[\mathcal{P}, L \otimes L] = 0$ which nullifies the numerator for $u_1 = u_2$. According to a theorem by Sophus Lie [**44**], for any Poisson bracket there exist local coordinates (X, x, c) such that X and x are canonical (2.1), and c are central, that is

$$\{c_j, c_k\} = \{c_j, X_k\} = \{c_j, x_k\} = 0.$$

To obtain a symplectic manifold which can serve as a phase space for a mechanical system, one needs thus to restrict the Poisson bracket onto a level manifold of its central (or Casimir) functions.

In the case of the bracket (2.14) the Casimir functions can be found easily. First, the leading coefficient $L^{(n)}$ provides 4 casimirs. More casimirs are given by the coefficients of the determinant $\det L(u)$. Being, generally speaking, a polynomial of degree $2n$, the determinant has $(2n+1)$ coefficients but its leading coefficient coincides with $\det L^{(n)}$, which gives us only $2n$ new casimirs. In total, we have $(2n+4)$ casimirs which corresponds to the level manifolds of dimension $4(n+1) -$

$(2n+4) = 2n$. To show that there are no more casimirs, it is sufficient to construct an example of a $2n$-dimensional symplectic leaf of the bracket (2.14).

The tool for constructing such examples is the *comultiplication property* of the bracket (2.14) mentioned in Section 2.2. It allows to build multidimensional symplectic leaves from simpler blocks. The simplest, 0-dimensional simplectic leaf of the bracket (2.14) is given by a constant matrix $L(u) \equiv K$. The next most natural choice is to take a linear polynomial in u with the unit matrix as the leading coefficient:

$$\ell^{\mathrm{XXX}}(u) = u\mathbf{1} + \mathbb{S}, \quad \mathbb{S} = \begin{pmatrix} S_3 & S_1 - \mathrm{i}S_2 \\ S_1 + \mathrm{i}S_2 & -S_3 \end{pmatrix}. \tag{2.20}$$

Substituting (2.20) for $L(u)$ into (2.14) we obtain for S_α the Poisson algebra isomorphic to the Lie algebra sl_2:

$$\{S_\alpha, S_\beta\} = -\mathrm{i}\sum_{\gamma=1}^{3} \varepsilon_{\alpha\beta\gamma} S_\gamma, \tag{2.21}$$

$\varepsilon_{\alpha\beta\gamma}$ being the standard antisymmetric tensor. The Poisson bracket (2.21) has the Casimir function $C = S_1^2 + S_2^2 + S_3^2$, and its generic symplectic leaves $C = \mathrm{const} \neq 0$ are 2-dimensional spheres.

Noting that, due to the fact that the r-matrix (2.15) depends only on the difference $(u_1 - u_2)$, the shift of the spectral parameter $u \mapsto u - c$ is an automorphism of the Poisson algebra (2.14). Therefore, taking a direct product of n copies of the triplets S_α restricted to the level surfaces $C = \rho_j^2$, $j = 1, \ldots, n$, we obtain a $2n$-dimensional symplectic leaf of the bracket (2.14) given by the product

$$L^{\mathrm{XXX}}(u) = K\ell_n^{\mathrm{XXX}}(u - c_n)\ldots\ell_1^{\mathrm{XXX}}(u - c_1). \tag{2.22}$$

Note that the number of parameters contained in the symplectic leaf is $(2n+4)$, where 4 comes from the constant matrix K and the rest from n casimirs ρ_j and n shifts c_j, $j = 1, \ldots, n$. The parameters are easily identified with $L^{(n)} = K$ and the zeroes of the determinant $\det L(u) = \det K \prod_j (u - c_j - \rho_j)(u - c_j + \rho_j)$. We are thus led to the conclusion that the constructed symplectic leaf is in fact the generic leaf for the bracket (2.14).

The Lax matrix (2.22) defines an integrable system known as the *inhomogeneous Heisenberg magnetic chain* [**13, 31**]. All other integrable models associated with the Poisson bracket (2.14) and the sl_2-invariant r-matrix (2.15) can be obtained from degenerations of the Lax matrix (2.22).

To describe some important degenerations of (2.22) let us parametrize the spin components S_α in (2.20) using a pair of canonical variables (X, x):

$$\ell^{\mathrm{XXX}}(u) = \begin{pmatrix} u + xX - \rho & -x^2X + 2\rho x \\ X & u - xX + \rho \end{pmatrix}. \tag{2.23}$$

Multiplying $\ell^{\mathrm{XXX}}(u)$ from the right by the diagonal matrix $\mathrm{diag}(1, -1/(2\rho))$ and performing the shift $u \mapsto u + \rho$ (note that these are legal operations which do not change the Poisson bracket (2.14)) we are capable to take the limit $\rho \to \infty$. The result is the Lax matrix for the so-called dimer-self-trapping (DST) model [**23**]

$$\ell^{\mathrm{DST}}(u) = \begin{pmatrix} u + xX & -x \\ X & -1 \end{pmatrix}. \tag{2.24}$$

Note that the determinant $\det \ell^{\mathrm{DST}}(u) = -u$ is linear in u. A further degeneration of the DST model produces the Toda lattice. To this end, one multiplies $\ell^{\mathrm{DST}}(u)$ from the right by the matrix $\operatorname{diag}(1, a^{-1})$ and, after making the substitutions $u \mapsto u - a$, $x \mapsto a\mathrm{e}^x$, $X \mapsto \mathrm{e}^{-x}(1 + a^{-1}X)$, obtains in the limit $a \to \infty$ the unimodular Lax matrix (2.10) for the Toda lattice.

More symplectic leaves can be obtained by applying the automorphism $\ell(u) \mapsto \check{\ell}(u) \equiv \ell^{-1}(-u)$ of the r-matrix Poisson algebra (2.14) to $\ell^{\mathrm{DST}}(u)$ and $\ell^{\mathrm{Toda}}(u)$. Up to a scalar factor we have:

$$\check{\ell}^{\mathrm{DST}}(u) \sim \begin{pmatrix} 1 & -x \\ X & u - xX \end{pmatrix}, \quad \check{\ell}^{\mathrm{Toda}}(u) \sim \begin{pmatrix} 0 & -\mathrm{e}^x \\ \mathrm{e}^{-x} & u - X \end{pmatrix} \tag{2.25}$$

(on $\ell^{\mathrm{XXX}}(u)$ the automorphism acts trivially: $\check{\ell}^{\mathrm{XXX}}(u) \sim \ell^{\mathrm{XXX}}(u)$).

CONJECTURE. *Any symplectic leaf $L(u)$ of the quadratic r-matrix Poisson bracket* (2.14) *which is polynomial in u can be decomposed (in a non-unique way, of course) into a product of a constant matrix K and linear matrix polynomials of the form $\ell^{\mathrm{XXX}}(u-c)$, $\ell^{\mathrm{DST}}(u-c)$, $\check{\ell}^{\mathrm{DST}}(u-c)$, $\ell^{\mathrm{Toda}}(u-c)$, $\check{\ell}^{\mathrm{Toda}}(u-c)$.*

In the case of a generic symplectic leaf the factorization (2.22) in terms of $\ell^{\mathrm{XXX}}(u-c)$ only should suffice. The difficult part is to analyze the degenerate cases when the leading coefficient $L^{(n)}$ is a degenerate matrix and/or the degree of $\det L(u)$ is less than $2n$. Hopefully, one of the readers will provide a proof soon.

One can find more information about the properties of the r-matrix Poisson bracket (2.14) in the papers [**13, 34, 33**] as well as in the lectures by Harnad and Reshetikhin in the present volume.

2.4. Bäcklund Transformation and its Properties. In this section we begin to study the Bäcklund transformation for the periodic Toda lattice. The Bäcklund transformation $\mathcal{B}_\lambda$, depending on a complex parameter λ, is defined as the mapping from the variables (X, x) to (Y, y) given implicitely by the equations

$$X_j = \mathrm{e}^{x_j - y_j} + \mathrm{e}^{y_{j+1} - x_j} - \lambda, \tag{2.26a}$$

$$Y_j = \mathrm{e}^{x_j - y_j} + \mathrm{e}^{y_j - x_{j-1}} - \lambda. \tag{2.26b}$$

The equations (2.26) are algebraic in the momenta and the exponents of coordinates. Resolving (2.26a) with respect to $\mathrm{e}^{y_{j+1}}$:

$$\mathrm{e}^{y_{j+1}} = \mathrm{e}^{x_j}(X_j + \lambda) - \mathrm{e}^{2x_j - y_j} \tag{2.27}$$

and iterating the equation (2.27) for $j = 1, 2, \ldots, n$, we finally arrive at a *quadratic* equation for e^{y_1}, which implies that the transformation $\mathcal{B}_\lambda$ is a *two-valued* algebraic function in terms of X, e^x. Fortunately, for all our purposes the simple implicit formulas (2.26) are sufficient.

Another attractive feature of the equations (2.26) is their *locality*: they involve only the variables with the indices differing by 0 and 1. Note that even for real λ resolving the equations (2.26) can produce complex values of Y and y. To avoid complications, we shall not make any attempt to study the reality conditions, and treat both (X, x) and (Y, y) as complex variables, in the spirit of *algebraic integrability* [**2**].

We start the list of properties of the Bäcklund transformation by noting its *canonicity*: the variables (Y_j, y_j) are canonical. This can be seen from the fact that the equations (2.26) can be written down in the form

$$X_j = \frac{\partial F_\lambda}{\partial x_j}, \quad Y_j = -\frac{\partial F_\lambda}{\partial y_j}, \tag{2.28}$$

where

$$F_\lambda(y;x) = \sum_{i=1}^{n} \left(\mathrm{e}^{x_j - y_j} - \mathrm{e}^{y_{i+1} - x_j} - \lambda(x_j - y_j) \right) \tag{2.29}$$

is the generating function [**5**] of the canonical transformation.

The next property is the *invariance of Hamiltonians*:

$$H_j(X,x) = H_j(Y,y), \quad j = 1, \ldots, n. \tag{2.30}$$

Though the invariance of physical Hamiltonian H (2.2) can be proved by a direct calculation [**42, 16**], to prove the invariance of the whole set of commuting Hamiltonians H_j we will need some more effective technique. The easiest way is to make use of the Inverse Scattering Method explained in Section 2.2. The invariance of H_j under $\mathcal{B}_\lambda$ is then equivalent to the invariance of the spectrum of $L(u)$, which implies that there exists an invertible matrix $M(u,\lambda)$ such that

$$M(u,\lambda)L(u;Y,y) = L(u;X,x)M(u,\lambda), \quad \forall u \in \mathbb{C}. \tag{2.31}$$

Such a matrix is called a *Darboux matrix*, and the tranformation of L given by (2.31) is called a *Darboux transformation* [**25**].

In our case, due to the factorization (2.9) of $L(u)$ into local Lax matrices $\ell_j(u)$, we can be more specific about the structure of the Darboux transformation. Setting $M_1(u,\lambda) \equiv M(u,\lambda)$, we introduce matrices $M_{j+1}(u,\lambda)$, $j = 1, \ldots, n-1$, inductively as

$$M_{j+1}(u,\lambda) = \ell_j(u;X_j,x_j)M_j(u,\lambda)\ell_j^{-1}(u;Y_j,y_j)$$

(note that for $j = n$, due to the periodicity $n+1 \equiv 1$ we recover the equality (2.31)). The global transformation (2.31) thus takes the form of the local *gauge transformation*

$$M_{j+1}(u,\lambda)\ell_j(u;Y_j,y_j) = \ell_j(u;X_j,x_j)M_j(u,\lambda). \tag{2.32}$$

The converse is also true: from (2.32) it follows that the spectrum of $L(u)$ is preserved. To prove the invariance of Hamiltonians (2.30) it is therefore sufficient to find the matrices M_j satisfying (2.32). Using the equations (2.26) it is easy to verify that (2.32) is satisfied with the following matrices [**16**]:

$$M_j(u,\lambda) = \begin{pmatrix} u - \lambda + \mathrm{e}^{y_j - x_{j-1}} & -\mathrm{e}^{y_j} \\ \mathrm{e}^{-x_{j-1}} & -1 \end{pmatrix}. \tag{2.33}$$

The two properties: canonicity and invariance of Hamiltonians constitute the definition of what is called an *integrable map* [**43**]. It can be considered as a discrete-time analog of an integrable hamiltonian flow. Veselov [**43**] has proved a discrete-time analog of Liouville's theorem which claims that in the action-angle variables (Φ,φ) any integrable map acts as a shift $\varphi_j \mapsto \varphi_j + \Omega_j(\Phi)$, or, speaking more precisely, as a collection of shifts due to the multivaluedness of algebraic mappings. Applying the theorem to the case of the Bäcklund transformation depending on a parameter λ, and noting that shifts on the Liouville torus commute, we obtain as an immediate consequence the *commutativity* of BT

$$\mathcal{B}_{\lambda_1} \circ \mathcal{B}_{\lambda_2} = \mathcal{B}_{\lambda_2} \circ \mathcal{B}_{\lambda_1}. \tag{2.34}$$

Note that a direct proof of the commutativity of BT is not a simple task, see for example [**42, 17**]. It is trivialized in our case entirely due to the fact that from the very beginning we are working in the hamiltonian context.

The last property in our list is *spectrality.* It was discovered rather recently [**22**], and the main motivation for its existence comes from the quantum case, see Section 3.3.

Let us introduce the quantity μ which is, in a sense, canonically conjugated to λ:

$$\mu \equiv \frac{\partial F_\lambda}{\partial \lambda} = \sum_{j=1}^{n} (x_j - y_j). \tag{2.35}$$

The spectrality of BT means that the pair $(\mathrm{e}^\mu, \lambda)$ lies on the spectral curve of the Lax matrix. Since $\det L(u) = 1$ this means that both e^μ and $\mathrm{e}^{-\mu}$ are eigenvalues of $L(\lambda)$

$$W(\mathrm{e}^{\pm\mu}, \lambda) \equiv \det\bigl(\mathrm{e}^{\pm\mu} - L(\lambda)\bigr) = 0 \tag{2.36}$$

(it does not matter if we take $L(\lambda; X, x)$ or $L(\lambda; Y, y)$ since they are isospectral).

The property of spectrality of BT still remains somewhat mysterious and certainly needs more research to uncover its algebraic and geometric meaning. The main drawback of the present definition is its being formulated in quite noninvariant terms of generating function F_λ.

To prove (2.36) it suffices to show that, say e^μ is an eigenvalue of the matrix $L(\lambda; Y, y)$. We shall construct explicitly the corresponding eigenvector ω_1:

$$L(\lambda; Y, y)\omega_1 = \mathrm{e}^\mu \omega_1. \tag{2.37}$$

From (2.33) it follows that $\det\bigl(M_j(u, \lambda)\bigr) = \lambda - u$. It is easy to see that for $u = \lambda$ the matrix $M_j(\lambda, \lambda)$ degenerates into a projector

$$M_j(\lambda, \lambda) = \begin{pmatrix} \mathrm{e}^{y_j} \\ 1 \end{pmatrix} \begin{pmatrix} \mathrm{e}^{-x_{j-1}} & -1 \end{pmatrix} \tag{2.38}$$

and, as a consequence, has the unique, up to a scalar factor, null-vector

$$\omega_j = \begin{pmatrix} \mathrm{e}^{x_{j-1}} \\ 1 \end{pmatrix}, \qquad M_j(\lambda, \lambda)\omega_j = 0. \tag{2.39}$$

Using the identity (2.31) with $M \equiv M_1$ we conclude that

$$M_1(\lambda, \lambda) L(\lambda; Y, y)\omega_1 = 0 \tag{2.40}$$

which, combined with the uniqueness of the null-vector ω_1 of M_1, implies that ω_1 is an eigenvector of $L(\lambda; Y, y)$. To determine the corresponding eigenvalue, we apply the same argument to the identity (2.32) obtaining the equality

$$M_{j+1}(\lambda, \lambda)\ell_j(\lambda; Y_j, y_j)\omega_j = 0$$

from which it follows that $\ell_j(\lambda; Y_j, y_j)\omega_j \sim \omega_{j+1}$. The direct calculation shows that

$$\ell_j(\lambda; Y_j, y_j)\omega_j = \mathrm{e}^{x_{j-1} - y_j}\omega_{j+1}. \tag{2.41}$$

It remains only to use the formulas (2.9) and (2.35) to arrive finally at (2.37).

An alternative variant of the proof, closer to what we shall use in the quantum case (see Section 3.5), is to introduce a gauge transformation with a triangular matrix N_j:

$$\widehat{\ell}_j \equiv N_{j+1}^{-1}\ell_j(\lambda; Y_j, y_j)N_j = \begin{pmatrix} \mathrm{e}^{y_j - x_{j-1}} & 0 \\ \mathrm{e}^{-y_j} & \mathrm{e}^{x_{j-1}-y_j} \end{pmatrix}, \quad N_j = \begin{pmatrix} 1 & \mathrm{e}^{x_{j-1}} \\ 0 & 1 \end{pmatrix} \tag{2.42}$$

(note that ω_j coincides with the second column of N_j).

The result, as expected, is

$$t(\lambda) = \operatorname{tr} \ell_n(\lambda) \dots \ell_1(\lambda) = \operatorname{tr} \widehat{\ell}_n \dots \widehat{\ell}_1 = \mathrm{e}^{\mu} + \mathrm{e}^{-\mu}. \tag{2.43}$$

We conclude this section with a remark on using BT for generating solitons, which is the main application of BT to the integrable nonlinear evolution equations [**14, 17, 42**]. We are following here the argument by Gaudin [**16**]. Let us apply the Bäcklund transformation (2.26) to the vacuum state $X_j = x_j = 0$. The equations (2.26) turn into

$$0 = \mathrm{e}^{-y_j} + \mathrm{e}^{y_{j+1}} - \lambda, \tag{2.44a}$$

$$Y_j = \mathrm{e}^{-y_j} + \mathrm{e}^{y_j} - \lambda. \tag{2.44b}$$

Concentrating on the first equation (the second equation describes the time evolution $Y_j = dy_j/dt$ with respect to the Hamiltonian H) we introduce the parametrization: $\lambda = 2\cosh\kappa$, $\mathrm{e}^{y_0} = \cosh(\alpha+\kappa)/\cosh\alpha$. The general solution can now be written as

$$\mathrm{e}^{y_j} = \frac{\cosh\bigl(\alpha + \kappa(j+1)\bigr)}{\cosh(\alpha+\kappa j)}. \tag{2.45}$$

In the case of the infinite lattice, when $j \in \mathbb{Z}$, the formula (2.45) describes a soliton solution. Note, however, that the solution (2.45) has different asymptotics $\mathrm{e}^{y_j} \to \mathrm{e}^{\pm\kappa}$ as $j \to \pm\infty$, thus satisfying boundary conditions different from those for the vacuum state. As a result, the energy and values of other integrals of motion for the soliton solution differ from those for the vacuum.

The situation is quite different in the periodic case. The periodicity condition $y_{n+1} = y_1$ can be satisfied in two ways. The first one leads to the quantization of the parameter: $\kappa n \in \pi\mathrm{i}\mathbb{Z}$ and is inacceptable if we want to keep λ free. Besides, in this way we get a complex solution for e^{y_j}. Another option is to fix the free parameter α by setting $\alpha = \pm\infty$, which gives us another vacuum state $\mathrm{e}^{y_j} = \mathrm{e}^{\pm\kappa}$ having the same values of Hamiltonians as the vacuum.

The fact that BT in the periodic case does not produce solitons and always preserves the integrals of motion may disappoint those accustomed to other usages of BTs. A merit of our variant is, however, that it has deep analogies in the quantum case, as we shall see further.

2.5. Duality. Besides the 2×2 Lax matrix $L(u)$ which we used until now there exists another, $n \times n$ Lax matrix $\mathcal{L}(v)$ for the Toda lattice which is dual to $L(u)$ in the sense that the corresponding spectral curves are equivalent up to interchanging the spectral parameters u and v

$$(-1)^{n-1}\det\bigl(u - \mathcal{L}(v)\bigr) = \det\bigl(v - L(u)\bigr). \tag{2.46}$$

Referring the reader to the paper [**1**] where the geometric meaning of the duality is elucidated, we present here a more elementary approach.

To produce the dual Lax matrix $\mathcal{L}(v)$ we take an eigenvector $\theta_1(u)$ of $L(u)$ corresponding to the eigenvalue v (for brevity, we will not mark the dependence on u in θ),

$$L(u)\theta_1 = v\theta_1 \tag{2.47}$$

and define θ_j by induction as

$$\theta_{j+1} = \ell_j(u)\theta_j, \quad j = 1, \ldots, n. \tag{2.48}$$

From (2.47) it follows that $\theta_{n+1} = v\theta_1$. The function $\theta_j(u)$, when properly normalized, is called the *Baker-Akhiezer function.* Denoting the components of the vector θ_j by φ_j and ψ_j, we write down (2.48) explicitly as

$$\begin{pmatrix} \varphi_{j+1} \\ \psi_{j+1} \end{pmatrix} = \begin{pmatrix} u + X_j & -\mathrm{e}^{x_j} \\ \mathrm{e}^{-x_j} & 0 \end{pmatrix} \begin{pmatrix} \varphi_j \\ \psi_j \end{pmatrix}.$$

Then, splitting the components and taking into account the quasiperiodicity condition $\theta_{n+1} = v\theta_1$ we arrive at the following linear equations for φ_j and ψ_j:

$$u\varphi_j = \varphi_{j+1} - X_j\varphi_j + \mathrm{e}^{x_j}\psi_j, \quad j = 1, \ldots, n-1 \tag{2.49a}$$

$$u\varphi_n = v\varphi_1 - X_n\varphi_n + \mathrm{e}^{x_n}\psi_n, \tag{2.49b}$$

$$\psi_{j+1} = \mathrm{e}^{-x_j}\varphi_j, \quad j = 1, \ldots, n-1 \tag{2.50a}$$

$$v\psi_1 = \mathrm{e}^{-x_n}\varphi_n. \tag{2.50b}$$

Eliminating ψ_j we obtain a second-order finite-difference equation for φ_j

$$u\varphi_1 = \varphi_2 - X_1\varphi_1 + \mathrm{e}^{x_1 - x_n}v^{-1}\varphi_n, \tag{2.51a}$$

$$u\varphi_j = \varphi_{j+1} - X_j\varphi_j + \mathrm{e}^{x_j - x_{j-1}}\varphi_{j-1}, \quad j = 2, \ldots, n-1 \tag{2.51b}$$

$$u\varphi_n = v\varphi_1 - X_n\varphi_n + \mathrm{e}^{x_n - x_{n-1}}\varphi_{n-1}, \tag{2.51c}$$

which can be rewritten as the linear problem for the vector Φ with the components φ_j in the matrix form:

$$\mathcal{L}(v)\Phi = u\Phi, \qquad \Phi = \begin{pmatrix} \varphi_1 \\ \ldots \\ \varphi_n \end{pmatrix}$$

where the matrix $\mathcal{L}(v)$, defined as

$$\mathcal{L}(v) = \begin{pmatrix} -X_1 & 1 & \ldots & 0 & v^{-1}\mathrm{e}^{x_{1n}} \\ \mathrm{e}^{x_{21}} & -X_2 & \ldots & 0 & 0 \\ \ldots & \ldots & \ldots & \ldots & \ldots \\ 0 & 0 & \ldots & -X_{n-1} & 1 \\ v & 0 & \ldots & \mathrm{e}^{x_{n,n-1}} & -X_n \end{pmatrix}, \qquad x_{jk} \equiv x_j - x_k, \tag{2.52}$$

is the dual Lax matrix we were looking for.

We leave the proof of the identity (2.46) as an exercise to the reader. For the r-matrix corresponding to the Lax matrix $\mathcal{L}(v)$ see [**23, 18**].

Similarly to the case of the 2×2 matrix $L(u)$, for $\mathcal{L}(v)$ there must also exist a Darboux matrix $\mathcal{M}$ intertwining $\mathcal{L}(v; X, x)$ and $\mathcal{L}(v; Y, y)$. The explicit expression for $\mathcal{M}$, like the one for $\mathcal{L}(v)$, can be found from the Baker-Akhiezer function. Let θ_j and $\tilde{\theta}_j$ refer, respectively, to $\mathcal{L}(v; X, x)$ and $\mathcal{L}(v; Y, y)$. Let us assume that θ_j

and $\tilde{\theta}_j$ are linked by the relation $\theta_j = M_j\tilde{\theta}_j$, which is obviously compatible with (2.48) and (2.32). Expanding $\theta_j = M_j\tilde{\theta}_j$ as

$$\begin{pmatrix}\varphi_j\\ \psi_j\end{pmatrix} = \begin{pmatrix} u-\lambda+\mathrm{e}^{y_j-x_{j-1}} & -\mathrm{e}^{y_j}\\ \mathrm{e}^{-x_{j-1}} & -1\end{pmatrix}\begin{pmatrix}\tilde{\varphi}_j\\ \tilde{\psi}_j\end{pmatrix},$$

taking its first line

$$\varphi_j = (u-\lambda+\mathrm{e}^{y_j-x_{j-1}})\tilde{\varphi}_j - \mathrm{e}^{y_j}\tilde{\psi}_j$$

and substituting $u\tilde{\varphi}_j = \tilde{\varphi}_{j+1} - Y_j\tilde{\varphi}_j + \mathrm{e}^{y_j-y_{j-1}}\tilde{\varphi}_{j-1}$ $\tilde{\psi}_j = \mathrm{e}^{-y_{j-1}}\tilde{\varphi}_{j-1}$ from $\tilde{\theta}_{j+1} = \ell_j(u;Y_j,y_j)\tilde{\theta}_j$, as well as $Y_j = \mathrm{e}^{x_j-y_j} + \mathrm{e}^{y_j-x_{j-1}} - \lambda$ from (2.26b), we obtain, after making the necessary correction for $j=n$ the following result:

$$\varphi_j = \tilde{\varphi}_{j+1} - \mathrm{e}^{x_j-y_j}\tilde{\varphi}_j, \quad j=1,\dots,n-1 \tag{2.53a}$$

$$\varphi_n = v\tilde{\varphi}_1 - \mathrm{e}^{x_n-y_n}\tilde{\varphi}_n \tag{2.53b}$$

or, in matrix form, $\Theta = \mathcal{M}\tilde{\Theta}$, with

$$\mathcal{M}(v) = \begin{pmatrix} -\mathrm{e}^{x_1-y_1} & 1 & \dots & 0 & 0\\ 0 & -\mathrm{e}^{x_2-y_2} & \dots & 0 & 0\\ \dots & \dots & \dots & \dots & \dots\\ 0 & 0 & \dots & -\mathrm{e}^{x_{n-1}-y_{n-1}} & 1\\ v & 0 & \dots & 0 & \mathrm{e}^{x_n-y_n}\end{pmatrix}. \tag{2.54}$$

By construction, we have

$$\mathcal{M}(v)\mathcal{L}(v;Y,y) = \mathcal{L}(v;X,x)\mathcal{M}(v). \tag{2.55}$$

Alternatively, one could introduce $\widetilde{M}_j \sim -M_j^{-1}$,

$$\widetilde{M}_j(u,\lambda) = \begin{pmatrix} 1 & -\mathrm{e}^{y_j}\\ \mathrm{e}^{x_{j-1}} & \lambda-u-\mathrm{e}^{y_j-x_{j-1}}\end{pmatrix},$$

such that

$$\widetilde{M}_{j+1}(u,\lambda)\ell_j(u;X_j,x_j) = \ell_j(u;Y_j,y_j)\widetilde{M}_j(u,\lambda)$$

and repeat the same calculation, starting from $\tilde{\theta}_j = \widetilde{M}_j\theta_j$. The result is $\tilde{\Theta} = \tilde{\mathcal{M}}\Theta$, with

$$\tilde{\mathcal{M}}(v) = \begin{pmatrix} 1 & 0 & \dots & 0 & -v^{-1}\mathrm{e}^{y_1-x_n}\\ -\mathrm{e}^{y_2-x_1} & 1 & \dots & 0 & 0\\ \dots & \dots & \dots & \dots & \dots\\ 0 & 0 & \dots & 1 & 0\\ 0 & 0 & \dots & -\mathrm{e}^{y_n-x_{n-1}} & 1\end{pmatrix} \tag{2.56}$$

satisfying

$$\tilde{\mathcal{M}}(v)\mathcal{L}(v;X,x) = \mathcal{L}(v;Y,y)\tilde{\mathcal{M}}(v). \tag{2.57}$$

Despite the fact that $\tilde{\mathcal{M}} \neq \mathcal{M}^{-1}$ the formulas (2.55) and (2.57) are compatible because of the the remarkable factorization of $\mathcal{L}(v)$:

$$\mathcal{L}(v;X,x) - \lambda\mathbf{1} = \mathcal{M}(v)\tilde{\mathcal{M}}(v), \quad \mathcal{L}(v;Y,y) - \lambda\mathbf{1} = \tilde{\mathcal{M}}(v)\mathcal{M}(v), \tag{2.58}$$

see [**3, 43**] for a discussion of the factorization as a mechanism for generating Bäcklund transformations.

In the above formulas v is, by definition, an eigenvalue of $L(u)$, so the pair (v,u) lies on the spectral curve $\det(v - L(u)) = 0$ of $L(u)$. When dealing with $L(u)$, it is convenient to take u as independent variable, and when dealing with $\mathcal{L}(v)$

respectively v. For the Bäcklund transformation it means in fact swapping the roles of λ and μ: the parameter μ becomes the independent numeric variable instead of λ. All the formulas defining BT remain the same but their interpretation changes: the equality (2.35) becomes a constraint for x and y rather than a definition of μ, whereas λ becomes a dynamical variable—a Lagrange multiplier for the constraint which can be determined from equations (2.26). The respective dual Bäcklund transformation $\tilde{\mathcal{B}}_\mu$ possesses all characteristic properties of BT which can be proven using the Lax matrix $\mathcal{L}(v)$ in the same manner as for $\mathcal{B}_\lambda$, see **[22]** for details.

2.6. General Construction of Bäcklund Transformation. As shown in Section 2.4, to any Bäcklund transformation $\mathcal{B}_\lambda$ there corresponds a Darboux matrix $M(u,\lambda)$ intertwining the corresponding Lax matrices, see formula (2.31). In practice, however, one usually does not know the BT *a priori*, and has to deal with the inverse problem: given $L(u)$ how to find admissible $M(u,\lambda)$ producing a BT. If one is not interested in the Hamiltonian properties of the transformation the usual strategy is to try some ansatz for $M(u,\lambda)$, say, as a low-degree polynomial in u. See the monograph **[25]** for a plentitude of examples.

In this section we shall restrict our attention to the integrable models generated by the quadratic Poisson algebra (2.14) with the SL(2)-invariant r-matrix (2.15), and address the following question: which $M(u,\lambda)$ are admissible, that is, produce canonical mapppings $\mathcal{B}_\lambda$?

Answer: It is sufficient that $M(u,\lambda)$ *as a smooth manifold* coincide with a symplectic leaf of the same quadratic Poisson bracket (2.14) as $L(u)$, the leading coefficients $M^{(m)}$ of $M(u,\lambda)$ and $L^{(n)}$ of $L(u)$ in u commute:

$$[M^{(m)}, L^{(n)}] = 0 \tag{2.59}$$

and also, $M^{(m)}L^{(n)} \neq 0$ (nondegeneracy condition).

Open problem: Are these conditions necessary?

The Bäcklund transformation $\mathcal{B}_\lambda$ constructed for the Toda lattice in Section 2.4 also fits our scheme. Indeed, the Darboux matrix $M(u,\lambda)$ given by (2.33) has, as a smooth manifold, the same structure as the local Lax operator (2.24) for the DST model. The parameter λ is introduced through the shift $u \mapsto u - \lambda$ which is an automorphism of the Poisson algebra. To elaborate, let $M(u,\lambda)$ be

$$M(u,\lambda; S, s) = \ell^{\mathrm{DST}}(u-\lambda; S, s) \equiv \begin{pmatrix} u-\lambda+sS & -s \\ S & -1 \end{pmatrix}. \tag{2.60}$$

As we shall see, the equation (2.31) allows then to determine S, s and, eventually, $L(u; Y, y)$ in terms of $L(u; X, x)$. Expand first (2.31) in powers of u using (2.19) and (2.60). The coefficient at u^{n+1} vanishes because of (2.59). The matrix element 21 of the coefficient at u^n gives the expression for S:

$$S = L_{21}^{(n-1)}(X, x). \tag{2.61}$$

To determine s, take again (2.31) and substitute $u = \lambda$. Multiplying the resulting matrix equality by the row-vector $(1, -s)$ and noting that at $u = \lambda$ the matrix $M(u,\lambda)$ degenerates:

$$M(\lambda,\lambda) = \begin{pmatrix} s \\ 1 \end{pmatrix} \begin{pmatrix} S & -1 \end{pmatrix}, \tag{2.62}$$

we obtain the quadratic equation for s:

$$L_{12}(\lambda; X, x) + s\big(L_{11}(\lambda; X, x) - L_{22}(\lambda; X, x)\big) - s^2 L_{12}(\lambda; X, x) = 0. \tag{2.63}$$

Expressing the variables S and s in terms of X and x one can, in principle, calculate the Poisson brackets for $L(u; Y, y)$ directly and verify that they have the same r-matrix form (2.14) as for $L(u; X, x)$, thus proving the canonicity of the transformation from (X, x) to (Y, y). See [**38**] where it is done in a slightly more general situation. The calculation by brute force, however, is not particularly instructive, and below, following [**39**], we present a quite simple and general proof. The construction we describe mimicks the construction of the quantum Q-operator described in Section (3.4).

Suppose that $M(u)$ is a symplectic leaf of the same Poisson algebra (2.14) as $L(u)$:

$$\{\overset{1}{M}(u), \overset{2}{M}(v)\} = [r_{12}(u-v), \overset{1}{M}(u)\overset{2}{M}(v)] \tag{2.64}$$

satisfying the condition (2.59). Note that $M(u)$ is by no means restricted to $\ell^{\mathrm{DST}}(u - \lambda; S, s)$ as above. Let $M(u)$ be parametrized by the canonical variables (S, s), and $L(u)$, respectively, by (X, x). The matrix $M(u)$ might contain one or more parameters λ which we neglect. Assuming the commutativity (2.59), consider the two products: $M(u)L(u)$ and $L(u)M(u)$. By virtue of the comultiplication property of the bracket (2.14) they both are symplectic leaves of the same bracket. Furthermore, due to the condition $M^{(m)}L^{(n)} = L^{(n)}M^{(m)} \neq 0$, they share the same values of casimirs described in Section 2.3, namely, the leading coefficient and determinant. In a generic situation, provided there is no accidental degeneration, which we shall assume, the equality of casimirs implies an isomorphism of the symplectic leaves, or, in other words, there should exist a canonical transformation $\mathfrak{L} : (X, x; S, s) \to (Y, y, T, t)$, determined from the equation

$$M(u; T, t)L(u; Y, y) = L(u; X, x)M(u; S, s). \tag{2.65}$$

Suppose that the canonical transformation $\mathfrak{L}$ has a generating function $F(t, y; s, x)$

$$X = \frac{\partial F}{\partial x}, \quad Y = -\frac{\partial F}{\partial y}, \quad S = \frac{\partial F}{\partial s}, \quad T = -\frac{\partial F}{\partial t} \tag{2.66}$$

(for simplicity, we omit the indices j in X_j, x_j, etc.)

Let us impose now the constraint

$$t = s, \quad T = S, \tag{2.67}$$

and note that on the constraint surface we have $M(u; T, t) = M(u; S, s)$, and therefore the equality (2.65) is transformed to the Darboux form (2.31). It remains to prove that the transformation $\mathfrak{L}$ remains canonical after being restricted on the constraint surface.

Suppose that one can resolve the equations $\partial F/\partial s + \partial F/\partial t = 0$ with respect to $s \equiv t$ and express X and Y from (2.66) as functions of (x, y).

PROPOSITION. *The resulting transformation $\mathcal{B} : (X, x) \to (Y, y)$ is canonical and is given by the generating function $\Phi(x, y) = F\big(s(x, y), y; s(x, y), x\big)$, such that*

$$X = \frac{\partial \Phi}{\partial x}, \quad Y = -\frac{\partial \Phi}{\partial y}. \tag{2.68}$$

PROOF. Let $|_{st}$ mean the restriction on the constraint manifold $s = t = s(x, y)$. The proof consists of two lines:

$$X = \frac{\partial \Phi}{\partial x} = \left.\frac{\partial F}{\partial x}\right|_{st} + \left.\frac{\partial F}{\partial s}\right|_{st} \frac{\partial s}{\partial x} + \left.\frac{\partial F}{\partial t}\right|_{st} \frac{\partial t}{\partial x} \tag{2.69}$$
$$= \left.\frac{\partial F}{\partial x}\right|_{st} + \frac{\partial s}{\partial x} \left.\left(\frac{\partial F}{\partial s} + \frac{\partial F}{\partial t}\right)\right|_{st}.$$

We observe now that

$$\left.\left(\frac{\partial F}{\partial s} + \frac{\partial F}{\partial t}\right)\right|_{st} = 0 \tag{2.70}$$

due to $S = T$, and, consequently, $X = \partial\Phi/\partial x$. Similarly, one establishes $Y = -\partial\Phi/\partial y$, thus completing the proof. □

In many applications the Lax matrix $L(u)$, as in the Toda case, is a monodromy matrix factorized into the product of local Lax matrices $\ell_j(u)$, see formula (2.9), having the same Poisson brackets (2.14) as $L(u)$:

$$\{\overset{1}{\ell}_i(u), \overset{2}{\ell}_j(v)\} = [r_{12}(u-v), \overset{1}{\ell}_i(u)\overset{2}{\ell}_j(v)]\delta_{ij}. \tag{2.71}$$

The similarity transformation (2.31) is replaced now with a gauge transformation (2.32) which ensures the preservation of the spectral invariants of $L(u)$.

The modification of the reduction procedure described above is quite straightforward. Supposing that $\ell_j(u)$ and $M_j(u)$ depend on local canonical variables we define first the local canonical transformations $\mathfrak{L}^{(i)} : (X_j, x_j; S_j, s_j) \to (Y_j, y_j, T_j, t_j)$ from the equations

$$M_j(u; T_j, t_j)\ell_j(u; Y_j, y_j) = \ell_j(u; X_j, x_j)M_j(u; S_j, s_j). \tag{2.72}$$

Let the corresponding generating functions be $f^{(j)}(t_j, y_j; s_j, x_j)$. Consider the direct product of n phase spaces $(X_j, x_j; S_j, s_j)$ and $(Y_j, y_j; T_j, t_j)$, respectively. The generating function

$$F := \sum_{j=1}^{n} f^{(j)}(t_j, y_j; s_j, x_j) \tag{2.73}$$

then determines the direct product $\mathfrak{L}$ of the local canonical transformations $\mathfrak{L}^{(j)}$.

Let us now impose the constraint

$$t_j = s_{j+1}, \quad T_j = S_{j+1}, \tag{2.74}$$

assuming periodicity $j+n \equiv j$. The proof of the canonicity of the resulting transformation $\mathcal{B} : (X, x) \to (Y, y)$ parallels the proof given previously. It remains to notice that after imposing the constraint (2.74) we have $M_j(u; T_j, t_j) = M_j(u; S_j, s_j)$ and obtain the equality (2.32).

It is convenient to represent the structure of the Bäcklund transformation graphically. Let the local transformation $\mathfrak{L}^{(j)}$ be depicted as a four-legged vertex (see Figure 1), each leg corresponding to a canonical pair like (X, x), etc. The arrows show the direction of the transformation.

The Bäcklund transformation $\mathcal{B}$ is then represented by the Figure 2, where the joint horizontal lines mark the constraints (2.74).

In conclusion to this section, a few general remarks. The proof of canonicity presented above is pretty general using only the comultiplication property of the

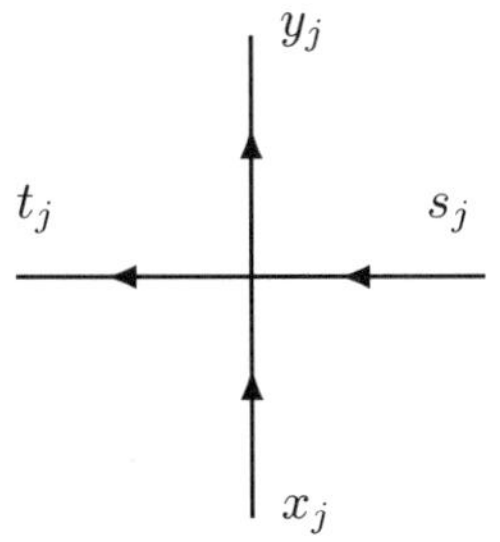

FIGURE 1. Local transformation

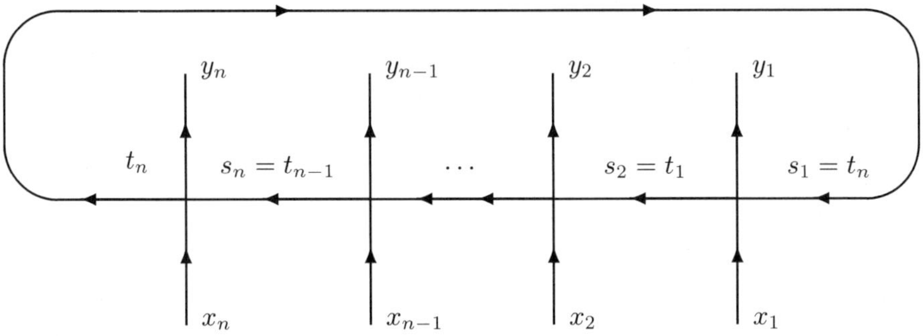

FIGURE 2. Composition of local transformations

quadratic r-matrix Poisson bracket (2.14). It covers thus all integrable models governed by the bracket (2.14) for any r-matrix, not necessarily the SL(2)-invariant one. The only thing one needs is to study the structure of symplectic leaves of the Poisson bracket and to choose some elementary matrices $M(u)$.

Note that the product of two M-matrices produces the composition of the corresponding Bäcklund transformations. Given the conjecture about the factorization of symplectic leaves from Section 2.3 is true, it implies that any BT is decomposable into elementary BTs corresponding to the above Lax matrices $\ell^{\mathrm{XXX}}(u-\lambda)$, $\ell^{\mathrm{DST}}(u-\lambda)$, $\check{\ell}^{\mathrm{DST}}(u-\lambda)$, $\ell^{\mathrm{Toda}}(u-\lambda)$, $\check{\ell}^{\mathrm{Toda}}(u-\lambda)$.

An interesting and as yet unsolved problem is how to deal with the spectrality property of BT within our construction. Our conjecture is that there is a spectrality identity $\det(\mathrm{e}^{\mu} - L(\lambda)) = 0$ with respect to any zero $u = \lambda$ of $\det M(u)$.

2.7. Application to the Toda Lattice. Let us demonstrate how the construction described in the previous section produces the Bäcklund transformation for the Toda lattice described in section 2.4. Substituting into the formula (2.72) the expressions (2.10) for $\ell_j(u)$ and (2.60) for $M_j(u,\lambda)$ we get

$$\begin{pmatrix} u-\lambda+t_jT_j & -t_j \\ T_j & -1 \end{pmatrix} \begin{pmatrix} u+Y_j & -\mathrm{e}^{y_j} \\ \mathrm{e}^{-y_j} & 0 \end{pmatrix} = \begin{pmatrix} u+X_j & -\mathrm{e}^{x_j} \\ \mathrm{e}^{-x_j} & 0 \end{pmatrix} \begin{pmatrix} u-\lambda+s_jS_j & -s_j \\ S_j & -1 \end{pmatrix}. \tag{2.75}$$

The system of equations obtained by equating the coefficients at powers of u has a unique solution:

$$Y_j = -\lambda + e^{x_j} s_j^{-1} + s_j S_j, \tag{2.76a}$$

$$e^{y_j} = s_j, \tag{2.76b}$$

$$T_j = e^{-x_j}, \tag{2.76c}$$

$$t_j = \lambda e^{x_j} - e^{2x_j} s_j^{-1} + e^{x_j} X_j. \tag{2.76d}$$

defining the local transformation $\mathfrak{L}_\lambda^{(j)}$. Strictly speaking, due to the degeneracy of the Lax matrix $\ell^{\mathrm{Toda}}(u)$, the proof of the canonicity of $\mathfrak{L}_\lambda^{(j)}$ given in Section 2.6 does not apply here directly because the transformation (2.76) does not possess a generating function in terms of (t, y, x, s). It is easy, however, to verify the canonicity by a straightforward calculation.

The equalities (2.76b) and (2.76c) allow to resolve the constraint (2.74) yielding $s_j = e^{y_j}$, $S_j = e^{-x_{j-1}}$ which, upon being substituted into (2.76a) and (2.76d), produce exactly the defining relations (2.26) for the Bäcklund transformation studied in the Section 2.4.

EXERCISE. Find what canonical transformations preserving the Hamiltonians of the Toda lattice are generated by the Darboux matrices $M(u) = \operatorname{diag}(1, a)$ and $M(u) = \ell^{\mathrm{Toda}}(u - \lambda)$.

3. Quantization

3.1. Quantum/Classical Correspondence. Here we give only a very brief account of the quantum mechanical notions we are going to use. For more information on the basics of quantum mechanics see any good textbook. See also Reshetikhin's lectures in this volume for references on deformation quantization.

The quantum observables are usually introduced as self-adjoint operators in a Hilbert space. In the limit of the classical Hamiltonian mechanics, as the deformation parameter $\hbar$ (Planck constant) goes to 0, the observables commute, and the next order term in $\hbar$ produces the Poisson bracket of the corresponding classical observables:

$$[\,\cdot\,,\cdot\,] = -i\hbar\{\,\cdot\,,\cdot\,\} + O(\hbar^2).$$

We shall work with the realization of the Hilbert space of quantum states as a space $\mathcal{L}_2(\mathbb{R}^n)$ of square integrable functions of canonical coordinates $(x_1, \dots, x_n)$. The corresponding canonical momenta are then quantized as the differentiation operators $X_j = -i\hbar\partial/\partial x_j$. Generally speaking, any operator Q in $L_2(\mathbb{R}^n)$ can be realized as an integral operator

$$Q : f(x) \mapsto \int dx_1 \dots dx_n \, \mathcal{Q}(y \mid x) f(x) \tag{3.1}$$

with the kernel $\mathcal{Q}(y \mid x)$ which possibly is a generalized function (distribution).

To the canonical transformations in classical mechanics (automorphisms of the Poisson algebra) there correspond in quantum mechanics the automorphisms of the associative operator algebra, that is, similarity transformations $A \mapsto QAQ^{-1}$ with unitary operators Q. The following beautiful formula [**15**]

$$Q(y \mid x) \sim \exp\bigl(i\hbar^{-1} F(y \mid x)\bigr), \quad \hbar \to 0 \tag{3.2}$$

gives the correspondence between the kernel $Q(y \mid x)$ of a unitary transformation and the generating function of the classical canonical transformation into which it turns in the classical limit. The formula (3.2) works for non-unitary transformations as well.

In what follows we shall occasionally use non-self-adjoint and non-unitary operators which corresponds in the classical case to working with complex rather than real manifolds.

3.2. Quantum Toda Lattice. From now on we shall drop $\hbar$ from our formulas assuming $\hbar = 1$. The commutative Hamiltonians of the periodic quantum Toda lattice are differential operators in $\mathcal{L}_2(\mathbb{R}^n)$. They are obtained from exactly the same formulas (2.2), (2.9), (2.10), (2.12) as in the classical case where one should substitute $X_j = -\mathrm{i}\partial_{x_j}$. The proof of their commutativity is based on the algebraic framework called *The Quantum Inverse Scattering Method*; see [**21, 35, 36**] for a detailed exposition of the method.

Starting with the quantum local Lax matrix $\ell_j(u)$

$$\ell_j(u) = \begin{pmatrix} u - \mathrm{i}\partial_{x_j} & -\mathrm{e}^{x_j} \\ \mathrm{e}^{-x_j} & 0 \end{pmatrix}$$

we observe that it satisfies a quadratic commutation relation

$$R_{12}(u_1 - u_2)\overset{1}{\ell}(u_1)\overset{2}{\ell}(u_2) = \overset{2}{\ell}(u_2)\overset{1}{\ell}(u_1)R_{12}(u_1 - u_2) \tag{3.3}$$

with

$$R_{12}(u) = u + \mathrm{i}\mathcal{P}_{12}. \tag{3.4}$$

As its classical counterpart (2.14), the relation (3.3) possesses the comultiplication property which implies that the monodromy matrix $L(u)$ given by (2.9) satifies the same relation

$$R_{12}(u_1 - u_2)\overset{1}{L}(u_1)\overset{2}{L}(u_2) = \overset{2}{L}(u_2)\overset{1}{L}(u_1)R_{12}(u_1 - u_2) \tag{3.5}$$

from which the commutativity of the Hamiltonians,

$$[t(u_1), t(u_2)] = 0,$$

follows immediately (see [**21, 35, 36**] for explanations).

The associative algebra given by the generators $L(u)$ and quadratic relations (3.5) is called yangian $\mathcal{Y}[gl_2]$. Its representations correspond in the classical limit to the symplectic leaves of the quadratic Poisson bracket (2.14). A convenient way of viewing the equality (3.5) is to treat it as a particular form of the quantum Yang-Baxter equation

$$R_{12}(u)R_{13}(u+v)R_{23}(v) = R_{23}(v)R_{13}(u+v)R_{23}(v), \tag{3.6}$$

which is considered as an operator equality in the tensor product $V_1 \otimes V_2 \otimes V_3$ of three linear spaces V_1, V_2 and V_3. Respectively, $R_{jk}(u)$ is an operator in the space $V_j \otimes V_k$ naturally embedded in $V_1 \otimes V_2 \otimes V_3$. For any pair V_j, V_k of the yangian $\mathcal{Y}[gl_2]$ moduli there exists R_{jk} such that the Yang-Baxter equation (3.6) holds for any triplet V_1, V_2, V_3.

In particular, for $V_1 = V_2 = V_3 = \mathbb{C}^2$ we have the YBE (3.6) for the R-matrix (3.4). The relation (3.5) also can be considered as a particular case of (3.6) for $V_1 = V_2 = \mathbb{C}^2$ (auxiliary spaces), $V_3 = \mathcal{L}_2(\mathbb{R}^n)$ (quantum space), $R_{13} = \overset{1}{L}$, $R_{23} = \overset{2}{L}$.

The commutativity of the quantum Hamiltonians being established, the next problem is to find an effective way of determining their joint spectrum. There are two known ways of approaching this problem: separation of variables [**35, 40, 19, 20**] and the Q-operator [**16, 28**]. Here we shall consider the latter approach.

3.3. Properties of the Q-Operator. The original idea of Baxter [**7, 8, 9, 10**] which enabled him to solve the XYZ spin chain was to construct a one-parameter family of operators Q_λ commuting with the Hamiltonians of the model

(3.7) $$[Q_\lambda, t(u)] = 0,$$

and hence sharing with $t(u)$ the common set of eigenvectors. Moreover, Q_λ must satisfy the *Baxter equation*

(3.8) $$Q_\lambda t(\lambda) = \Delta_+(\lambda) Q_{\lambda+\mathrm{i}} + \Delta_-(\lambda) Q_{\lambda-\mathrm{i}}$$

where $\Delta_\pm(\lambda)$ are scalar functions determined by the parameters of the model. Note that in the left-hand-side of (3.8), the order Qt or tQ is not important because of the commutativity (3.7). Applying the Baxter equation (3.8) to a common eigenvector of Q_λ and $t(\lambda)$ one can replace the operators in (3.8) by their eigenvalues. The resulting finite-difference equation of second order for the eigenvalues of Q_λ, considered in an appropriate functional class, allows then to determine the spectrum of $t(\lambda)$.

Baxter succeeded in constructing a Q-operator for the XYZ spin chain as a trace of a monodromy matrix

(3.9) $$Q_\lambda = \mathrm{tr}_V\, \mathbb{L}_n(\lambda) \ldots \mathbb{L}_1(\lambda),$$

constructed with a specially chosen auxiliary space V. Graphically, the structure of Q_λ is represented by the same Figure 2 which we used in the classical case. The horizontal lines correspond to the auxiliary space, the vertical ones—to the quantum space. Each vertex represents an operator $\mathbb{L}_j(\lambda)$. The commutativity (3.7) is then guaranteed by the Yang-Baxter equation, and the only problem is to choose a V which would produce the Baxter equation (3.8). Later, Gaudin and Pasquier [**28**] have constructed a Q-operator for the quantum periodic Toda lattice by giving an explicit expression for its kernel $\mathcal{Q}_\lambda(y \mid x)$ as an integral operator. They have also noticed that the classical limit of the similarity transformation $Q_\lambda(\,\cdot\,)Q_\lambda^{-1}$ is exactly the Bäcklund transformation studied in the previous sections.

Below we reproduce the result by Gaudin and Pasquier. Our approach combines their integral operators technique with Baxter's original idea of constructing a Q-operator as a trace of a monodromy matrix (3.9).

Note that such properties of the Bäcklund transformation as the invariance of Hamiltonians (2.30) and spectrality (2.43) are the classical counterparts of such properties of the Q-operator as, respectively, commutativity (3.7) and the Baxter equation (3.8). The former being obvious, we comment only on the latter. Observing that the shift operators $\lambda \mapsto \lambda \mp \mathrm{i}$ are expressed as $\exp(\mp\mathrm{i}\partial_\lambda) = \exp(\pm\mu)$, where μ is the canonical momentum conjugate to λ, we can rewrite (3.8) in the form $t(\lambda) = \Delta_+(\lambda)\mathrm{e}^{-\mu} + \Delta_-(\lambda)\mathrm{e}^{\mu}$ which gives (2.43) in the classical limit (for the Toda lattice $\Delta_\pm \equiv 1$).

3.4. Q-Operator for the Toda Lattice. We shall construct the Q-operator as the trace of the monodromy matrix (3.9) taking for the auxiliary space V the space $\mathbb{C}[s]$ of polynomials in the variable s. The corresponding representation of the yangian $\mathcal{Y}[gl_2]$ is then realized as the Lax operator of the quantum DST model

$$M(u,\lambda) = \begin{pmatrix} u - \lambda - \mathrm{i}\partial_s & -s \\ -\mathrm{i}\partial_s & -1 \end{pmatrix}, \tag{3.10}$$

compare with the formula (2.60) for the classical case.

To prove the commutativity (3.7) it is sufficient to establish the identity

$$M(u,\lambda)\ell(u)\mathbb{L}_\lambda = \mathbb{L}_\lambda \ell(u) M(u,\lambda) \tag{3.11}$$

which can be considered as a variant of the YBE (3.6) with the following layout of spaces: $V_1 = \mathbb{C}^2$, $V_2 = \mathbb{C}[s]$, $V_3 = \mathcal{L}_2(\mathbb{R}^n)$. We shall use (3.11) as the equation for determining $\mathbb{L}_\lambda$. Rewriting (3.11) as the system of equations for the kernel $\mathbb{L}_\lambda(t, y \mid s, x)$ of $\mathbb{L}_\lambda$,

$$\begin{aligned} &\begin{pmatrix} u - \lambda - \mathrm{i}t\partial_t & -t \\ -\mathrm{i}\partial_t & -1 \end{pmatrix} \begin{pmatrix} u - \mathrm{i}\partial_y & -\mathrm{e}^y \\ \mathrm{e}^{-y} & 0 \end{pmatrix} \mathbb{L}_\lambda(t, y \mid s, x) \\ &\qquad = \begin{pmatrix} u + \mathrm{i}\partial_x & -\mathrm{e}^x \\ \mathrm{e}^{-x} & 0 \end{pmatrix} \begin{pmatrix} u - \lambda + \mathrm{i} + \mathrm{i}s\partial_s & -s \\ \mathrm{i}\partial_s & -1 \end{pmatrix} \mathbb{L}_\lambda(t, y \mid s, x), \end{aligned} \tag{3.12}$$

we obtain a solution, unique up to a scalar factor,

$$\mathbb{L}_\lambda(t, y \mid s, x) \sim \delta(s - \mathrm{e}^y) \exp\bigl(\mathrm{i}t\mathrm{e}^{-x} - \mathrm{i}\mathrm{e}^{x-y} + \mathrm{i}\lambda(x - y)\bigr). \tag{3.13}$$

From (3.9) we get the formula for the kernel of Q_λ:

$$\mathcal{Q}_\lambda(y \mid x) = \int ds_n \cdots \int ds_1 \prod_{j=1}^{n} \mathbb{L}_\lambda(s_{j+1}, y_j \mid s_j, x_j). \tag{3.14}$$

Due to the delta-function factor in (3.13), the integration over s_j in (3.14) reduces to the substitution $s_j = \mathrm{e}^{y_j}$. Finally, we have

$$\mathcal{Q}_\lambda(y \mid x) = \prod_{j=1}^{n} \exp\bigl(\mathrm{i}\mathrm{e}^{y_{j+1} - x_j} - \mathrm{i}\mathrm{e}^{x_j - y_j} + \mathrm{i}\lambda(x_j - y_j)\bigr). \tag{3.15}$$

Note that $\mathcal{Q}_\lambda(y \mid x) = \exp(-\mathrm{i}F_\lambda(y \mid x))$, where $F_\lambda(y \mid x)$ is the generating function (2.29) of the classical BT, that is, the semiclassical formula (3.2) is exact in our case. This is an accidental peculiarity of the Toda lattice which usually does not hold for other models.

In [**28**] another version of the Q-operator is used which differs from (3.15) by the shift $y_j \mapsto y_j + \mathrm{i}\pi/2$:

$$\check{\mathcal{Q}}_\lambda(y \mid x) = \prod_{j=1}^{n} \exp\bigl(-\mathrm{e}^{y_{j+1} - x_j} - \mathrm{e}^{x_j - y_j} + \lambda(x_j - y_j)\bigr) \tag{3.16}$$

which, in operator terms, corresponds to multiplying Q_λ by the factor $\exp(-\pi H_1/2)$. The kernel (3.16) is more convenient for analytical study since it rapidly decreases along the real axis in x_j.

3.5. Baxter's Equation. The commutativity (3.7) being already established, it remains to prove for our Q_λ the Baxter equation (3.8). We reproduce here the proof by Gaudin and Pasquier [**28**] which parallels the proof for the classical case given at the end of Section 2.4, see formula (2.43).

First, note that the kernel (3.15) factorizes as

$$\mathcal{Q}_\lambda(y \mid x) = \prod_{j=1}^{n} w_j(\lambda) \tag{3.17}$$

into factors

$$w_j(\lambda) = \exp\bigl(\mathrm{i}e^{y_j - x_{j-1}} - \mathrm{i}e^{x_j - y_j} + \mathrm{i}\lambda(x_{j-1} - y_j)\bigr). \tag{3.18}$$

Applying then $t(\lambda)$ to the kernel $\mathcal{Q}_\lambda(y \mid x)$ and using (2.9) we observe that each $\ell_j(\lambda; -\mathrm{i}\partial_{y_j}, y_j)$ acts locally only on $w_j(\lambda)$ and obtain

$$t(\lambda)\mathcal{Q}_\lambda(y \mid x) = \operatorname{tr}\bigl(\ell_n(\lambda)w_n(\lambda)\bigr)\dots\bigl(\ell_1(\lambda)w_1(\lambda)\bigr) = Q_\lambda(y \mid x)\operatorname{tr}\tilde{\ell}_n\dots\tilde{\ell}_1, \tag{3.19}$$

where

$$\tilde{\ell}_j \equiv \ell_j(\lambda)\ln w_j(\lambda) = \begin{pmatrix} e^{y_j - x_{j-1}} + e^{x_j - y_j} & -e^{y_j} \\ e^{-y_j} & 0 \end{pmatrix}. \tag{3.20}$$

After that we can use the triangular gauge transformation $\hat{\ell}_j \equiv N_{j+1}^{-1}\tilde{\ell}_j N_j$ with N_j and the resulting matrix $\widehat{\ell}_j$ given by the same formulas (2.42) as in the classical case. Noticing then that

$$\frac{w_j(\lambda + \mathrm{i})}{w_j(\lambda)} = e^{y_j - x_{j-1}}, \qquad \frac{w_j(\lambda - \mathrm{i})}{w_j(\lambda)} = e^{x_{j-1} - y_j} \tag{3.21}$$

we obtain the required result,

$$t(\lambda)Q_\lambda = Q_{\lambda+\mathrm{i}} + Q_{\lambda-\mathrm{i}}.$$

Similarly, for the modified kernel (3.16) one obtains

$$t(\lambda)Q_\lambda = \mathrm{i}^n Q_{\lambda+\mathrm{i}} + \mathrm{i}^{-n} Q_{\lambda-\mathrm{i}}. \tag{3.22}$$

The Toda Hamiltonians $\{H_j\}_{j=1}^n$ enter the Baxter equation (3.22) through the generating function $t(u) = u^n + H_1 u^{n-1} + \dots + H_n$. Their eigenvalues are determined by the condition that the finite-difference equation (3.22) possesses a solution Q_λ which is holomorphic and rapidly decreases along the real axis. For a detailed analysis of the equation (3.22) see [**28, 19, 20**].

4. Conclusion

We have discussed here only the most elementary properties of the Bäcklund transformation and the Q-operator using the sole example of Toda lattice. For further reading see [**11, 12, 19, 20, 22, 23, 24, 29, 30, 40, 41**].

5. Acknowledgments

I am grateful to the University of Montreal and CRM for the hospitality and the opportunity to put together these lectures. My deep thanks are addressed also to my coauthor Vadim Kuznetsov the collaboration with whom provided material for these lectures.

References

1. M. R. Adams, J. Harnad, and J. Hurtubise, *Dual moment maps to loop algebras*, Lett. Math. Phys. **20** (1990), 294–308.
2. M. Adler and P. van Moerbeke, *Algebraic integrable systems: A systematic approach*, Academic Press, Boston, 1989.
3. M. Adler and P. van Moerbeke, *Toda-Darboux maps and vertex operators*, Internat. Math. Res. Notices **10** (1998), 489–511.
4. A. Antonov and B. Feigin, *Quantum group representations and Baxter equation*, Phys. Lett. B **392** (1997), 115-122.
5. V. I. Arnold, *Mathematical methods of classical mechanics*, 2nd edition, Springer, Berlin, 1989.
6. O. Babelon and C-M. Viallet, *Hamiltonian structures and Lax equations*, Phys. Lett. B **237** (1990), 411–416.
7. R. J. Baxter, *Partition function of the eight-vertex lattice model*, Ann. Physics **70** (1972), 193–228.
8. R. J. Baxter, *One-dimensional anisotropic Heisenberg chain*, Ann. Physics **70** (1972), 323–337.
9. R. J. Baxter, *Eight-vertex model in lattice statistics and one-dimensional anisotropic Heisenberg chain. III. Eigenvectors of the transfer matrix and Hamiltonian*, Ann. Physics **76** (1973), 48–71.
10. R. J. Baxter, *Exactly solved models in statistical mechanics*, Academic Press, London, 1983, chapters 9–10.
11. V. Bazhanov, S. Lukyanov, and A. Zamolodchikov, *An integrable structure of conformal field theory. II. Q-operator and DDV equation*, Commun. Math. Phys. **190** (1997), 247–278.
12. S. E. Derkachov, *Baxter's Q-operator for the homogeneous XXX spin chain*, J. Phys. A **32** (1999), 5299–5316.
13. L. D. Faddeev and L. A. Takhtajan, *Hamiltonian methods in the theory of solitons*, Springer, Berlin, 1987.
14. H. Flaschka and D. W. McLaughlin, *Some comments on Bäcklund transformations, canonical transformations, and the inverse scattering method*, Bäcklund Transformations (R. M. Miura, ed.), Lecture Notes in Math., vol. 515, Springer, Berlin, 1976, pp. 253–295.
15. V. A. Fock, *On canonical transformation in classical and quantum mechanics*, Vestnik Leningrad. Univ. Ser. I Mat. Meh. **19** (1959), 67 (Russian).
16. M. Gaudin, *La fonction d'onde de Bethe*, Masson, Paris, 1983.
17. R. Hirota and J. Satsuma, *A simple structure of superposition formula of the Bäcklund transformation*, J. Phys. Soc. Japan **45** (1978), 1741–1750.
18. M. Jimbo, *Quantum R matrix for the generalized Toda system*, Comm. Math. Phys. **102** (1985), 527–547.
19. S. Kharchev and D. Lebedev, *Integral representation for the eigenfunctions of quantum periodic Toda chain*, hep-th/9910265.
20. S. Kharchev and D. Lebedev, *Eigenfunctions of* $\mathrm{GL}(N,\mathbb{R})$ *Toda chain: the Mellin-Barnes representation*, Pis'ma Zh. Èksper. Teoret. Fiz. **71** (2000), 338-343.
21. V. A Korepin, N. M. Bogoliubov, and A. G. Izergin, *Quantum inverse scattering method and correlation functions*, Cambridge University Press, 1993.
22. V. B. Kuznetsov and E. K. Sklyanin, *On Bäcklund transformations for many-body systems*, J. Phys. A **31** (1998), 2241–2251.
23. V. B. Kuznetsov, E. K. Sklyanin, and M. Salerno, *Quantum Bäcklund transformation for DST dimer model*, J. Phys. A **33** (2000), 171–189.
24. V. B. Kuznetsov and P. Vanhaecke, *Bäcklund transformations for finite-dimensional integrable systems: a geometric approach*, nlin/0004003.
25. V. B. Matveev and M. A. Salle, *Darboux transformations and solitons*, Springer, Berlin, 1991.
26. P. van Moerbeke, *The spectrum of Jacobi matrices*, Invent. Math. **37** (1976), 45–81.
27. F. W. Nijhoff and V. Papageorgiou, *Lattice equations associated with the Landau-Lifshitz equations*, Phys. Lett. A **141** (1989), 269–274.
28. V. Pasquier and M. Gaudin, *The periodic Toda chain and a matrix generalization of the Bessel function recursion relations*, J. Phys. A **25** (1992), 5243–5252.

29. G. P. Pron'ko, *On The Baxter's Q-operator for the XXX spin chain*, hep-th/9908179; Commun. Math. Phys. (to appear).
30. G. P. Pron'ko, *On Baxter Q-operators for Toda chain*, nlin.SI/0003002.
31. G. R. W. Quispel, F. W. Nijhoff, H. W. Capel, and J. van der Linden, *Linear integral equations and nonlinear difference-difference equations*, Physica A **125** (1984), 344–380.
32. A. G. Reyman and M. A. Semenov-Tian-Shansky, *Integrable systems* II, chap. 2 in Dynamical Systems VII (V. I. Arnold and S. P. Novikov, eds), Encyclopaedia Math. Sci., vol. 16, Springer, Berlin, 1994.
33. M. A. Semenov-Tian-Shansky, *What is classical r-matrix?*, Funct. Anal. Appl. **17** (1983), 259–272.
34. M. A. Semenov-Tian-Shansky, *Dressing transformations and Poisson-Lie group actions*, Publ. Res. Inst. Math. Sci. **21** (1985), 1203–1221.
35. E. K. Sklyanin, *The quantum Toda chain*, Non-Linear Equations in Classical and Quantum Field Theory (N. Sanchez, ed.), Lecture Notes in Phys., vol. 226, Spinger, Berlin, 1985, pp. 196–233.
36. E. K. Sklyanin, *Quantum inverse scattering method. Selected topics*, Quantum Group and Quantum Integrable Systems (Mo Lin Ge, ed.), Nankai Lectures Math. Phys., World Scientific, Singapore, 1992, pp. 63–97.
37. E. K. Sklyanin, *Separation of variables. New trends*, Progr. Theoret. Phys. Suppl. **118** (1995), 35–60.
38. E. K. Sklyanin, *Canonicity of Bäcklund transformation: r-matrix approach.* I, solv-int/9903016.
39. E. K. Sklyanin, *Canonicity of Bäcklund transformation: r-matrix approach.* II, solv-int/9903017.
40. F. A. Smirnov, *Structure of matrix elements in quantum Toda chain*, math-ph/9805011.
41. F. A. Smirnov, *Dual Baxter equations and quantization of affine Jacobian*, math-ph/0001032.
42. M. Toda, *Theory of nonlinear lattices*, Springer, Berlin, 1981.
43. A. P. Veselov, *Integrable maps*, Russian Math. Surveys **46** (1991), no. 5, 1–51.
44. A. Weinstein, *The local structure of Poisson manifolds*, J. Differential Geom. **18** (1983), 523–557.

Steklov Mathematical Institute at St. Petersburg, Fontanka 27, St. Petersburg 191011, Russia

E-mail address: sklyanin@euclid.pdmi.ras.ru

Centre de Recherches Mathématiques
CRM Proceedings and Lecture Notes
Volume **26**, 2000

Universality of the Distribution Functions of Random Matrix Theory

Craig A. Tracy and Harold Widom

ABSTRACT. We give a brief overview of some recent developments in random matrix theory. The focus is on various scaling limits and the associated limiting distribution functions. These limiting distributions are expressible in terms of solutions to integrable differential equations of the Painlevé type. The universality of these limiting laws is discussed.

1. Random Matrix Models

In probability theory and statistics a common first approximation to many random processes is a sequence $X_1, X_2, X_3, \dots$ of independent and identically distributed (iid) random variables. Let F denote their common distribution. To motivate the material below, we take these random variables and construct a particularly simple $N \times N$ random matrix,

$$\operatorname{diag}\bigl(X_1(\omega), X_2(\omega), \dots, X_N(\omega)\bigr).$$

The order statistics are the eigenvalues ordered

$$\lambda_1 \le \lambda_2 \le \dots \le \lambda_N,$$

and the distribution of the largest eigenvalue, $\lambda_{\max}(N) = \lambda_N$, is

$$\begin{aligned}\operatorname{Prob}(\lambda_{\max}(N) \le x) &= \operatorname{Prob}(X_1 \le x, \dots, X_N \le x)\\ &= F(x)^N.\end{aligned}$$

Since the distribution F is arbitrary, we observe that so too is the distribution of the largest eigenvalue of an $N \times N$ random matrix. However, one is really interested in limiting laws as $N \to \infty$. That is, we ask if there exist constants a_N and b_N such that

$$\frac{\lambda_{\max}(N) - a_N}{b_N} \tag{1.1}$$

1991 *Mathematics Subject Classification.* 60G70, 60G55, 47B99, 34M55.

First-named author supported in part by the National Science Foundation under grant DMS–9802122.

Second-named author supported in part by the National Science Foundation under grant DMS–9732687.

This is the final form of the paper.

converges in distribution to a nontrivial limiting distribution function G. In the present situation a complete answer is provided by

THEOREM 1.1. *If* (1.1) *converges in distribution to some nontrivial distribution function* G, *then* G *belongs to one of the following forms*:

(1) $e^{-e^{-x}}$ *with support* $\mathbf{R}$.
(2) $e^{-1/x^{\alpha}}$ *with support* $[0,\infty)$ *and* $\alpha > 0$.
(3) $e^{-(-x)^{\alpha}}$ *with support* $(-\infty, 0]$ *and* $\alpha > 0$.

This theorem is a model for the type of results we want for nondiagonal random matrices.

A random matrix model is a probability space $(\Omega, \mathcal{P}, \mathcal{F})$ where Ω is a set of matrices. Here are some examples

- Circular Unitary Ensemble (CUE, $\beta = 2$)
  - $\Omega = \mathcal{U}(N) = N \times N$ unitary matrices.
  - $\mathcal{P} =$ Haar measure.
- Gaussian Orthogonal Ensemble (GOE, $\beta = 1$)
  - $\Omega = N \times N$ real symmetric matrices.
  - $\mathcal{P} =$ unique[1] measure that is invariant under orthogonal transformations and the matrix elements (say on and above the diagonal) are iid random variables.
- Gaussian Unitary Ensemble (GUE, $\beta = 2$)
  - $\Omega = N \times N$ hermitian matrices.
  - $\mathcal{P} =$ unique measure that is invariant under unitary transformations and the real and imaginary matrix elements (say on and above the diagonal) are iid random variables.
- Gaussian Symplectic Ensemble (GSE, $\beta = 4$)
  - $\Omega = 2N \times 2N$ Hermitian self-dual matrices.[2]
  - $\mathcal{P}=$ unique measure that is invariant under symplectic transformations and the real and imaginary matrix elements (say on and above the diagonal) are iid random variables.

Expected values of random variables $f : \Omega \to \mathbf{C}$ are computed from the usual formula

$$E_{\Omega}(f) = \int_{\Omega} f(M)\, d\mathcal{P}(M).$$

If $f(M)$ depends only on the eigenvalues of $M \in \Omega$, then one can be more explicit:

- CUE (Weyl's Formula)

$$E_{\mathcal{U}(N)}(f) = \frac{1}{N!(2\pi)^N} \int_{-\pi}^{\pi} \cdots \int_{-\pi}^{\pi} f(\theta_1, \ldots, \theta_N) \times \prod_{1 \le \mu < \nu \le N} \left|\Delta\left(e^{i\theta_1}, \ldots, e^{i\theta_N}\right)\right|^2 d\theta_1 \ldots d\theta_N,$$

- Gaussian Ensembles ($\beta = 1, 2, 4$):

[1]Uniqueness is up to centering and a normalization of the variance.

[2]Identify the $2N \times 2N$ matrix with the $N \times N$ matrix whose entries are quaternions. If the quaternion matrix elements satisfy $\overline{M}_{ji} = M_{ij}$ where the bar is quaternion conjugation, then the $2N \times 2N$ matrix is called Hermitian self-dual. Each eigenvalue of a Hermitian self-dual matrix has multiplicity two.

$$E_{N\beta}(f) = c_{N\beta} \int_{-\infty}^{\infty} \cdots \int_{-\infty}^{\infty} f(x_1, \ldots, x_N) |\Delta(x_1, \ldots, x_N)|^\beta e^{-\beta/2 \sum x_j^2} \, dx_1 \ldots dx_N,$$

where $c_{N\beta}$ is chosen so that $E_{N\beta}(1) = 1$ and $\Delta(x_1, \ldots, x_N) = \prod_{1 \le i < j \le N}(x_i - x_j)$. The factor $e^{-\beta/2 \sum x_j^2}$ explains the choice of the word "gaussian" in the names of these ensembles. A commonly studied generalization of these gaussian measures is to replace the sum of quadratic terms appearing in the exponential with $\sum V(x_i)$ where V is, say, a polynomial (with the obvious restrictions to make the measure well-defined).

Choosing $f = \prod_i (1 - \chi_J(x_i))$, χ_J the characteristic function of a set $J \subset \mathbf{R}$, we get the important quantity[3]

$$E_{N\beta}(f) = E_{N\beta}(0; J) := \text{probability no eigenvalues lie in } J,$$

and in the particular case $J = (t, \infty)$

$$F_{N\beta}(t) := \text{Prob}(\lambda_{\max} \le t) = E_{N\beta}(0, J).$$

The level spacing distribution[4] is expressible in terms of the mixed second partial derivative of $E_{N\beta}\big(0; (a,b)\big)$ with respect to the endpoints a and b.

2. Fredholm Determinant Representations

Though $E_{N\beta}(0; J)$ are explicit N-dimensional integrals, these expressions are not so useful in establishing limiting laws as $N \to \infty$. What turned out to be very useful are Fredholm determinant representations for $E_{N\beta}(0; J)$. In 1961 M. Gaudin proved for $\beta = 2$ (using the newly developed orthogonal polynomial method of M. L. Mehta) that $E_{N2}(0; J) = \det(I - K_{N2})$ where K_{N2} is an integral operator acting on J whose kernel is of the form

$$\frac{\varphi(x)\psi(y) - \psi(x)\varphi(y)}{x - y}, \tag{2.1}$$

with $\varphi(x) = c_N e^{-x^2/2} H_N(x)$, $\psi(x) = c_N e^{-x^2/2} H_{N-1}(x)$, and $H_j(x)$ are the Hermite polynomials.[5] For $\beta = 1$ or 4, generalizing F. J. Dyson's 1970 analysis of the n-point correlations for the circular ensembles, it follows from work by Mehta the following year that the square of $E_{N\beta}(0; J)$ again equals a Fredholm determinant, $\det(I - K_{N\beta})$, but now the kernel of $K_{N\beta}$ is a 2×2 matrix.[6]

3. Scaling Limits (Limiting Laws)

3.1. Bulk Scaling Limit. Let $\rho_N(x)$ denote the density of eigenvalues at x and pick a point x_0, independent of N with $\rho_N(x_0) > 0$. We scale distances so that resulting density is one at x_0, $\xi := \rho_N(x_0)(x - x_0)$, and we call the limit

$$N \to \infty, \quad x \to x_0, \quad \text{such that } \xi \text{ is fixed},$$

[3] This quantity has an obvious extension to other random matrix models.

[4] Let the eigenvalues be ordered. The conditional probability that given an eigenvalue at a, the next one lies between s and $s + ds$ is called the level spacing density.

[5] In the case of random matrix models corresponding to general potential V, $\varphi(x) = c_N e^{-V(x)/2} p_N(x)$ and $\psi(x) = c_N e^{-V(x)/2} p_{N-1}(x)$ where $p_j(x)$ are the orthogonal polynomials associated with weight function $w(x) = e^{-V(x)}$. It is in this generalization that we see the close relation between the general theory of orthogonal polynomials and random matrix theory.

[6] See **[37]** for elementary proofs of these facts.

the bulk scaling limit. For $\beta = 2$,

$$E_{N2}(0; J) \to E_2(0; J) = \det(I - K_2)$$

where the integral operator K_2 (acting on $L^2(J)$) has as its kernel (the sine kernel)

$$\frac{1}{\pi} \frac{\sin \pi(\xi - \xi')}{\xi - \xi'}.$$

(We use the same symbol J to denote the scaled set J.) Furthermore,

$$p_2(s) = -\frac{d^2}{ds^2} E_2\big(0; (0, s)\big)$$

is the (limiting) level-spacing density for GUE; known as the Gaudin distribution.[7] We observe that the limiting kernel is translationally invariant and independent of x_0.

3.2. Edge Scaling Limit. In the gaussian ensembles, the density decays exponentially fast around $2\sigma\sqrt{N}$; perhaps surprisingly, it is also the case that

$$\lim_{N\to\infty} \frac{\lambda_{\max}(N)}{\sqrt{N}} = 2\sigma, \quad \text{a.s.} \tag{3.1}$$

where σ is the standard deviation of the off-diagonal matrix elements. (In the normalization we've adopted, $\sigma = 1/\sqrt{2}$.) If we introduce the scaled random variable $\hat{\lambda}$ through

$$\lambda_{\max} = 2\sigma\sqrt{N} + \frac{\sigma\hat{\lambda}}{N^{1/6}},$$

then

$$\text{Prob}(\lambda_{\max} \le t) = \text{Prob}(\hat{\lambda} \le s) \to F_\beta(s) \quad \text{as } N \to \infty,$$

where $t = 2\sigma\sqrt{N} + \sigma s/N^{1/6}$. For $\beta = 2$,

$$F_2(s) = \det(I - K_{\text{Airy}}),$$

where K_{Airy} has kernel of the form (2.1) with $\varphi(x) = \text{Ai}(x)$, $\psi(x) = \text{Ai}'(x)$ and $J = (s, \infty)$. (See, e.g., [**37**] for the $\beta = 1, 4$ results.)

4. Connections with Integrable Systems

4.1. Bulk Scaling Limit. In 1980 M. Jimbo, T. Miwa, Y. Môri, and M. Sato [**21**] expressed the Fredholm determinant of the sine kernel in terms of a solution to a certain system of integrable differential equations.[8] In the simplest case of a single interval, $J = (0, s)$, the differential equation is a particular case of Painlevé V (P_V)[9] and the Fredholm determinant is given by

$$\det(I - \lambda K_2) = \exp\left(\int_0^{\pi s} \frac{\sigma(x; \lambda)}{x}\, dx\right),$$

$$(x\sigma'')^2 + 4(x\sigma' - \sigma)(x\sigma' - \sigma + (\sigma')^2) = 0,$$

$$\sigma(x; \lambda) \sim -\frac{\lambda}{\pi} x, \quad \text{as } x \to 0.$$

---

[7] For the analogous $\beta = 1, 4$ results, see, e.g., [**27**] or [**37**].

[8] A simplified proof of their results can be found in [**38**].

[9] The differential equation below is the sigma representation of P_V.

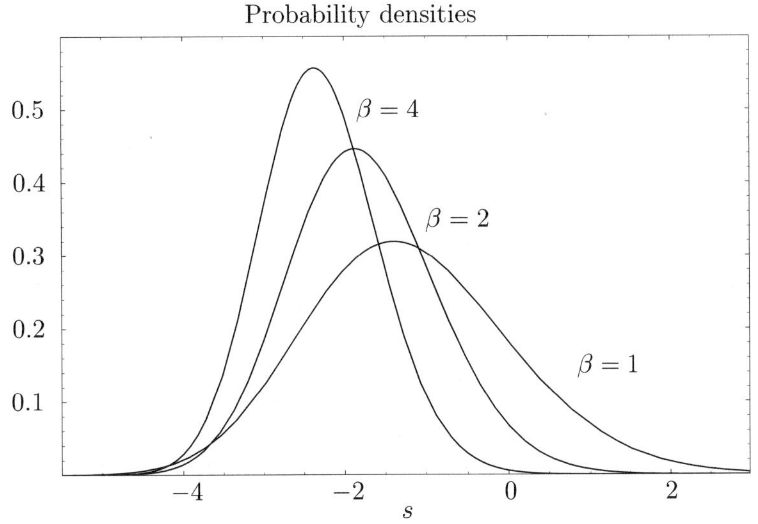

FIGURE 1. Densities for the scaled largest eigenvalues, $f_\beta(s)$.

For $\beta = 1, 4$ and $J = (0, s)$, $E_\beta\big(0; (0, s)\big)$ can also be expressed in terms of the same function $\sigma(x; 1)$. A down-to-earth application of these Painlevé representations (and using the known asymptotics of $\sigma(x; 1)$) is that one can easily produce graphs of the level spacing densities $p_\beta(s)$.[10]

4.2. Edge Scaling Limit. The limiting distributions (edge scaling) of the largest eigenvalue, $F_\beta(s)$, can also be expressed in terms of Painlevé functions—this time P_{II} [**39, 40**]:

$$F_1(s)^2 = \exp\Big(-\int_s^\infty q(x)\,dx\Big) F_2(s), \tag{4.1}$$

$$F_2(s) = \exp\Big(-\int_s^\infty (x-s) q(x)^2\,dx\Big), \tag{4.2}$$

$$F_4(s/\sqrt{2})^2 = \cosh^2\Big(\frac{1}{2}\int_s^\infty q(x)\,dx\Big) F_2(s), \tag{4.3}$$

where q satisfies the Painlevé II equation

$$q'' = xq + 2q^3$$

with boundary condition $q(x) \sim \mathrm{Ai}(x)$ as $x \to \infty$.[11] The graphs of the densities $f_\beta(s) = dF_\beta(s)/ds$ are in Figure 1.

[10] Without the Painlevé representations, the numerical evaluation of the Fredholm determinants is quite involved.

[11] That a solution q exists and is unique follows from the representation of the Fredholm determinant in terms of it. Independent proofs of this, as well as the asymptotics as $x \to -\infty$, were given by S. Hastings and J. McLeod, P. Clarkson and McLeod and by P. Deift and X. Zhou.

4.3. Generalizations. Both the sine kernel and the Airy kernel are of the form (2.1). Kernels of this form arise in many problems in integrable systems; indeed, so much so that A. Its, A. Izergin, V. Korepin and V. Slavnov [**20**] in 1990 initiated a general analysis of these kernels. The following theorem [**41**], which applies to a wide class of $\beta = 2$ random matrix ensembles, gives the general situation:

THEOREM 4.1. *Let* $J = \bigcup_{j=1}^{m}(a_{2j-1} - a_{2j})$ *be a union of open intervals. Define* $\tau(a) = \det(I - K)$ *where* K *is an integral operator acting on* $L^2(J)$ *whose kernel is of the form* (2.1) *where* φ *and* ψ *are assumed to satisfy*

$$\frac{d}{dx}\begin{pmatrix}\varphi\\ \psi\end{pmatrix} = \Omega(x)\begin{pmatrix}\varphi\\ \psi\end{pmatrix}$$

with $\Omega(x)$ *a* 2×2 *matrix with zero trace and rational entries in* x.

Then $\partial/\partial a_j \log\det(I - K)$ *are expressible polynomially in terms of solutions to a total system of partial differential equations* (a_j *are the independent variables*). *The differential equations are given explicitly in terms of the coefficients of the rational functions appearing in* $\Omega(x)$.

4.4. Historical Comments. The first connection between Toeplitz/Fredholm determinants and Painlevé functions was established in 1973–77 in work of T. T. Wu, B. M. McCoy, E. Barouch and the first author concerning the scaling limit of the 2-point functions of the 2D Ising model of statistical mechanics. The Painlevé function that arose was P_{III}. This work was subsequently generalized by Sato, Miwa and Jimbo to n-point functions and, more generally, holonomic quantum fields. The Kyoto School then took up the problem of the density matrix of the impenetrable Bose gas and it was in this context that they discovered that the Fredholm determinant of the sine kernel is related to P_V.

A crucial simplification of the Kyoto School work, as it applies to random matrix theory, was made by Mehta in 1992 [**28**]. This last work inspired the commutator methods introduced by the present authors in the period 1993–96. Since then both Riemann-Hilbert methods of Deift, Its, Zhou and others (see, e.g. [**15**]); and Virasoro methods of M. Adler, T. Shiota, P. van Moerbeke, and others (see, e.g. [**1**]), have played an increasingly important role in the development of random matrix theory. The connection of these methods with the isomonodromy method has been clarified by J. Palmer [**32**] and J. Harnad [**19**].

Space does not permit us to discuss the interesting connections between random matrices and Szegö type limit theorems. See E. Basor [**7**] for connections with linear statistics and the review papers [**8, 42**] for some related historical comments.

5. Universality

5.1. Universality of Gaussian Ensembles in Random Matrix Models.

5.1.1. *Invariant Measures,* $\beta = 2$. As briefly mentioned above, a widely studied class of random matrix models is defined by the replacement of the gaussian potential, x^2, by the general potential $V(x)$. For the weight functions most studied, the parameter N is put into the exponent so that the weight function becomes $e^{-NV(x)}$. For different V's, the limiting density $\rho_V(x)$ can be quite different. It may be supported on many distinct intervals, and it may vanish at interior points of its support. In the gaussian case, the limiting density is the Wigner semicircle law: $\rho_W(x) = 2/\pi\sqrt{1-x^2}$. Heuristic arguments suggest that the behavior exhibited by the Wigner law—that ρ is positive on the interior of its support and vanishes like a square root at endpoints—is the typical behavior for ρ_V. The bulk scaling limit and edge scaling limit are defined in analogous ways to the gaussian cases. To establish universality of these scaling limits, one must show (for $\beta = 2$ ensembles) that the scaled kernels approach the sine kernel and the Airy kernel, respectively. The potential $V(x) = (t/2)x^2 + (g/4)x^4$ ($g > 0$, $t < 0$) is an example of a "two interval" potential. Indeed, for this important example P. Bleher and A. Its [**9**] proved precisely this statement of universality. (See their paper for related work in the orthogonal polynomial literature as well as the physics literature.) Recently, building on work of [**16**], A. Kuijlaars and K. McLaughlin [**26**] have shown this behavior is generic for real analytic V satisfying $\lim_{x\to\infty} V(x)/\log|x| = +\infty$.

In the physics literature, M. Bowick, E. Brézin [**12**] and others have argued (for $\beta = 2$ ensembles) that if ρ_V vanishes faster than a square root, then the corresponding edge scaling limit will result in nonAiry universality classes. The resulting new kernels will have form (2.1) and the theory developed in [**41**] will apply, but there remains much to be understood in these cases.

For $\beta = 1, 4$, the situation is more complicated due to the structure of K_V [**37, 45**], and the "universality" theorems are not so general.

5.2. Noninvariant Measures: Wigner Ensemble. The Wigner ensembles are defined by requiring that the matrix elements on and above the diagonal in either the real symmetric case or the complex hermitian case are independent and identically distributed random variables. It is only in the case when the distribution is gaussian is the measure invariant. One usually assumes, as we do here, that all moments of the common distribution function exist. It was Wigner himself who showed that the limiting density of states is the Wigner semicircle. Subsequently several authors—culminating in a theorem by Z. Bai and Y. Lin clarifying which moments need exist—showed (3.1) continues to hold for the Wigner ensembles.

It should be noted that because the measure is noninvariant, the nongaussian Wigner ensembles do not, as far as we understand, have Fredholm determinant representation for their distribution functions. This means, for one, that the methods of integrable systems are not directly applicable to Wigner ensembles. It is therefore particularly important, as A. Soshnikov [**36**] recently proved, that in the edge scaling limit the Wigner ensembles are in the same universality class as the gaussian models. In particular, the limiting distribution of the scaled largest eigenvalue is given by $F_1(s)$ for real symmetric Wigner matrices and by $F_2(s)$ for complex hermitian Wigner matrices.

5.3. Examples from Physics. A second type of universality, and the one first envisioned by Wigner in the context of nuclear physics, asserts in Wigner's words [**46**]

> Let me say only one more word. It is very likely that the curve in Figure I [an approximate graph of $p_1(s)$] is a universal function. In other words, it doesn't depend on the details of the model with which you are working.

The modern version of this asserts that for a classical, "fully" chaotic Hamiltonian the corresponding quantum system has a level spacing distribution equal to $p_\beta(s)$ in the bulk. (The symmetry class determines which ensemble.) This quantum chaos conjecture, due to O. Bohigas, M. Giannoni and C. Schmit [**10**], has been a guiding principle for much subsequent work, though it is the authors' understanding that it remains a conjecture. A particularly nice numerical example supporting this conjecture is M. Robnik's work [**35**] on chaotic billards. The reader is referred to the recent review article [**18**] for further numerical examples that support this conjecture. It should be noted that there are examples from number theory where the conjecture fails. Thus, as it has been said, the conjecture is undoubtedly true except where it is demonstrably false.

5.3.1. *Aperiodic Tiling Adjacency Matrix.* The discovery of quasicrystals has made the study of statistical mechanical models whose underlying lattice is quasiperiodic of considerable interest to physicists. In particular, in order to understand transport properties, tight binding models have been defined on various quasiperiodic lattices. One such study by Zhong *et al.* [**47**] defined a simplified tight binding model for the octagonal tiling of Ammann and Beenker. This quasiperiodic tiling consists of squares and rhombi with all edges of equal lengths (see Figure 2) and has a D_8 symmetry around the central vertex. On this tiling the authors take as their Hamiltonian the adjacency matrix for the graph with free boundary conditions. The largest lattice they consider has 157,369 vertices. The matrix splits into ten blocks according to the irreducible representations of the dihedral group D_8. For each of these ten independent subspectra, they compare the empirical distribution of the normalized spacings of the consecutive eigenvalues with the GOE level spacing density $p_1(s)$. In Figure 2 we have reproduced a portion of their data for one such subspectrum together with p_1.

5.4. Spacings of the Consecutive Zeros of Zeta Functions. Perhaps the most surprising appearance of the distributions of random matrix theory is in number theory. Analytical work by H. Montgomery and extensive numerical calculations by A. Odlyzko on the zeros of the Riemann zeta function have given convincing evidence that the normalized consecutive spacings follow the Gaudin distribution, see Figure 3. Recent results of Z. Rudnick and P. Sarnak are also compatible with the belief that the distribution of the spacings between zeros, not only of the Riemann zeta function, but also of quite general automorphic L-functions over $\mathbf{Q}$, are all given by this Montgomery-Odlyzko Law. In their landmark book [**25**], N. Katz and P. Sarnak establish the Montgomery-Odlyzko Law for wide classes of zeta and L-functions over finite fields.

5.5. Random Matrix Theory and Combinatorics. The last decade has seen a flurry of activity centering around connections between combinatorial probability of the Robinson-Schensted-Knuth (RSK) type on the one hand and random matrices and integrable systems on the other. From the point of view of probability theory, the quite surprising feature of these developments is that the methods came from Toeplitz determinants, integrable differential equations of the Painlevé type and the closely related Riemann-Hilbert techniques as they were applied and refined in random matrix theory. Using these techniques new, and apparently quite universal, limiting laws have been discovered. The earliest signs of these connections can be found in the work of A. Regev [**34**] and I. Gessel [**17**]. Here, however, we introduce this subject by examining a certain card game of D. Aldous and P. Diaconis [**3**], called patience sorting.

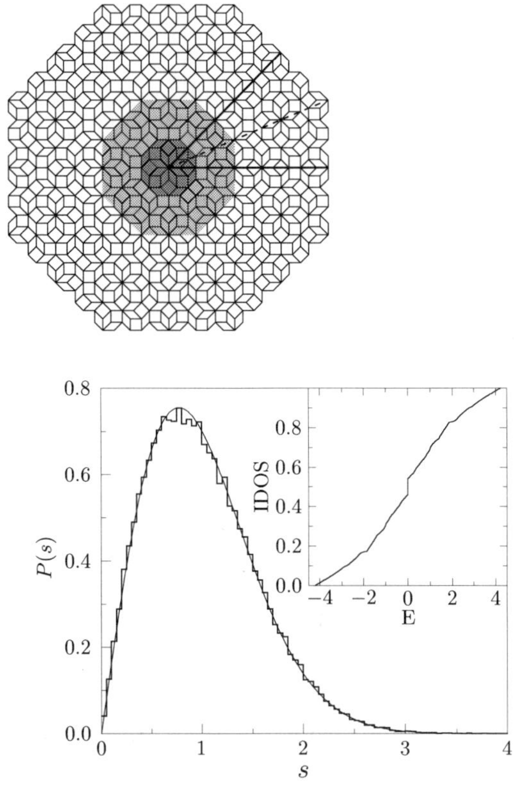

FIGURE 2. Data for nearest neighbor normalized spacings of eigenvalues of the adjacency matrix for a quasiperiodic octagonal tiling are plotted together with the GOE level spacing density $p_1(s)$. Data are from one independent subspectrum of a D_8-symmetric octagonal patch of a tiling with 157,369 vertices. Courtesy of Zhong *et al.* [**47**].

5.5.1. *Patience Sorting and Random Permutations.* Our deck of cards is labeled $\{1, 2, \ldots, N\}$ and we order the cards with their natural ordering. Shuffle the deck of cards and

- Turn over the first card.
- Turn over the second card. If it is of higher rank, start a new pile to the right of the first card. Otherwise place the second card on top of the first card.
- Turn over the third card. If it is of higher rank than either the first or the second card, start a new pile to the right of the second card. Otherwise place the third card on top of the card of higher rank. If both first and second are of higher rank, place the third card on the smaller ranked card. (That is, play cards as far as possible to the left.)

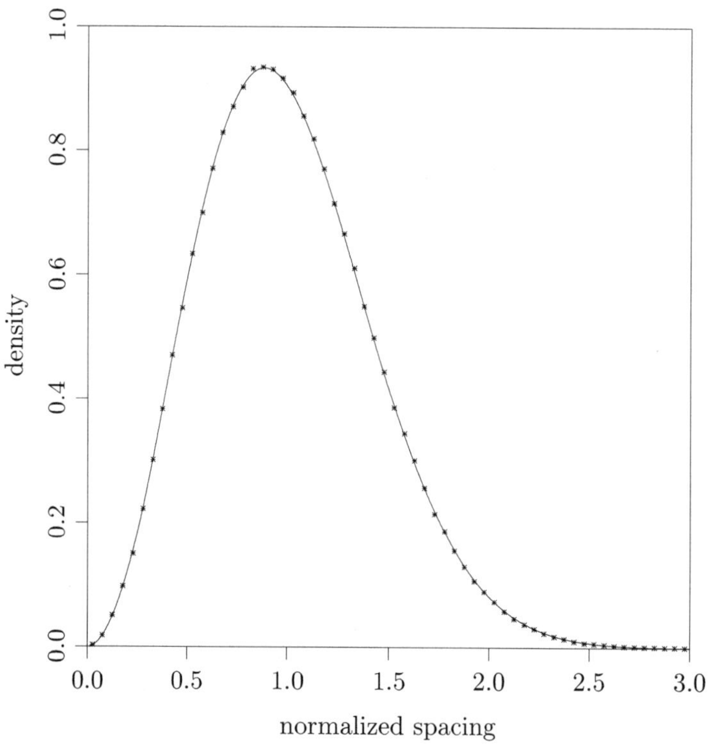

FIGURE 3. Data for nearest neighbor spacings among 1,041,600 zeros of the Riemann zeta function near the 2×10^{20}-th zero are plotted together with the GUE spacing density. Courtesy of Andrew Odlyzko [**29**].

- Continue playing the game, playing cards as far left as possible, until all the cards are turned over.

The object of the game is to end with a small number of piles. Let $\ell_N(\sigma)$ equal the number of piles at the end of the game where we started with deck $\sigma = \{i_1, i_2, \ldots, i_N\}$. Clearly, $1 \le \ell_N(\sigma) \le N$, but what are some typical values for a shuffled deck? Starting each time with a newly shuffled deck of $N = 52$ cards, the computer played patience sorting 100,000 times. Here are the statistics for ℓ_{52}:

- Mean = 11.56 (11.00).
- Standard Deviation = 1.37 (1.74)
- Skewness = 0.33 (0.22)
- Kurtosis Excess = 0.16 (0.09)
- Sample Range = 7 to 19 (Probability 0.993)

where the numbers in parentheses are the asymptotic predictions (as the number of cards tends to infinity) of the theory of J. Baik, P. Deift and K. Johansson [**4**] to be described below.

A shuffled deck of cards, $\sigma = \{i_1, i_2, \ldots, i_N\}$, is a permutation of $\{1, 2, \ldots, n\}$, and so we think of the shuffled deck as a random permutation. A moment's reflection will convince the reader that $\ell_N(\sigma)$ is equal to the length of the longest increasing subsequence in the permutation σ. As a problem in random permutations, determining the asymptotics of $E(\ell_N)$ as $N \to \infty$ is called Ulam's Problem. In the 1970's A. Vershik and S. Kerov and independently B. Logan and L. Shepp showed $E(\ell_N) \sim 2\sqrt{N}$ with important earlier work by J. Hammersley. Hammersley's analysis introduced a certain interacting particle system interpretation. This was developed by Aldous and Diaconis [**2**] who in 1995 gave a "soft" proof of this result using hydrodynamic scaling arguments from interacting particle theory.

Introducing the exponential generating function

$$\sum_{N\geq 0} \mathrm{Prob}(\ell_N \leq n)\frac{t^N}{N!},$$

Gessel showed that it is equal to $D_n(t)$, the determinant of the $n \times n$ Toeplitz determinant with symbol $e^{\sqrt{t}(z+z^{-1})}$. (Recall that the i, j entry of a Toeplitz matrix equals the $i - j$ Fourier coefficient of its symbol.) It is in this work of Gessel and subsequent work of Odlyzko *et al.* [**30**] and E. Rains [**33**], that the methods of random matrix theory first appear in RSK type problems.[12]

Starting with this Toeplitz determinant representation, Baik, Deift and Johansson [**4**], using the steepest descent method for Riemann-Hilbert problems [**14**], derived a delicate asymptotic formula for $D_n(t)$ which we now describe. Introduce another parameter s and suppose that n and t are related by $n = [2t^{1/2} + st^{1/6}]$. Then as $t \to \infty$ with s fixed one has

$$\lim_{t\to\infty} e^{-t} D_n(t) = F_2(s)$$

where $F_2(s)$ is the distribution function (4.2). Using a dePoissonization lemma due to Johansson [**22**], these asymptotics led Baik, Deift and Johansson to the limiting law

$$\lim_{N\to\infty} \mathrm{Prob}\left(\frac{\ell_N - 2\sqrt{N}}{N^{1/6}} < s\right) = F_2(s).$$

Since the work of Baik, Deift and Johansson, several groups have extended this connection between RSK type combinatorics and the distribution functions of random matrix theory. The aforementioned result is equivalent to the determination of the limiting distribution of the number of boxes in the first row in the RSK correspondence $\sigma \leftrightarrow (P, Q)$. In [**5**] the same authors show that the limiting distribution of the number of boxes in the *second* row is (when centered and normalized) distributed as the *second* largest scaled eigenvalue in GUE [**37**]. They then conjectured that this correspondence extends to all rows. This conjecture was recently proved by A. Okounkov [**31**] using topological methods and by A. Borodin, A. Okounkov and G. Olshanski [**13**] and Johansson [**24**] using analytical methods.

Placing restrictions on the permutations σ (that they be fixed point free and involutions), Baik and Rains [**6**] have shown that the limiting laws for the length of the longest increasing/decreasing subsequence are now the limiting distributions F_1 and F_4 for the scaled largest eigenvalue in GOE and GSE, see (4.1) and (4.3). Generalizing to signed permutations and colored permutations the present authors

[12]Gessel [**17**] does not mention random matrices, but in light of well-known formulas in random matrix theory relating Toeplitz determinants to expectations over the unitary group, we believe it is fair to say that the connection with random matrix theory begins with this discovery. See, however, Regev [**34**].

and Borodin [**43, 11**] showed that the distribution functions of the length of the longest increasing subsequence involve the same F_2.

Johansson [**23**] showed that the shape fluctuations of a certain random growth model, again appropriately scaled, converges in distribution to F_2. (This random growth model is intimately related to certain randomly growing Young diagrams.) In subsequent work, Johansson [**24**] showed that the fluctuations in certain random tiling problems (related to the Arctic Circle Theorem) are again described by F_2. Finally, Johansson [**24**] and the present authors [**44**] have considered analogous problems for random *words* and have discovered various random matrix theory connections.

Acknowledgments

The authors have benefited from conversations with A. Its and it is a pleasure to acknowledge this. The first author thanks J. Harnad and P. Winternitz for their invitation to speak at the workshop *Integrable Systems: From Classical to Quantum*.

References

1. M. Adler, T. Shiota, and P. van Moerbeke, *Random matrices, Virasoro algebras, and noncommutative KP*, Duke Math. J. **94** (1998), 379–431.
2. D. Aldous and P. Diaconis, *Hammersley's interacting particle process and longest increasing subsequences*, Probab. Theory Related Fields **103** (1995), 199–213.
3. D. Aldous and P. Diaconis, *Longest increasing subsequences: from patience sorting to the Baik-Deift-Johansson theorem*, preprint.
4. J. Baik, P. Deift, and K. Johansson, *On the distribution of the length of the longest increasing subsequence of random permutations*, preprint (math.CO/9810105).
5. J. Baik, P. Deift, and K. Johansson, *On the distribution of the length of the second row of a Young diagram under Plancherel measure*, preprint (math.CO/9901118).
6. J. Baik and E. M. Rains, *The asymptotics of monotone subsequences of involutions*, preprint (math.CO/9905084).
7. E. L. Basor, *Distribution functions for random variables for ensembles of positive Hermitian matrices*, Comm. Math. Phys. **188** (1997), 327–350.
8. E. L. Basor, *Connections between random matrices and Szegö limit theorems*, Spectral Problems in Geometry and Arithmetic (T. Branson, ed.), Amer. Math. Soc., Providence, 1999, pp. 1–7.
9. P. Bleher and A. Its, *Semiclassical asymptotics of orthogonal polynomials, Riemann-Hilbert problem, and universality in the matrix model*, Ann. of Math. **150** (1999), 185–266.
10. O. Bohigas, M. J. Giannoni, and C. Schmit, *Characterization of chaotic quantum spectra and universality of level fluctuation laws*, Phys. Rev. Lett. **52** (1984), 1–4.
11. A. Borodin, *Longest increasing subsequences of random colored permutations*, Electron. J. Combin., **6**(1) (1999), # R12.
12. M. J. Bowick and E. Brézin, *Universal scaling of the tail of the density of eigenvalues in random matrix models*, Phys. Lett. B **268** (1991), 21–28.
13. A. Borodin, A. Okounkov, and G. Olshanski, *Asymptotics of Plancherel measures for symmetric groups*, preprint (math.CO/9905032).
14. P. A. Deift and X. Zhou, *A steepest descent method for oscillatory Riemann-Hilbert problems: Asymptotics for the MKdV equation*, Ann. of Math. **137** (1993), 295–368.
15. P. A. Deift, A. R. Its, and X. Zhou, *A Riemann-Hilbert approach to asymtptotic problems arising in the theory of random matrix models, and also in the theory of integrable statistical mechanics*, Ann. of Math. **146** (1997), 149–235.
16. P. Deift, T. Kriecherbauer, K. T-R. McLaughlin, S. Venakides, and X. Zhou, *Uniform asymptotics for polynomials orthogonal with respect to varying exponential weights and applications to universality questions in random matrix theory*, preprint.

17. I. M. Gessel, *Symmetric functions and P-recursiveness*, J. Combin. Theory Ser. A **53** (1990), 257–285.
18. T. Guhr, A. Mueller-Groeling, and H. A. Weidenmueller, *Random-matrix theories in quantum physics: common concepts*, Phys. Rep. **299** (1998), 189–425.
19. J. Harnad, *On the bilinear equations for Fredholm determinants appearing in random matrices*, preprint (solv-int/9906004).
20. A. R. Its, A. G. Izergin, V. E. Korepin, and N. A. Slavnov, *Differential equations for quantum correlation functions*, Internat. J. Modern Phys. B **4** (1990), 1003–1037.
21. M. Jimbo, T. Miwa, Y. Môri, and M. Sato, *Density matrix of an impenetrable Bose gas and the fifth Painlevé transcendent*, Physica D **1** (1980), 80–158.
22. K. Johansson, *The longest increasing subsequence in a random permutation and a unitary random matrix model*, Math. Res. Lett. **5** (1998), 63–82.
23. K. Johansson, *Shape fluctuations and random matrices*, Comm. Math. Phys. **209** (2000), 437–476.
24. K. Johansson, *Discrete orthogonal polynomial ensembles and the Plancherel measure*, preprint (math.CO/9906120).
25. N. M. Katz and P. Sarnak, *Random matrices, Frobenius eigenvalues, and monodromy*, Amer. Math. Soc., Providence, 1999.
26. A. B. J. Kuijlaars and K. T-R. McLaughlin, *Generic behavior of the density of states in random matrix theory and equilibrium problems in the presence of real analytic external fields*, to appear in Comm. Pure Appl. Math.
27. M. L. Mehta, *Random matrices*, 2nd ed., Academic Press, San Diego, 1991.
28. M. L. Mehta, *A non-linear differential equation and a Fredholm determinant*, J. Physique I **2** (1992), 1721–1729.
29. A. M. Odlyzko, *On the distribution of spacings between zeros of the zeta function*, Math. Comp. **48** (1987), 273–308.
30. A. M. Odlyzko, B. Poonen, H. Widom, and H. S. Wilf, *On the distribution of longest increasing subsequences in random permutations*, unpublished notes.
31. A. Okounkov, *Random matrices and random permutations*, preprint (math.-CO/9903176).
32. J. Palmer, *Deformation analysis of matrix models*, Physica D **78** (1994), 166–185.
33. E. M. Rains, *Increasing subsequences and the classical groups*, Electron. J. Combin. **5** (1998), #R12.
34. A. Regev, *Asymptotic values for degrees associated with strips of Young diagrams*, Adv. Math. **41** (1981), 115–136.
35. M. Robnik, *Improved energy level statistics for a family of billiards with analytic boundaries*, J. Phys. A **25** (1992), 3593–3602.
36. A. Soshnikov, *Universality at the edge of the spectrum in Wigner random matrices*, Comm. Math. Phys. **207** (1999), 697–733.
37. C. A. Tracy and H. Widom, *Correlation functions, cluster functions and spacing distributions for random matrices*, J. Statist. Phys. **92** (1998), 809–835.
38. C. A. Tracy and H. Widom, *Introduction to random matrices*, Geometric and Quantum Aspects of Integrable Systems (G. F. Helminck, ed.), Lecture Notes in Phys., vol. 424, Springer-Verlag, Berlin, 1993, pp. 103–130.
39. C. A. Tracy and H. Widom, *Level-spacing distributions and the Airy kernel*, Comm. Math. Phys. **159** (1994), 151–174.
40. C. A. Tracy and H. Widom, *On orthogonal and symplectic ensembles*, Comm. Math. Phys. **177** (1996), 727–754.
41. C. A. Tracy and H. Widom, *Fredholm determinants, differential equations and matrix models*, Comm. Math. Phys. **163** (1994), 33–72.
42. C. A. Tracy and H. Widom, *Asymptotics of a class of Fredholm determinants*, Spectral Problems in Geometry and Arithmetic (T. Branson, ed.), American Mathematical Society, Providence, 1999, pp. 167–174.
43. C. A. Tracy and H. Widom, *Random unitary matrices, permutations and Painlevé*, Comm. Math. Phys. **207** (1999), 665–685.
44. C. A. Tracy and H. Widom, *On the distributions of the lengths of the longest monotone subsequences in random words*, preprint (math.CO/9904042).
45. H. Widom, *On the relation between orthogonal, symplectic and unitary ensembles*, J. Statist. Phys. **94** (1999), 347–363.

46. E. P. Wigner, *Distribution of neutron resonance level spacing*, Columbia University Report CU–175, 1957; Reprinted, The Collected Works of Eugene Paul Wigner, Part A, Vol. II, Springer-Verlag, Berlin, 1996, pp. 337–338.
47. J. X. Zhong, U. Grimm, R. A. Römer, and M. Schreiber, *Level-spacing distributions of planar quasiperiodic tight-binding models*, Phys. Rev. Lett. **80** (1998), 3996–3999. See also, U. Grimm, R. A. Römer, M. Schreiber, J. X. Zhong, *Universal level-spacing statistics in quasiperiodic tight-binding models*, preprint (cond-mat/9908063).

Department of Mathematics, Institute of Theoretical Dynamics, University of California, Davis, CA 95616, USA
E-mail address: tracy@itd.ucdavis.edu

Department of Mathematics, University of California, Santa Cruz, CA 95064, USA
E-mail address: widom@cats.ucsc.edu